AF606775

SAP® Integration Suite

SAP PRESS is a joint initiative of SAP and Rheinwerk Publishing. The know-how offered by SAP specialists combined with the expertise of Rheinwerk Publishing offers the reader expert books in the field. SAP PRESS features first-hand information and expert advice, and provides useful skills for professional decision-making.

SAP PRESS offers a variety of books on technical and business-related topics for the SAP user. For further information, please visit our website: *www.sap-press.com*.

Bilay, Gutsche, Krimmel, Stiehl
SAP Cloud Platform Integration: Comprehensive Guide (3rd Edition)
2020, 906 pages, hardcover and e-book
www.sap-press.com/5077

Bönnen, Jegadeesan, Mary, Vij
SAP API Management
2020, 365 pages, hardcover and e-book
www.sap-press.com/4928

Acharya, Debelic, Joshi, Dhawan
ABAP in the Cloud: Development and Operations with SAP BTP, ABAP Environment
2021, 316 pages, hardcover and e-book
www.sap-press.com/5236

Acharya, Bajaj, Dhar, Ghosh, Lahiri
SAP Cloud Platform: Cloud-Native Development
2019, 489 pages, hardcover and e-book
www.sap-press.com/4713

Bönnen, Drees, Fischer, Heinz, Strothmann
SAP Gateway and OData (3rd Edition)
2019, 841 pages, hardcover and e-book
www.sap-press.com/4724

Christopher Aron, Piyush Gakhar, Shilpa Vij

SAP® Integration Suite

Editor Megan Fuerst
Acquisitions Editor Hareem Shafi
Copyeditor Melinda Rankin
Cover Design Graham Geary
Photo Credit iStockphoto.com: 1254825733/© shulz
Layout Design Vera Brauner
Production Hannah Lane
Typesetting III-satz, Husby (Germany)
Printed and bound in the United States of America, on paper from sustainable sources

ISBN 978-1-4932-2134-9

1st edition 2021

Library of Congress Cataloging-in-Publication Data
Names: Aron, Christopher, author. | Gakhar, Piyush, author. | Vij, Shilpa, author.
Title: SAP integration suite / Christopher Aron, Piyush Gakhar, Shilpa Vij.
Description: 1st edition. | Bonn ; Boston : Rheinwerk Publishing, 2021. | Includes index.
Identifiers: LCCN 2021020802 | ISBN 9781493221349 (hardcover) | ISBN 9781493221356 (ebook)
Subjects: LCSH: Business--Data processing. | Integrated software. | SAP HANA (Electronic resource)
Classification: LCC HD30.2 .A755 2021 | DDC 658.4/038--dc23
LC record available at https://lccn.loc.gov/2021020802

Contents at a Glance

Dear Reader,

Bread baking is quite complex. I recently learned that if you want to master the perfect loaf, there are a lot of questions to ask, and a lot of things to buy: Bleached, unbleached, or whole wheat flour? Regular or instant yeast? Kneading by hand or with an electric mixer? Sheet pan or baking stone?

If you're like me, your kitchen may be entirely unprepared for this new hobby. My foray into bread baking required several online purchases and trips to different grocery stores, searching for everything from 00 flour for pizza dough, to molasses powder for bagel dough, to rice flour to keep the doughs from sticking. When it came to tools, I needed a new rolling pin, a wicker proofing basket, a dough scraper—the list goes on. Eventually, I achieved a fully stocked pantry containing everything I might want to integrate into a given bake.

Think of SAP Integration Suite as the fully stocked pantry for your organization's landscape. Prepackaged or custom integrations? Hybrid or cloud? SAP or third party? SAP Integration Suite combines all of the ingredients (or, capabilities) you might need, from Cloud Integration to API Management and beyond. Your guide to using all of these capabilities to create a seamlessly integrated enterprise landscape? This book!

What did you think about *SAP Integration Suite*? Your comments and suggestions are the most useful tools to help us make our books the best they can be. Please feel free to contact me and share any praise or criticism you may have.

Thank you for purchasing a book from SAP PRESS!

Megan Fuerst
Editor, SAP PRESS

meganf@rheinwerk-publishing.com
www.sap-press.com
Rheinwerk Publishing · Boston, MA

Contents

Foreword

In today's fast-paced world, digital transformation has accelerated across industries and businesses globally. Only enterprises that can adapt and execute business transformation strategies quickly can sustain growth. Emerging technologies, digital platforms, and new business models have exponentially fast-tracked the growth and profitability of organizations across regions. Furthermore, the COVID-19 pandemic has accelerated this transition as companies worldwide are engaging with their customers and their supply chains entirely digitally now. To stay ahead of the game, enterprises need to adjust to evolving ecosystem expectations, offer seamless customer experiences, and rapidly improve business operations. Enterprises can do this by intelligently integrating diverse and heterogenous applications, processes, and data.

SAP helps enterprises reinvent themselves as intelligent enterprises. With an integrated and modular suite of applications for many industries, a business network that revolutionizes how businesses collaborate, and a business technology platform, SAP helps enterprises become profitable, resilient, and sustainable best-run businesses. SAP Integration Suite is a vital pillar of SAP Business Technology Platform. SAP Integration Suite helps enterprises to integrate end-to-end business processes across a heterogeneous and hybrid landscape.

The authors Shilpa Vij, Piyush Gakhar, and Christopher Aron are knowledgeable practitioners in the integration space, helping our customers simplify and accelerate enterprise integration. This book results from their ongoing engagements with customers and partners and deep understanding of real-world use cases. This book serves as a ready reference for your integration teams to help them solve any enterprise integration challenges to support your digital transformation.

Use SAP Integration Suite to help you reimagine and automate your key business processes. I hope you have an enjoyable read and great success in your journey to becoming an intelligent enterprise. We are always ready to help you so that you can stay up-to-speed and innovate in this ever-changing digital age.

Michael Paul Weingartner
President, SAP Business Technology Platform Core
Technology & Innovation

Preface

Welcome to our book on SAP Integration Suite. Integration has always been an important part of customer IT landscapes. In the age of the cloud, hybrid heterogenous landscapes have only exacerbated the need for seamless, flexible, andf easy-to-use integration capabilities, as companies seek to enable digitally enabled business processes to leverage data from multiple sources. We want to provide readers in an integration-centric role a comprehensive guide to SAP Integration Suite that offers a combination of useful perspective and hands-on guidance so that they can optimize their own landscape and simplify integrations.

In this book, we've sought to provide a comprehensive guide. This book starts with an introduction that provides a solid grounding in the underlying business context of integration, as well as an initial overview of SAP Integration Suite and how each element can provide you with the tools you need to be successful. The rest of the book is focused on providing a series of deep dives for each functional area, with guidance on how you can best utilize these capabilities to achieve your integration goals.

Who This Book Is For

This book is for anyone interested in learning about SAP Integration Suite and how to take advantage of its industry-leading capabilities within their landscape. Customers, partners, consultants, technology leaders, and process owners from all business areas (sales, marketing, finance, human resources, technology, operations, and service) who are planning to evaluate or implement SAP Integration Suite solutions will find this book a great starting point, with integration-related planning, implementation, and optimization in mind.

We've designed the book to combine both high-level and hands-on content for all key capabilities of SAP Integration Suite. Not only have we featured technology components and best practices, but we have also provided guidance on related aspects, such as operations, governance, security, and monitoring. This book can therefore readily serve as a reference guide for implementing SAP Integration Suite cloud solutions. This book doesn't include exhaustive detail on every integration scenario (e.g., scenarios focused in depth on specific industries or lines of business), but it should provide sufficient guidance on how to implement SAP Integration Suite in a reliable and successful manner.

How This Book Is Organized

We've organized this book around a gradual introduction to SAP Integration Suite that provides business and technical context. Then we go through each major feature of SAP Integration Suite in turn.

We recommend reading this book sequentially from Chapter 1 on—but if you prefer, you can directly go to any chapter and start reading about its topic. For example, if you're interested in learning about using prepackaged integration content, you can start reading Chapter 3 without reading previous chapters. Now, let's review what is covered in each chapter of this book:

- **Chapter 1: Introduction**
  This chapter provides an introduction to typical IT landscapes and the common problems encountered that SAP Integration Suite can help solve. It explores the need to work smart in the area of integration to accelerate implementations and simplify ongoing management, and it discusses how SAP Integration Suite can help optimize the resulting business performance.
- **Chapter 2: SAP Integration Suite at a Glance**
  This chapter focuses on providing an overview of the SAP Integration Suite's capabilities. You will learn about operations, governance, security, and monitoring. This chapter also provides guidance on API best practices, multicloud landscapes, and how to maximize the impact of integration patterns and orchestration.
- **Chapter 3: Prepackaged Integrations**
  SAP offers out-of-the-box, prebuilt integrations across the value chain processes such as lead-to-cash, recruit-to-retire, design-to-operate, and source-to-pay and across SAP and non-SAP systems. This chapter gives you an overview of SAP API Business Hub and teaches you how to search, discover, and consume integrations and application programming interfaces (APIs).
- **Chapter 4: Open APIs**
  This chapter gives an overview of how an organization can leverage API approaches to cover hybrid integration needs. It showcases how the API Management capability provides you with end-to-end capabilities for designing, developing, and implementing an enterprise-wide API framework, including out-of-the-box policies, and teaches you how to use them.
- **Chapter 5: Designing Integration Scenarios**
  If you prefer a hands-on approach, we recommend Chapter 5. This chapter teaches you how to design, deploy, and monitor your first integration flow. You will also learn how the Cloud Integration capability helps to simplify application-to-application (A2A), business-to-business (B2B), business-to-government (B2G), cloud-to-cloud, and cloud-to-on-premise integrations in a low-code approach.
- **Chapter 6: AI-Driven and B2B Integration**
  This chapter provides a deep dive into the Integration Advisor, SAP Integration

Suite's artificial intelligence (AI)-based capability that helps you to simplify and streamline the implementation flow of your B2B, A2A, and B2G integration processes. You'll also understand how a crowd-based machine learning approach helps you create integration content easily.

- **Chapter 7: Third-Party Integration**
You can simplify third-party (non-SAP) connectivity through hub-based, one-to-many integration to many integration hubs such as customer relationship management (CRM), ERP, cloud storage, social media, and so on with more than 160 open connectors. This chapter provides an overview and hands-on experience with discovering, authenticating, and consuming prebuilt connectors with the Open Connectors capability.
- **Chapter 8: Event-Driven Integration**
As more and more systems support event-based updates, asynchronous architectures are becoming more common. In this chapter, you will learn how to leverage SAP Event Mesh as an event hub and how to benefit from a decoupled architecture.
- **Chapter 9: Future Developments**
This chapter closes the book by discussing recent innovations that enhance integration strategies and the future potential of SAP Integration Suite, including its roadmap.
- **Appendix A: Solution Blueprints**
In this appendix, additional information is provided about example customer implementations and their associated architectures.
- **Appendix B: Advisory Program**
In this appendix, additional information is provided about different channels of engagement that are available for a supportive approach to kick-starting your integration journey.

Acknowledgements

As mentioned previously, a book such as this one that covers so many different areas and the latest functional and technical innovations of SAP Integration Suite is a true team effort. Several people made especially large contributions to the book, and I would like to offer my sincere appreciation to and thank these leaders and experts: Jonathan Fung, Elijah Martinez, and Karsten Strothmann. Thanks to my close family (Carrie, Katie, and Colin) for their support during the creative process, providing me with lots of love throughout the COVID-19 pandemic. I would also like to thank my parents for all the love and personal investment they made in my formative years—which, despite my own efforts, seems to have worked out in the end!

—Christopher Aron

This book is very special to me as it was written during the pandemic, in a different, challenging situation putting family and work together. I would like to thank my family for their support, as it takes long hours to think, write, and review. In addition, I would like to express gratitude towards my management and our prestigious customers, partners, and integration community, who are always sources of great inspiration. At the outset, I offer my gratitude to our development and product management team members, who are full of energy, have a state-of-the-art technology skillset, and work tirelessly to create a best-in-class product to support our customers' integration journeys. I hope this book will bring value to readers and help them get familiar with SAP Integration Suite, a strategic solution offering from SAP.

—Piyush Gakhar

I offer credit to our customers, our ecosystem, and our ever-growing community of integration experts and integration blackbelts for adapting SAP Integration Suite, as they are the force behind inceptions such as the seamless composition of end-to-end business processes across heterogeneous and hybrid landscapes. Thank you for always inspiring us. This book is another milestone as we continue on in our mission of simplifying integration to accelerate the community's journey to be a complete intelligent enterprise. Thank you to my teams at SAP for being a continuous source of inspiration and tirelessly working to simplify integration and accelerate customers' transition to an intelligent enterprise. I also express my gratitude to Udo Paltzer for helping me with critical and valuable feedback on early drafts.

—Shilpa Vij

We would also like to thank Megan Fuerst and Hareem Shafi at Rheinwerk Publishing for their guidance and input throughout the creative and editorial stages of this publication.

Conclusion

We hope that you find value in reading this book. We have tried to provide a comprehensive and useful overview of SAP Integration Suite. We hope that after reading this book, you feel empowered and motivated to engage confidently in any SAP Integration Suite–related project that presents itself. We know that SAP Integration Suite has successfully helped thousands of organizations around the world, and we would like to add your experience to this community of integration specialists. We hope our book is successful in providing a foundational knowledge of SAP Integration Suite. Ultimately, we hope you can leverage this book as an important guide in your own success, enabling you to build on your own knowledge. So, without any further comment, let's get started with the first chapter, looking at the business and technology dynamics that are raising the visibility and importance of integration strategy and execution.

Chapter 1
Introduction

SAP Integration Suite enables businesses to easily share and exchange valuable data with both internal and external stakeholders. This chapter explains why SAP Integration Suite is considered an essential building block for decision-makers to unlock enterprise data and processes for real-time consumption and interaction.

SAP Integration Suite provides an efficient and scalable building block approach that facilitates quick implementation through plug-and-play content, optimization through ongoing management of the integration content, repeatability through tools such as the API Management capability and Open Connectors, plus different technical tools to cover all the different data sources and transactions that need to be covered in your integration landscape.

The key to SAP Integration Suite is its ability to unlock data access and usage within and outside of your organization. As the commercial world becomes increasingly digital, connected, and real time, it is important that both your legacy and new landscape initiatives support your company's business objectives. The technology is more mature than ever before to empower you to achieve seamless connectivity between your users and the desired data sources. By leveraging SAP Integration Suite, you'll have the most effective solution thanks to increased productivity, more automation, and less maintenance.

In this chapter, we'll set the stage for the book by introducing the intelligent enterprise and the SAP Integration Suite. We'll establish how the customer context and maturity plays an important part, examine how SAP's intelligent enterprise vision can help customers, and walk through how SAP Integration Suite is relevant to heterogenous landscapes and future trends. We'll also provide an overview of key integration concepts such as application programming interfaces (APIs), SAP Integration Suite's capabilities, and complementary solutions.

1.1 Customer Context

Mastering the integration discipline as a best practice for optimal business performance and agility has been growing in momentum over the past fifty years as the underlying enterprise landscapes and component systems have consistently evolved to become more heterogenous, distributed, and complex. In this section, we'll establish the growing need for integration and dive deeper into the customer context when it comes to integration maturity and planning your new integration landscape.

1.1.1 The Evolving Integration Landscape

Enterprise integration as a discipline has its origins in the 1960s, when IBM launched its first general purpose mainframe (System/360). Specific areas of integration interest were first documented by specific industries in the 1970s and 1980s, with two early examples being computer-integrated manufacturing and electronic trading. Enterprise-wide IT integration and engineering concepts began as an area of focus in the 1990s, in line with the transformation of the workplace through personal computers, enterprise applications, and the internet.

More recently, digital transformation, cloud computing platforms, consumerization of IT, and mobility have expanded the definition of integration to cover these new areas. Looking forward, intelligent technologies such as machine learning, robotic process automation, and edge and artificial intelligence will further push the boundaries of integration concepts and strategies. As integration becomes more important and intertwined with successful business outcomes, the breadth and depth of stakeholder interest will increase, with multiple organization disciplines, functions, and personas getting involved.

The reason seamless integration has become more important to so many different organization contributors is that the benefits and capabilities of a connected IT landscape have increased exponentially as organizations have seen their underlying business models become more directly tied to software-derived capabilities. Virtually every business now demands flexible data access, real-time data use, and optimal decisions. Customer experience and stakeholder empowerment are directly tied to the ability of an IT organization to provide a robust data management strategy, which hinges on comprehensive integration flows between users and applications.

The integration architecture may be ultimately owned by IT, but the majority of business stakeholder initiatives are ultimately tied to IT project speed and outcome directly, with virtually everyone needing accurate, real-time consumable information. The current paradigm in most organizations around the globe is that the vast majority of key company initiatives require the need for underlying integration capabilities but do not necessarily directly call that out—and if they do, it tends to be specific to the initiative at hand and therefore takes more of a tactical bent than a strategic point of view.

As we enter the third decade of the current millennium, companies around the world are still struggling with the basic premise of seamless, ubiquitous, and secure data access as companies face unprecedented volume demands and heterogeneity within their IT landscapes. By taking a step back and evaluating their wider integration needs, companies will be able to see the bigger picture. By looking at integration strategically, companies can put in place flexible platforms that will help fulfill their broader needs for fast delivery, seamless connectivity, and real-time decision making across all pressing company initiatives and avoid a piecemeal approach.

Based on its State of Enterprise SaaS Management survey, Zylo now estimates the average enterprise has over 600 applications (see *https://zylo.com/blog/saas-stats-it-strategy/*). This number of applications shows how diverse and unique modern IT landscapes are. This level of heterogeneity is unprecedented, and the average number of applications leveraged by businesses is expected to increase moving forward. Company operations are only going to become increasingly digital as they update their business processes and structures to achieve current and future objectives.

Integration concepts have matured over time. In the 1960s, system integration was the main area of focus, but in the 1970s, application integration became the new area of concentration. In the 1980s, process integration started to become popular, and since the 2000s a lot of organizations have started considering integration through an enterprise integration lens.

The first two decades of this millennium have been dominated by the growth of the internet; the adoption of cloud-based resources by businesses around the world; and the emergence of seamless, real-time, self-service-oriented transactions made by smartphone-empowered consumers. Consumer experiences and expectations have of course crossed over into the business landscape. The need of many enterprises therefore is to manage an increasingly fluid hybrid, heterogenous landscape that combines on-premise system information with cloud properties that are leveraged on a constant basis, all while delivering exceptional and satisfying user experiences.

1.1.2 Integration Maturity

Frequently, IT staff within companies recognize how taking a strategic approach to integration can ultimately help their organization achieve its broader goals within a much more attractive time frame than would happen in the paths of investment currently being adopted by their organizations. An issue commonly reported by these IT stakeholders is the struggle to tie the benefits of a comprehensive integration strategy to the needs and objectives of currently funded or approved priority projects. So the question is this: How can a company start realizing the importance of a robust and flexible integration platform in achieving their current and near-future corporate goals?

Another common circumstance is one in which customers have a very mature, on-premise-oriented landscape. How do they efficiently migrate their integration layer to the cloud as they lift and shift the rest of their IT landscape to the cloud? How does a company leverage the latest and greatest in cloud-based integration technologies, combined with the latest and greatest in approaches and best practices, all while ensuring that legacy integration assets, knowledge, and processes are leveraged?

Our recommendation as a first step is to start identifying where your organization currently sits with regard to integration, and then map a path toward a state in which the organization's approach to integration is both mature and mastered. A great way of doing this is leveraging SAP's complementary SAP Integration Solution Advisory Methodology, which will provide you with tools and guidance so that you can break the mapping of your current approach to integration down into a few simple steps.

Essentially, the methodology helps you map in detail how you are currently leveraging integration technology across your organization. This mapping includes on-premise and cloud assets, as well as integration flow types (process, data, Internet of Things [IoT], user, and more). The methodology also provides guidance for identifying best practices and mapping an ideal integration landscape. Figure 1.1 shows the methodology, visualizing the steps involved in terms of moving from a tactical to a strategic approach to integration.

As shown in Figure 1.1, customers can increase their integration maturity level by moving from an ad hoc integration practice to a more systematic method of integration. Based on established integration standards, customers can also define areas for self-service integration—for example, by allowing citizen integrators to implement simple integration scenarios on their own within the context of overall integration governance.

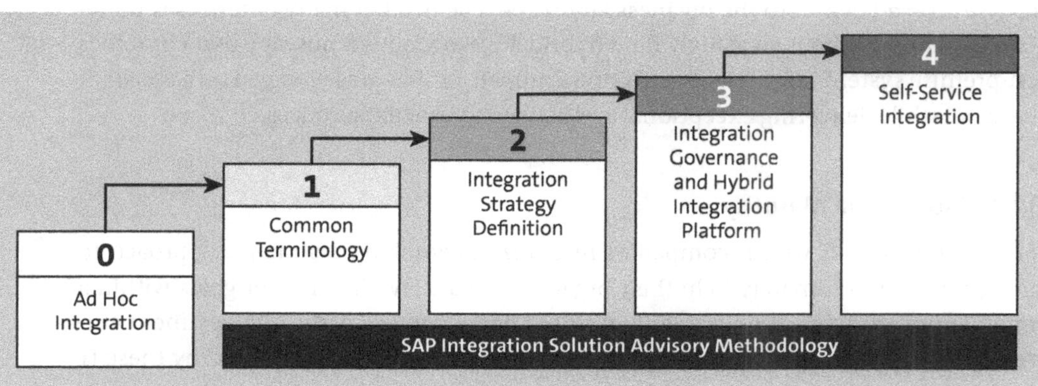

Figure 1.1 Integration Maturity Levels when Applying SAP Integration Solution Advisory Methodology

An *integration center of excellence* (ICoE) is a key milestone in recognizing the importance of integration in terms of fulfilling company objectives. The ICoE is a key enabler and accelerator of integration-related initiatives. Typically, an ICoE will consist of team members with developer and architecture-centric skill sets. These ICoE personas typically work in partnership with line of business (LoB) stakeholders, identifying and solving for their needs. The ICoE is responsible for standardizing the company's broader approach to data access and use in order to ultimately drive standards, break down silos, and help leverage repeatable building blocks based on common approaches.

However, a recent study by the Americas' SAP Users' Group (ASUG) found that only 8% of organizations have established an ICoE (see *http://s-prs.co/v532600*). The minority of companies that have established an ICoE shine as example for others to follow and provide proof of the associated benefits. An ICoE is unlikely to come to fruition organically, so conscious identification and associated planning is required. Ultimately, cross-departmental sign-off is going to be needed with key stakeholder support.

The ICoE is therefore primed for early majority adoption, following in the footsteps of the early adopters. The same ASUG survey reveals that only 55% of companies leverage internal resources for integration. This means that just under half of this survey's population are outsourcing a critical capability to their trusted partners. As the benefits of integration become more apparent through self-evaluation and alignment of company goals, this percentage will ultimately shift in the direction of internal ownership and stewardship. For companies adopting internal resources for integration but without an ICoE in place, a review of current approaches is needed and an assessment for how to master and ultimately get in front of company needs for data use.

Another study, sponsored by SAP and conducted by Technology Business Research (TBR; see *http://s-prs.co/v532601*), notes that just over half of the survey respondents believe the current integration technology available to an organization is not equipped to support their integration needs of the future. This is not surprising given the rapid changes taking place in the current economy. The winds of change relating to digitalization are only going to increase over the next few years as technology becomes more embedded in day-to-day operations and processes.

1.1.3 Planning Your Level of Maturity

For some companies, an ICoE will not be the desired outcome. Centralizing and standardizing data access and usage will not be the right approach for achieving business success. For example, companies may have regional variances, a subsidiary model in which independence and unique approaches are valued, or some other approach in which separation makes sense and is optimal.

Therefore, SAP's methodology is only a guide. The good news is the methodology is very adaptable, with over 1,000 organizations currently leveraging and customizing it. The methodology provides a flexible blueprint for identifying how integration topics are currently being addressed by your organization, and it provides optional steps for realizing where you want to take your organization in terms of optimal data access and usage by various stakeholders.

Ultimately, the guidance from SAP is, at a minimum, to cease viewing integrations as individual ad hoc needs. Doing so results in a potentially unlimited number of point-to-point integrations that add complexity to a landscape, which slows underlying performance and inhibits new innovations and approaches due to the need for constant care and attention.

It is true that the underlying systems have come a long way in terms of data access and standards-based formats. However, it is very beneficial to add a common access layer, which ultimately reduces the amount of ongoing management needed and improves performance through associated logic and best practices. The underlying systems are taxed less in terms of supporting requests, and common needs can leverage the same building blocks, which frees up integration resources to work on more value-adding activities.

Examples of building blocks that are beneficial to an organization are common prepackaged integration content, standardized system access governance, common performance key performance indicators (KPIs) and associated tooling, landscape blueprints, and communication of best practices. These steps outline common approaches and best practices for the broader enterprise audience that enable an organization to function at a high level while supporting data access and usage needs. As an independent organization, you can assess and implement the levels of integration processes and governance that make sense for your company goals.

1.1.4 Defining Your Steps to an Integration Platform

Once you have documented both your organization's current and desired positions as they relate to integration concepts and approaches, the next step logically is to specify how you are going to get there in terms of technology, resources, and timing. We recommend against a big bang approach. Instead, it is better to break the organizational journey down into a series of smaller projects. This has the advantage of reducing the level of effort and time needed to secure visible wins and organizational benefits.

By specifying and achieving quick wins, you will be able to provide stakeholders with evidence of project importance and therefore gain ongoing support for the overall project. You can also capture stakeholder feedback along the way to ensure optimal results and organizational alignment. Examples of quick wins could be standardized access for a couple of key systems, providing associated governance and improved application response times.

In terms of technology, the previously cited ASUG survey demonstrated that the quality of integration tools and capabilities is the number one factor in achieving integration benefits, followed by an organization's internal integration alignment. Therefore, it is important to consider the scope and fit of the technology vendor being chosen. You want to pick an integration solution that is focused on helping you achieve your business objectives. In other words, look for solution tooling that makes your life easier by providing prepackaged capabilities and building block features for rapid implementation and achievement of best practices.

Second, internal alignment is important. Even in the early stages of an integration initiative, you want to get key stakeholders to buy into a better way to approach organizational access and usage of data in order to drive dramatic changes in business processes and operating models. Quick wins in proving a strategic approach to integration provide dramatic business results, which will help build momentum for the phased implementation that has been charted.

One way to keep the integration phases small is to start working with one LoB or business unit at a time. Implementation can be done very quickly, so the sprints can be only a few weeks in duration to achieve the desired results.

Now that we've focused on getting support for *what* we are doing in terms of integration transformation, it is equally important, as mentioned earlier, to also focus on *why* we need to be conducting a transformative approach to the integration topic. In general, your company is going to have a slate of projects that are competing for funding. It is important to tie the integration project directly back to the expected business benefits as they relate to these initiatives.

Because taking a strategic approach to integration will have universal benefits in terms of data access, data usage, and therefore business impact, the exercise should be rewarding. A seamless integration platform will enable these business initiatives to achieve real-time visibility and action in a fast, scalable, and secure manner. Further, the integration platform will help with business initiatives that will require funding next year and the year after.

Internal stakeholders may at first be reluctant to step aside for a strategic effort, or to incur additional costs for their projects in order to fund a cross-project initiative. Therefore, the ask for an integration platform needs to be done in such a way that internal stakeholders are educated on the near-, mid- and long-term benefits of their motivations and understand that the integration platform can be implemented quickly, in bite-sized pieces. It's important to emphasize that a building block-based approach will provide more self-service opportunities for them and foster a shared services approach, in which additional support can be provided to their initiatives because the central IT integration team won't be bogged down with low-value "keeping the lights on" types of activities.

As mentioned previously, it's important to select quick wins to show immediate business value, and there should be obvious opportunities in which a building block approach to integration can help one of your key business stakeholders with their business needs and therefore help secure their support.

1.2 Intelligent Enterprise Vision

An *intelligent enterprise* can be defined as an organization that applies advanced technologies and best practices within agile, integrated business processes. This helps it to be more resilient, profitable, and sustainable and to become a best-run business.

In this section, we'll look a little deeper at the associated benefits of the intelligent enterprise, why it merits your attention, and how integration is integral to the whole concept. First, let's examine our general definition of the intelligent enterprise.

1.2.1 What Is an Intelligent Enterprise?

Our intelligent enterprise definition includes a few concepts that we will expand upon:

- **New technologies and approaches**
  First, consider the application of new technologies and approaches. Change within an organization is typically hard. This is because individuals or groups of individuals are accustomed to doing their jobs and performing their day-to-day responsibilities a certain way. There are normally a few reasons that business processes are structured the way they are, typically shaped and iterated upon over time. Adding new technologies or changing the current approach can involve risks, a learning curve for those involved, and possibly a change in responsibilities for team members, to name a few concerns.

  In terms of integration, new technologies could be related to the core tooling leveraged for achieving seamless integration flows, or they could involve harnessing a new intelligent technology such as machine learning to get recommendations on optimal integration flow construction. Or the technology could be more material, such as harnessing new data sources as part of an integration flow or business process that hasn't been leveraged previously. For example, the new data could be sourced from robotic process automation (RPA) bots or IoT devices.

  New approaches could take the form of a new ICoE team, as we discussed previously, and its impacts on personnel and organizational structure. We see a lot of SAP users adopting digital integration hubs as a new way to implement rapid change in existing business constructs by providing a 360-degree view of a business with heterogenous integration at its core and a dynamic user experience on top. This provides users with real-time visibility into the business, which in turn enables rapid contextual decision-making and thus leads to optimal business performance.

- **Best practices**
  The next aspect of the intelligent enterprise that we want to expand upon is the achievement of best practices. These best practices can be in the realm of an industry, an LoB, or a specific business process. In terms of integration, one way to achieve best practices is to leverage prepackaged integration content. Prepackaged integration content is created and maintained by subject matter experts for a given domain. By taking advantage of this content, your organization will benefit from optimal performance both now and in the future as it ensures best practices and takes advantage of new improvements as they become available.

  As mentioned, with an ICoE, the prepackaged construct also ensures that an organization is able to benefit from best practices across teams, which leads to consistent performance and standardization across an organization. That in turn eliminates data silos, duplication, and inefficiencies.

- **Agile business processes**
  With digitization, businesses are moving faster. Whether they want it to or not, the pace of business is increasing. Customer expectations are also rising in line with new technologies and efficiencies that are available to them. These rising expectations can take the form of a personalized experience or simply the ability to make changes based on variances in their needs. As a result, an intelligent enterprise needs to support agile business processes. For this to be possible, a robust integration layer is needed that can efficiently scale and accommodate business needs without overwhelming the IT team that supports these capabilities. By leveraging integration building blocks, changes can be made in real time without having to reengineer or maintain the entire technology stack.

  As business accommodations and changes are made, the supporting integration landscape needs to be resilient. Long-term performance cannot be compromised in terms of the underlying systems, even though data requests and usage are increasing. More often than not, increased performance without reduced capabilities is desired from these legacy applications as use cases broaden across the organization. This is another opportunity in which a digital integration hub can help by providing a database in a sidecar configuration to persist a subdataset for faster response times and delivery to users. This can offer an end-to-end "turbo" effect for providing data to users and shortening the length of round-trip requests.

- **Optimization of profitability**
  Getting more from your IT infrastructure has been a consistent theme so far in terms of the reasons you should prioritize integration efforts within your organization. A related aspect we haven't discussed so far is the ability to impact your company's bottom line by increasing profitability. This can be done in two ways. First, optimizing your approach will lower your total cost of ownership; second, new business models and business processes will be possible, which opens a path to new revenue streams and competitive differentiation.

- **Sustainability**
  The last aspect of the intelligent enterprise we would like to shine a light on is sustainability. This aspect can take many forms in the domain of integration. Sustainability can be enhanced by deriving more value from existing applications, extending their natural life, and increasing utilization rates. Another way that integration helps an intelligent enterprise increase its sustainability is by facilitating the digitization and transformation of many legacy analog processes that required paper, conversations, and duplication of steps. By streamlining and optimizing business processes, integration can help minimize a company's physical impact on the environment.

Figure 1.2 depicts the intelligent enterprise and shows the layers of capability that a fully integrated landscape can empower. From top to bottom, integration empowers business processes, connected applications, and the use of technology to optimize performance. The overview shows how the intelligent enterprise consists of technology, application, and process layers. SAP Business Technology Platform (SAP BTP) is the key piece of the technology layer, consisting of database, data management, analytics, extension, and integration capabilities. The intelligent suite and industry cloud solutions are able to harness SAP BTP capabilities within the application layer. Key business processes can then be optimized by providing enterprises with the ability to seamlessly connect with their business networks.

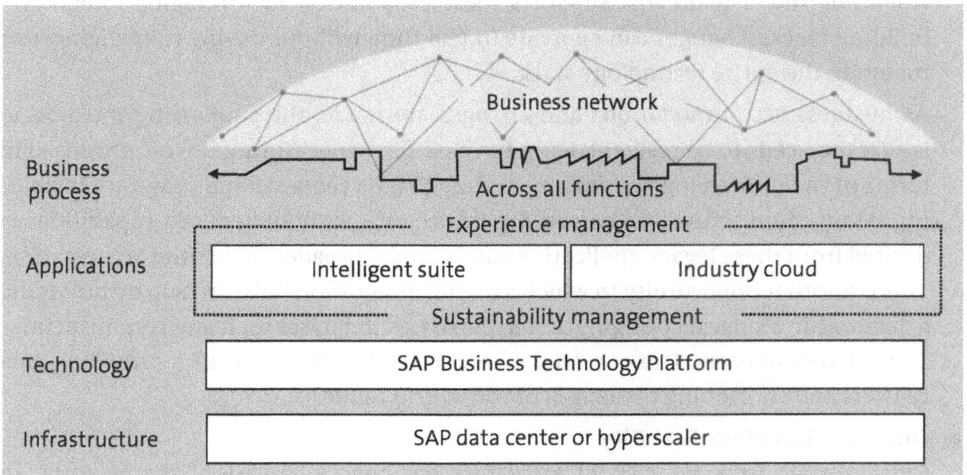

Figure 1.2 Intelligent Enterprise Overview

1.2.2 Benefits

Now that we've established what characterizes an intelligent enterprise, let's examine the benefits of these qualities. The following are beneficial features of an intelligent enterprise:

- **Intelligent suite**
  The intelligent suite is a logical collection of enterprise applications organized by LoB. Although LoBs often operate independently, sometimes there is a need to share data across applications to keep the organization in alignment. For example, it's quite common for the human resources LoB to be the system of record for employee data and the financial system or enterprise resource planning (ERP) application to maintain departmental budgets. Therefore, when costs are incurred by an individual, the ERP system will need access to the latest organization chart to identify which department or team the employee costs should be associated with. There are also

master processes—for example, hire-to-retire or lead-to-cash—that involve many subprocesses, which impact multiple applications with the intelligent suite.

In terms of integration, SAP provides prepackaged integrations oriented to these master processes (and their subprocesses) so that organizations can easily implement them across both SAP and third-party applications, with SAP maintaining the integration content on a regular basis. This helps achieve seamless end-to-end business processes that are based on best practices and optimized for efficiency and performance. This helps an intelligent enterprise seamlessly connect their customers, employees, and partners across their organizational pillars, ensuring real-time access to information relating to products, services, budget, finance, and more, which empowers the 360-degree aspects of a digital business.

- **Industry cloud**
  Integrating with industry clouds helps organizations apply industry-related expertise and intellectual property in unique ways to achieve a competitive advantage. Frequently, these industry clouds need to be seamlessly connected with the core lines of business to ensure optimal performance. By leveraging industry clouds in concert with your core applications, your fully integrated organization will be able to combine leading-edge industry best practices and thus leverage intelligent technologies to ensure that you achieve a competitive advantage in the broader marketplace.
- **SAP BTP**
  Next is the role SAP BTP plays in terms of access to industry-leading data management and analytical capabilities that help optimize data intelligence and associated decision-making. Companies need to both leverage transactional data from applications and glean insights from diverse sources of data through data integration and analytics. Looking at datasets helps companies glean insights from historical experiences and helps with trend analysis to identify new opportunities or predict future patterns for long-term performance.
- **Connected business networks**
  As we have briefly touched on so far, an intelligent enterprise is not solely focused on internal integration needs. In fact, an intelligent enterprise is able to harness both internal and external aspects as part of its optimization. A business's or customer's 360-degree point of view by definition is going to include both internal and external aspects in terms of coverage. Therefore, a company needs to ensure its business network is seamlessly connected: allowing for real-time changes and adjustments based on stakeholder need.

  For example, a supplier can provide updates to its production schedule that are either positive or negative; customers desire the ability to gain visibility into their current orders, making changes as their needs require. The business network can be more than procurement related, with other flexible value chains coming into play—

for example, contingent workforce sources or an auxiliary supplier/partner that is needed on an occasional basis.

- **Experience management**
  Last is the aspect of experience management. Intelligent enterprises need to track, understand, and address their customer needs. These needs can change over time, either as part of a long-term trend or on a short-term basis in response to particular events or transactions. Intelligent enterprises need to integrate these data points of sentiment and feeling into their digitized business processes if they want to successfully capture and exploit these insights for the benefit of their business.

A recent SAP-sponsored study conducted by Oxford Economics found that two-thirds of respondents are on a path to connecting traditional operations data with human behavior-related data such as emotions and attitudes (see *http://s-prs.co/v532602*).

As you can see, the intelligent enterprise touches all aspects of a modern business. Integration as an enabler is ever present in each of these areas. Therefore, a strategic approach to integration needs to be prioritized and implemented for your business to ultimately benefit from the concepts advocated as part of the intelligent enterprise. By implementing the building blocks of integration, you will be able to achieve seamless data flows and access both within and outside of your organization. You'll be better equipped to respond to market and customer needs through leveraging integration accelerators that enable new business processes and/or business initiatives to be executed and iterated on.

It's not just the data scientists and analysts within your organization that need accurate, real-time data; it's a mandatory need for every persona within your organization. Implementing a deliberate integration layer within your organization will enable you to provide the right data to the right stakeholder at the right time, in a format that is relevant to their usage and execution of an unlimited number of decisions and actions. This will help your organization reach new levels of productivity and efficiency. This platform will enable this speed of action to happen in a scalable, reliable, and impactful manner. This seamless fabric of data access will enable your company to successfully execute its most important initiatives with agility and speed.

Conversely, if your organization does not adopt a strategic approach to integration, then your organization will run the risk of becoming slower to respond, more expensive, and ultimately overwhelmed. The economic forces of today and tomorrow are not kind to companies that are inefficient, expensive, or out of touch with customer needs.

1.2.3 Value of Integration

Summarizing the value of integration starts with access to the right data at the right time by the desired stakeholder. The value then expands upon this starting point to provide an organization with a platform for optimizing its business operations based on efficient distribution of data across the organization in real time, which empowers efficient business processes and decision-making. Based on seamless data access, an organization can refine existing and implement new business processes with the confidence that real-time sequences can occur, automation can be implemented with confidence, and adjustments can be made to meet performance expectations and deliverables. After the COVID-19 pandemic, the time is right to push for the digitization of company data, with 70% of corporate boards accelerating digital business initiatives in the wake of the economic disruption.

Real-time seamless integration across an organization facilitates the aggressive pursuit of digital channels. Gartner cites in recent research that companies that increase their use of digital channels are 3.5 times more likely to be market leaders (see *http://s-prs.co/v532603*). The benefits of digital interactions with customers are multifaceted. Companies can establish self-service capabilities with customers that encourage them to spend more, develop deeper relations, and enable 24/7 interactions, all while minimizing human costs. Plus, in the previously cited Oxford Economics survey, 52% of respondents said that improving the flow of information is their top priority in terms of operational improvements.

The differing levels of digital investment and innovation by firms is also creating a data divide, with heavy investors holding positive points of view about their future business prospects, while those that don't invest in digital initiatives are more cautious in their outlook. The digital investors see their advantages as taking the form of faster decision-making, improved communication, and more opportunity for cross-team collaboration. To back up this concept of a digital divide, a recent survey by Board (*https://board.com*) found that just over half of companies (54%) are making important business decisions based on data and insights, while "gut" decisions are still being made by a significant portion of companies (45%). A majority of companies (57%) still rely on spreadsheets to aid their decision-making, despite more modern and reliable tools now being available (see *http://s-prs.co/v532604*).

Areas of impact for an integrated approach to business operations called out by the Oxford Economics survey include enhanced human experiences for both customers and employees, reduced costs and spend, increased speed, and a redesign of how an organization operates and the ability to make better-informed decisions.

The achievement of seamless data flows across an organization ultimately enables the unified entity to achieve higher levels of operation that deliver on the company's broader strategy. Similar to the ASUG survey, the vast majority of companies surveyed are not fully integrated (only 25% of Oxford Economics respondents reported being completely integrated). The Oxford Economics survey summed it up well:

> *"Executives must: interconnect and integrate. Organizations must view operations as though all business relationships—internal and external—are part of the same*

network. Build this philosophy into all decision making, nurture collaborative processes, and reap the benefits of interconnectivity."

SAP Integration Suite, with its broad technical capabilities, building block approach, and easy-to-use interface, is the tool to help you achieve the benefits of this interconnectivity.

1.3 Hybrid and Heterogeneous Landscapes

Now that we've covered the value of the intelligent enterprise in terms of integration, let's start reviewing how this concept applies to typical company IT landscapes. We'll discuss current landscape trends and data access considerations.

1.3.1 What Are Hybrid and Heterogeneous Landscapes?

A hybrid heterogenous landscape combines two aspects of modern IT architecture. The first is the combination of both on-premise and cloud elements, with the former being fully managed by the enterprise. The latter is typically a combination of infrastructure as a service (IaaS), platform as a service (PaaS), and software as a service (SaaS). These different approaches to cloud services provide customers with varying degrees of control and responsibility.

The second aspect of a hybrid heterogenous landscape is the unprecedented variety of applications being leveraged and managed. As mentioned previously, the average company is managing around 600 applications in its landscape. This heterogeneity is expected to increase over time as more specialized applications come to market. These applications are expected to have specific industry focuses or LoB functionalities.

Gartner has started articulating a concept it calls the *composable enterprise*. Gartner posits that we are now in the fourth generation of ERPs, with future application experiences being made up of smaller components and business units being encouraged to participate in the design of their application needs (see *http://s-prs.co/v532605*).

Deloitte, in a similar vein, promotes the concept of a *kinetic enterprise*, with an emphasis on how enterprises can focus their resources on building to evolve, no longer focusing their efforts on traditional values such as building to last (see *http://s-prs.co/v532606*).

So, in terms of these application trends and the resulting impact on a company's integration strategy, there is a need for a robust and flexible integration strategy that can easily plug new applications into a broader governance framework. The data stored within these applications needs to be accessible to the rest of the enterprise landscape to ensure agility in business processes and adaptation to customer needs.

1.3.2 Future Landscape Trends

It is generally accepted that enterprise landscapes will become more heterogenous and hybrid over time, with additional elements being added, such as IoT-centered edge capabilities, 5G mobile clouds, and metro area capabilities. These elements will be adopted by the majority of enterprises over the next few years.

As we discussed earlier in this chapter, every company will look to operate more efficiently and intelligently through the use of data, analytics, and applications working together in harmony. This makes the case for strategic application and data integration become more urgent as the resulting analytics will only be as insightful as the information made available. Companies will also seek to provide personalized services to customers, employees, and other stakeholders through increased use of location insights and situational data, allowing for behavioral analysis and results.

From an integration point of view, this will require increased real-time visibility and connection in order for the company to execute its value proposition. Security concepts will also evolve to become more dynamic, enabling stakeholder access to any associated transaction on any device. The security aspects will need to be robust, but underlying this concept is the assumption that the landscape will be fully integrated to support these objectives.

1.3.3 Data Access and Usage

The expectations with integration excellence are that there will no longer be data silos. Organizations will proactively and systematically break down the barriers to information sharing, such as eliminating departmental nonstandard data protocols by implementing enterprise-wide data management and data-tracking initiatives, system interoperation programs (such as API-based communication), and more encompassing digitalization and RPA-based initiatives.

Just as companies were incentivized to establish chief innovation officer (CIO) and chief digital officer (CDO) functions, it's conceivable that companies will expand these roles or create new ones that focus on ensuring a company has seamless access to accurate real-time data. Therefore, it's conceivable that for some companies, investing in a chief data officer or chief integration officer role will be of importance to the broader organization. Because integration is solidly within the scope of the chief information officer's domain, it's a great way to reinvigorate the role of the IT teams within an organization as more than just a cost control arm of the chief financial officer (CFO). Instead, the CIO can help steer the broader organization toward its technology and digital goals.

1.4 Why Simplify Integration?

In certain circumstances, customers have invested lots of time, money, and resources into building out and optimizing their current integration landscape. In some cases, this investment directly contributes to a company's core business operations and therefore represents core business value. In these situations, it can justifiably make sense to weigh the value of the intellectual property that has been established and refined over time. But in most cases, there will be clear advantages of evaluating your current integration approach, discovering its high costs of ongoing maintenance and subpar underlying performance when compared to the latest and greatest approaches to integration, which offer best practices and building blocks to facilitate rapid implementation.

In this section, we'll explore integration approaches: traditional, forward-looking, and building block-based.

1.4.1 Traditional Approaches

In Jim Collins's 1994 book *Built to Last: Successful Habits of Visionary Companies*, he outlines a concept based on studying several industry-leading companies: to achieve visionary status over multiple decades with superior market performance, companies need to focus internally on creating lasting organization value (clock building as opposed to time-telling) versus investing in externally based efforts. He wrote *(www.jimcollins.com/concepts/genius-of-the-and.html)*:

> *"A visionary company doesn't simply balance between preserving a tightly held core ideology and stimulating vigorous change and movement; it does both to the extreme."*

This long-term mindset has definitely stood the test of time for a number of companies' features. American Express's stock price has increased over 1800% since 1994, with IBM, Johnson and Johnson, Proctor and Gamble, 3M, and Walmart all witnessing three-figure percentage growths in their stock prices. Therefore, building to last as a concept is not at all bad as a corporate strategy and has shown to stand the test of time. But is building to last the best strategy in the digital era?

Companies need to be wise with their investments: they cannot build to last in every area of their businesses, or they will find themselves spread too thin and outmaneuvered by their competition. Instead, companies need to prioritize their investments in areas of key differentiation, finding where the return on investment will be the highest. It is a widely held opinion that continually investing in on-premise IT infrastructure and applications that take time, money, and resources to implement, maintain, and optimize is not the best practice, due to unnecessary costs, landscape complexity, and detrimental effects on performance, which impact core company objectives. In turn, investing in a sophisticated integration landscape to manage these IT assets and applications is not an optimal strategy. Instead, you need to pick an integration strategy that leverages those vendor strengths that ultimately make your life easier and leave you free and clear to solve bigger and more important issues.

1.4.2 Forward-Looking Approaches

In 1998, William L. Miller and Langdon Morris published *Fourth Generation R&D: Managing Knowledge, Technology, and Innovation*. The book articulates the 12 concepts for coordinating innovation in an effective and optimized manner. Miller and Morris describe the first three generations of research and development being performed in a linear manner, with severe challenges along the way in terms of implementation of innovation. Miller and Morris articulate the need for a chief innovation officer to help identify opportunities, coordinate resources, and nurture innovation projects. Proctor and Gamble (one of the companies highlighted in Jim Collins's *Built to Last* book) was one of the first companies to establish an innovation factory in the 2000s. Established companies have done the same, like 144-year-old AT&T, which has established several innovation foundries focused on specific areas of innovation (including network, software, AI and data science, and cybersecurity) and responsible for helping AT&T establish 500 partnerships per year and file an average of five patents per day.

This focus on deliberate, controllable innovation is a recognition that the status quo (building to last) will not be enough to ensure a company's long-term survival in the digital age. One hundred years ago, the average life expectancy of an S&P 500 company was close to 100 years; ten years ago, the life expectancy of a corporation had shrunk to less than 15 years. In the 2020s, the life expectancy of a corporation will be even shorter (for more information, see *http://s-prs.co/v532607*).

Therefore, it is vital to a company's long-term survival prospects that it gets out of the mindset of building to last and instead focuses its efforts on building to innovate, which will in turn ensure the company's long-term relevance and increase its chances of survival.

This concept of technology and software being enablers and not drivers is a key point. In the previously referenced SAP-sponsored study performed by TBR (*http://s-prs.co/v532608*), conducted in the summer of 2020, 85% of respondents saw the value of integration as high/medium-high in their organizations, but a significant number of respondents reported finding integration funding to be a struggle. This was typically because although integration was viewed as important, other projects had higher status or IT was busy establishing new revenue channels. This points to the need for highlighting integration strategy as a key enabler of building to innovate.

By facilitating seamless access to accurate real-time data for all stakeholders, integration enables the digital transformation aspirations of an organization. Also, by upgrading an organization's integration capabilities in such a way as to free an IT team from day-to-day management and optimization, a company will be better positioned to pursue additional innovation-based projects.

1.4.3 Building Block Approach

Integration building blocks aren't necessarily a new concept. The idea of leveraging integration building blocks has been an accepted part of an optimized integration approach for over a decade. However, what's interesting is the evolving nature and associated benefits of these building blocks. Integration building blocks can typically be classified as integration patterns, access patterns, and lifecycle patterns.

When it comes to *integration patterns*, we're describing functionality such as mediation actions, sequences, and accessing end points. An example of a building block in this area would be prepackaged integration content that is made available to integration users. This prepackaged content typically provides best practices for completing a certain transaction between two applications. The prepackaged content will include aspects relating to the recommended sequences and associated end points. An integration user can very quickly access and deploy this integration content for connecting a commonly used scenario between two applications. The prepackaged content is often maintained and updated by the integration vendor freeing up the integration user to focus on more value-adding activities. This content can also be used as a starting point for connecting two systems, whereby enhancements can be added or adjusted according to a company's needs.

Integration teams can also leverage an integration platform to build custom integration patterns specific to an organization, thereby creating building blocks that are internal in nature. Internal users can discover and leverage this integration content quickly to meet data access needs that may be held by multiple groups within an organization (e.g., the need to look up customer address details). Building blocks can also be leveraged for accessing external data needs that span different business units or stakeholder personas (e.g., looking up a customer's credit score). These last use cases (internal and external access) fall under the *access pattern* building block category.

The use cases for data access and usage can be sorted into a few main categories. The first is portals, used when stakeholders desire a mash-up or subset of data that may come from one or more underlying applications. This portal use case could be for an internal user seeking to get a comprehensive view of a customer status or an external customer seeking access to their account information and a way to make changes. Integration building blocks that can be implemented in this area can include enabling stakeholders to have access to various applications via a common abstracted access layer, such as an API-based catalog that abstracts the underlying variance in and complexity of accessing desired applications. Policies can also be implemented for this abstracted layer that can protect the underlying applications—for example, normalizing the peaks and valleys of data requests. Portal templates and low-code capabilities can also be exposed to citizen developers so they can self-service basic information needs.

A second common use case for data access and use is the need for data integration. Data integration covers the following disciplines:

- **Data processing and orchestration**
  Data processing and orchestration involves identifying data sources, implementing a data preparation layer (connections, automation, and storage management), preprocessing, enabling discovery, and ultimately data management.
- **Data governance**
  Data governance involves identifying who changes data, what data is changing, and who is using the associated data. The key step for most organizations is ultimately to help stakeholders (data scientists, business analysts, developers, and subject matter experts) to access data seamlessly in an efficient manner.
- **Data replication**
  Data replication involves connecting to and replicating data across multiple databases residing in different business units. The need to replicate data can vary, but some common examples include the need to provide a business unit with its own subset of data to improve response times and visibility, to protect source systems from being overwhelmed by frequent customer or user requests, and to perform in-depth business analysis for insights into a particular issue or opportunity.

As with previous areas of integration topics, the key for you as the owner of the integration fabric within your organization is to leverage the building blocks that make your life easier. Building blocks in the data integration area can take the form of machine learning–based models, automation of retraining, and maintenance and retirement of legacy integration solutions. All of these are tools that overcome data silos and complexity and operationalize machine learning lifecycles.

For data processing and orchestration, building blocks can provide launchpad-based views for machine learning tooling, data preparation, metadata exploration, and pipeline modeling. The key is helping your organization make the leap from experimentation to deployment. Once deployed, data integration tools need to manage data relevance and avoid becoming out-of-date or going stale—for example, avoiding *model drift*, in which the results no longer match the underlying business operations and therefore the related insights no longer correspond.

Building blocks also make it easier to identify and manage data sources, with tooling providing an easy-to-navigate preparation layer. This preparation layer helps onboard and maintain source connections and storage assets. Building blocks are available to refine and enrich these data sources for interested business users. Building blocks also can help catalog, publish, and enable user self-discovery. Next to consider are data automation tools that accelerate the development and refinement of machine learning models: build and train, deploy and validate, automate sharing, and perform ongoing management. This building block example ties to the *lifecycle pattern* category mentioned previously.

For data replication, building blocks can take the form of scaled-back data integration-embedded tooling within database and data warehouse solutions. Building blocks can also take the form of connectors for SAP and third-party sources, ABAP-specific adapters, message-based communication, machine learning operations, custom adapters, data lakes, and data caches.

Another building block in the data integration area is machine learning-based models that can be used to overcome data silos and complexity. To optimize machine learning-based models, IT teams need to factor the related amount of data flow, sensitivity of data, latency of delivery, and model complexity. IT teams can help manage these machine learning-based models and help ensure they move from ideation to production, capturing insights for optimization and maintenance activities.

Additional examples of integration building blocks include edge preparation and processing of data, API catalog hubs, microservices catalog hubs, business process-related prepackaged content, B2B libraries, and end-to-end tooling for the API Management capability.

Like all areas of technology, data integration continues to evolve in order to better serve the needs of businesses and their employees. Within data integration, one area of growth to evaluate is the use of graph databases. Graph databases store the native relationships of underlying objects, which facilitates faster queries and facilitates contextual displays. The use of graph technology is adding a semantic layer that offers a unified, consolidated view of data across the organization. This will be a rapid growth area for many companies in the coming years.

1.4.4 Integration Center of Excellence

The primary role of an ICoE is to aid the broader organization in optimizing the associated integration layer to achieve business objectives. This goal is achieved through a combination of direct and empowered indirect actions. The ICoE can play an important consultancy type of role in helping the broader company understand and approach integration-related projects more strategically.

The ICoE can help set benchmarks for where an organization currently sits with regard to current integration capabilities, help map an efficient effective path to an optimized state, and act as a trusted advisor with business stakeholders to make progress with associated data, systems, and processes.

ICoEs typically include a variety of skillsets and personas. Not only are integration experts part of the ICoE, but developer and architects skillsets are also needed to enable understanding and informed guidance for business stakeholders working with ICoE building blocks and assets. For example, a mobile developer within the organization will need to leverage integration assets such as API catalogs and associated APIs. With an ICoE including a developer persona itself, the ICoE can talk in the developer's own terms and ensure the required assets and guidance are in place.

Ultimately, any internal stakeholder should be able to effectively navigate existing integration assets for their own usage or efficiently make requests for new integration assets that may be needed to fill gaps. Internal stakeholders should be able to quickly understand the associated business benefits of approaching integration in a strategic manner and be able to witness a successful implementation directly and quickly.

1.5 APIs and the Intelligent Enterprise

Integration can help enterprises by delivering seamless end-to-end processes across multiple applications and business areas, providing unique experiences and facilitating real-time connectivity across business partners and customers. To achieve this wide range of capabilities, it's important for the associated integration layer to incorporate flexibility and adaptability while also ensuring security and scalability measures are in place. One method that should be employed is the utilization of APIs. APIs can form an abstraction layer between data-rich applications and the business users seeking to use them. APIs by their nature provide an opportunity for organizations to standardize their data access policies and techniques. These APIs can be used by multiple stakeholders in a standard way while also achieving optimum performance from the underlying applications.

In this section, we'll introduce key API-based concepts as we set the stage for an API-first landscape. We'll also introduce best practices, including how to ensure easy API access and usage.

1.5.1 Establishing an API-First Architecture

There are typically three levels of APIs within an API-first architecture:

1. **System APIs**
   At the base level are system APIs, which are typically tied to a specific application, database, or subsystem. Alternatively, a system API can be built around a specific domain or business capability. System APIs are therefore leveraged for accessing systems of record or data sources.
2. **Process APIs**
   The second type of APIs typically leveraged by organizations is the process API. Process APIs are leveraged for orchestrating and working with the data exposed by system APIs.
3. **User experience-oriented APIs**
   The third type of APIs typically leveraged by companies is the user experience-oriented API. These APIs are typically consumed by a channel leveraged by a user—for example, mobile apps, a portal, or another type of consumption.

For all of these API flavors, it's important to have well-defined specifications with strong supporting documentation. It's important that the API specifications provide clarity and help support standardized approaches for access and associated policies (e.g., the transformation of data to a common format or the need to cleanse data and the ability to support error handling). API specifications can also include best practices, such as how to add encryption when sensitive data is being shared outside an organization. Finally, it's important to factor in aspects related to API lifecycles: How will they be managed and updated once in use, and how will updates changes and retirements be handled?

Beyond the design, development, and publishing of APIs, you also need to factor how APIs will be discovered, utilized, controlled, and monitored. How can APIs be structured to facilitate reusability and scaling? How can API policies be implemented in such a way as to optimize usage but at the same time protect and support the underlying landscape elements?

1.5.2 Enterprise API Hubs

A common method to optimize an API-first strategy is to leverage an API hub. In an enterprise, the API hub can provide a marketplace for both internal and external stakeholders. The API hub is a location for developers, product managers, IT, creators, partners, and customers to gain access to a catalog of APIs. The API hub typically provides guidance on how to leverage the associated APIs in the form of documentation, code snippets, and examples, as well as the ability to test API usage. For the enterprise team that manages the API program, the API hub helps with the administration of the APIs, providing a means to manage and govern visibility into the API community and the ability to connect with users.

A readily available example of an API hub is the SAP API Business Hub at *api.sap.com*, developed by SAP to support the SAP community of API users. In the SAP API Business Hub, both SAP and third-party APIs are cataloged. The SAP API Business Hub is a rich example of integration content. Not only are APIs cataloged, but the community also gains access to other types of content, such as events, processes, and integrations.

Figure 1.3 shows the SAP API Business Hub home page. SAP API Business Hub provides prepackaged integration content that is structured around SAP and third-party applications, content types, and themes. The content types include API-based, process-based, and event-based types, plus workflow packages and more. Each package provides content relevant to a specific scenario, along with associated documentation. We'll discuss SAP API Business Hub further in Chapter 3.

Depending on your company's needs, you need to structure an API hub in such a way as to optimize its usage for your main stakeholders. You may have a good mix of professional, low-code, and citizen developer personas. Accordingly, each stakeholder type will have different needs and different uses for the API hub. For example, professional developers will be focused on technology aspects, whereas a citizen developer will be entering your API hub with a solutions-oriented point of view.

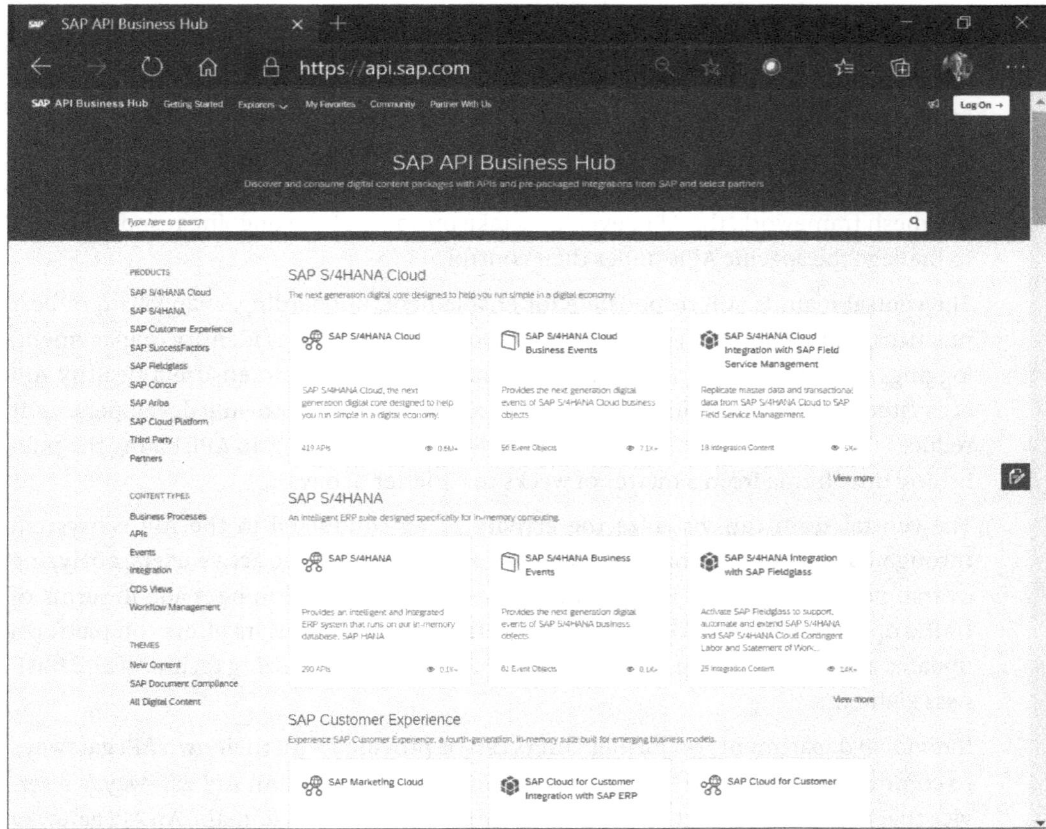

Figure 1.3 SAP API Business Hub

API hubs provide an opportunity to standardize data access and usage across your organization. In the past, system usage may have been built up to be opaque within your organization. Centralizing API calls through a single API hub provides you with the ability to track, monitor, and possibly monetize API usage. It also serves as an opportunity to provide developers with a consistent experience, with clear guidance and best practices that ensure optimal performance and help developers efficiently obtain the information they need. This frees them up to work on more value-adding activities; for example, a mobile developer can spend more time on user interface and application functionality.

You can also use an API hub to standardize API usage, data standards, and communication policies to optimize underlying system performance, eliminate data silos, and achieve seamless usage across your organization. An API hub such as SAP API Business Hub is therefore a great asset for your ICoE team to help to achieve your strategic objectives.

1.5.3 Establishing an API Mesh

One of the potential downsides of a centralized API hub is the tendency of a centralized IT capability to become burdened with lots of requests, governance, prioritization, and, as a result, potential bottlenecks. One potential option is to take a less centralized approach. The central team can take a services mindset to provide the core API platform capabilities, ensuring it's scalable and secure, while working with the API producers to take on some of the development capabilities in a self-service manner. The company still benefits from a centralized discovery and use ecosystem, but under an API mesh framework, the API developers take on some of the ongoing responsibilities to manage the specific APIs under their control.

The central team is still responsible for enablement, onboarding, evangelism of best practices, and taking on shared services responsibilities, such as identity management, logging, security, standards, testing, governance, and more, to ensure a healthy API ecosystem. These centralized services are of great benefit to API developers as it reduces their level of effort in terms of building and publishing an API, taking the publishing time frame from a matter of weeks to a matter of days.

The central team can visualize the activity being conducted in the API ecosystem through dashboards, seeing what's connected and who are the active users, analyzing overall performance, and identifying where improvements can be made in terms of traffic optimization and API usage. The central team can focus its efforts on platform updates and on enhancing developer and user value by optimizing technical and business elements.

Individual departments or business units can be provided with their own API gateways to communicate with the central API management platform. An API gateway is a service that provides the single entry point for API requests into a domain. An API gateway encapsulates the domain's internal architecture and provides a secure API for users. It also usually includes authentication, monitoring, and rate-limiting capabilities.

In Figure 1.4, the end-to-end lifecycle elements of API management with the API Management capability of SAP Integration Suite are shown. An enterprise-grade API management solution needs to include all the functionality you need for the different stages, starting with design and going through development, management, monitoring, and engagement.

Figure 1.4 API Management Capability End-to-End Lifecycle

As mentioned previously, in the context of an ICoE, the team managing the API management platform needs to offer capabilities to support this technology through all its key stages, but also should support skillsets that are similar to those of the consumers of the resulting APIs—that is, mobile developers, architects, and LoB IT teams. Figure 1.5 shows how the different personas work with the different layers of the API Management capability.

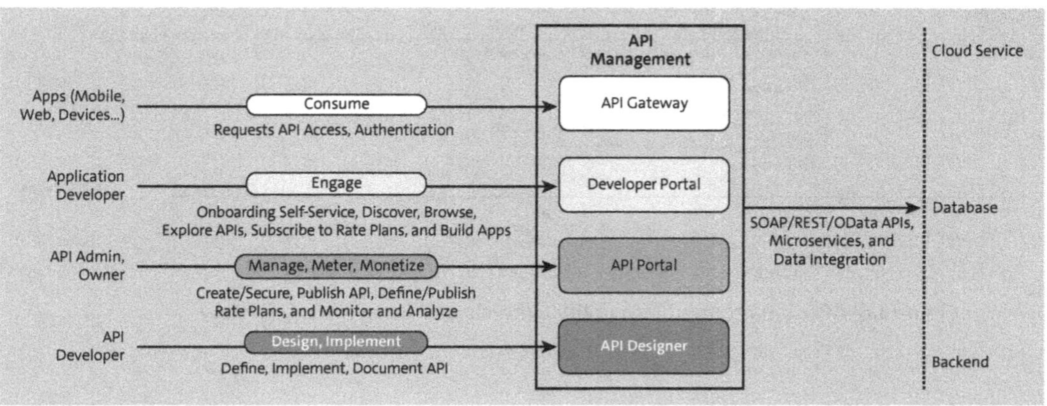

Figure 1.5 API Management Personas

The foundational layer of the API Management capability is the API designer layer, in which the API developer conceives, writes, and implements the functional API. Next is the management layer, where the API administrator can catalog, manage, and put in place policies relating to API usage in the API portal. The next level is the engagement layer of API Management, where application developers can gain access to and leverage the API through the developer portal. The top layer is the consumption layer, where associated applications leverage the API via the API gateway to access and utilize specific pieces of information.

Figure 1.6 shows the lifecycle aspects of APIs in the context of the API Management capability combined with personas, underlying source systems, and resulting use cases that cover an unlimited number of digital interactions both inside and outside an organization.

The API Management capability provides end-to-end abilities, starting with design and moving through implementation, management, analysis, monetization, and engagement. At each stage of the API lifecycle, comprehensive capabilities are made available through the API designer, API portal, API gateway, and SAP API Business Hub, ultimately serving the entire spectrum of use cases (mobile, marketplace, partners, and IoT). The API Management capability can provide these abilities universally across on-premise and cloud-based environments.

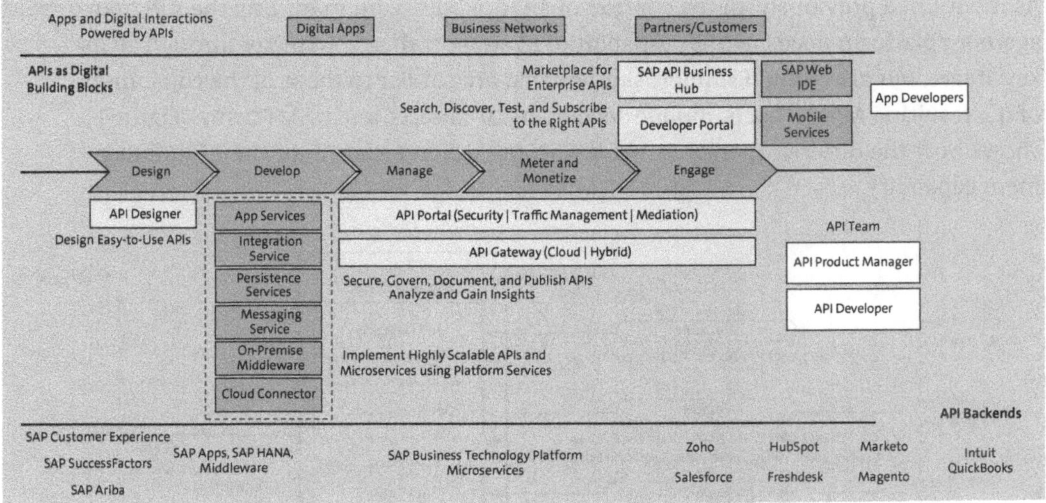

Figure 1.6 API Management Capability Lifecycle and Ecosystem Overview

1.6 SAP Integration Suite

So far, we've been exploring aspects of integration as they relate to your current circumstance. We've covered where your organization sits with regard to integration and why it matters to evaluate your current approach versus what's possible by taking a new approach and how you can have a rapid impact on your organization. Now we'll start exploring how SAP Integration Suite can provide you with real value and ensure you realize the associated benefits of a strategic approach to integration.

SAP has a deep heritage in the integration space, starting with SAP Exchange Infrastructure, which was launched at the end of 2002 (a precursor to SAP Process Integration and SAP Process Orchestration). The idea was to simplify how SAP connected with other business systems and applications to enable sharing of information between SAP and third-party systems.

Further Resources

For more information about SAP Process Integration and SAP NetWeaver, you can visit the following blog posts:

- *http://s-prs.co/v532609*
- *http://s-prs.co/v532610*

SAP went on to introduce its first cloud offering in 2012 with SAP HANA Cloud Integration, which was subsequently renamed for SAP Cloud Platform Integration and then, in 2020, SAP Integration Suite. From its inception, the primary driver for SAP Integration

Suite has been to harness the lessons learned from previous product generations and ultimately to provide users with the best possible combination of functionality and ease of implementation.

One of SAP Integration Suite's guiding capabilities has been to leverage prepackaged process integration content. SAP Integration Suite therefore features a centrally managed catalog of plug-and-play integration content for its user base (which we'll discuss in detail in Chapter 3). This prepackaged content is continually updated and managed by SAP, enabling users to benefit from best practices and optimal performance. This prepackaged approach enables users to quickly implement their needed integration flows, with the added benefit of being able to make changes or build their own integration flows as needed.

Over time, SAP Integration Suite has grown to include many different types of integration. This is representative of the different integration technologies desired by the typical organization. In the SAP Integration Solution Advisory Methodology developed by SAP that we discussed in Section 1.1, different technologies have been developed to accommodate the different needs of the underlying data characteristics and business needs. For example, process integration as a category typically handles transactional exchanges between applications, whereas data integration tooling tends to focus on batch migration or replication of large amounts of data. SAP has developed end-to-end capabilities around event-based architectures, API-led strategies, and specialist solutions around IoT or e-documents. Grouping these technologies is logical as every organization requires them in order to service internal and external needs—plus, in certain situations, it's important to leverage these complementary capabilities in combination.

Now, let's begin our journey into SAP Integration Suite with an overview of its key capabilities and offerings.

1.6.1 Integration Types and Data Sources

Within most organizations, there is a need to gain access to and view data sets residing in different sources. Typically, users are seeking a complete or unified view of the combined data set. For example, a company could want to see a set of monthly results sourced from separate and distinct databases. There is also a lot of use of data integration technologies in relation to big data. Other data repositories include data warehouses and cloud data stores. Machine learning is another area where large amounts of data need to be leveraged and analyzed and insights gleaned through the resulting models.

There are two main methods of extracting data from data sources: extract, transform, load (ETL); and extract, load, transform (ELT). These methods migrate large amounts of raw data that needs to be transformed in order for it to be meaningful. This transformation involves activities like sorting, joining, reformatting, filtering, merging, and aggregating. These migrations may occur on a repeating basis and be tracked as data pipelines. ETL and ELT essentially help break down data silos and help data scientists glean insights across broader data sets (business intelligence). ETL is the more traditional approach, with ELT taking off in recent times. ETL requires more interaction at each stage of the data transformation. ELT leaves the hands-on activities until the end, taking advantage of modern trends in enterprise computing centered on cloud concepts (e.g., scalability, low storage costs, and in-memory processing).

The common example of monthly reporting across multiple data sets leveraging data integration capabilities is an area where additional integration capabilities may be needed. For example, data analysts may wish to supplement these data sets with additional sources. These sources could be application-based transactions; external sources such as weather, traffic, and social media; and other types that reflect environmental factors that may influence business performance not captured from the original data sets.

According to IDC, 75% of organizations will be leveraging external sources in combination with internal sources by 2021 (*http://s-prs.co/v532611*). These additional sources may be leveraged via process-based integration flows (where logic is needed) or possibly by a straightforward API call. It's also possible, with the growth of event-based architectures, that certain types of events may be subscribed to in order to capture a comprehensive point of view of the business.

1.6.2 Event-Based Integration

When we discuss data-, process-, and API-based integration flows, we are typically referencing point-to-point integrations. Although prepackaged content and building blocks can make the operation and maintenance of these integration flows easier, they fundamentally require a central integration team to oversee, manage, and optimize over time. This can lead to delays, increased governance, and cost. Therefore, there have been efforts in the IT world to augment these integration efforts with an event-based approach. Point-to-point integrations are still needed, but an event-based approach works well with especially chatty applications for which the underlying business involves a lot of transactions.

Figure 1.7 shows how event-driven innovation is leveraged by the intelligent enterprise in terms of powering situation awareness, real-time decisions, automation, and dynamic ecosystems.

The event-based approach is a decentralized concept. The concept is based around multiple systems subscribing to an event bus. These subscribing systems can specify the types of events they are interested in receiving alerts on and therefore ignore the events they are not interested in.

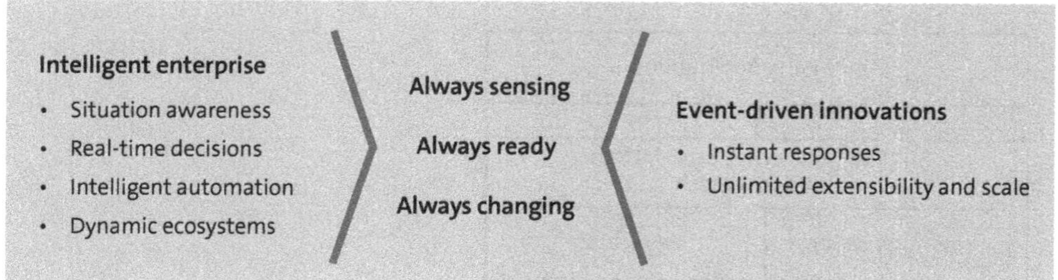

Figure 1.7 Event-Driven Innovations and the Intelligent Enterprise

Figure 1.8 shows how SAP Event Mesh can operate as an event bus, receiving event messages from publishing applications and sending event messages to subscribing applications.

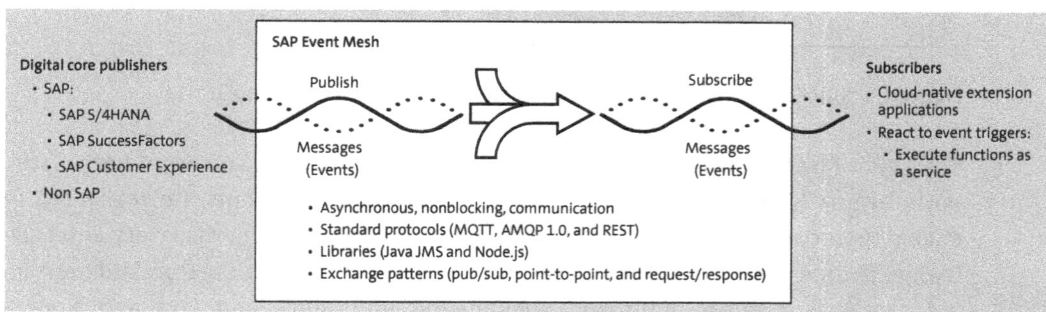

Figure 1.8 SAP Event Mesh as an Event Bus

At the same time, there are networked systems that are publishing the subscribed events. *Pub/sub* is the common shorthand for this type of integration flow. The messaging network captures which systems are interested in certain events and ensures the events are delivered to the subscribing systems.

The key capabilities of SAP Event Mesh, shown in Figure 1.9, cover both an event and message broker compatible with SAP and third-party sources connecting applications and services at scale seamlessly. Events are captured by the event broker, and then messages are sent with summary details by the messaging broker. SAP Event Mesh supports high-availability clusters and provides elastic scalability. Messaging protocols supported include AMQP, MQTT, and REST.

Another approach for event-based architectures is not to pursue a pub/sub type of structure, possibly when real-time visibility is not required, and to instead implement an event stream. With the event stream, events are logged and recorded for all events. A client system won't receive the stream of events in real time but instead can at any time access the log and replay events in order from the stream to identify an order of events for analysis.

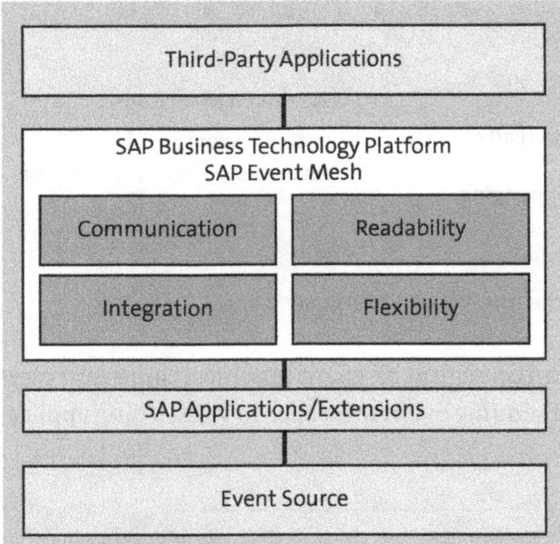

Figure 1.9 SAP Event Mesh Key Messaging Components

SAP Event Mesh provides both an event and a message broker. The event broker supports events from both SAP and third-party applications. Events are registered to enable their consumption and to provide a mechanism for event discovery and cataloging. The message broker supports asynchronous communication, standard messaging protocols, point-to-point and queue-based messaging, and Java and Node.js messaging clients. We'll discuss event-driven integration and SAP Event Mesh in more detail in Chapter 8.

1.6.3 Data Intelligence

We've established that the growth of data is exponential. Statista estimates that by 2024, 149 zetabytes will have been generated globally based on a compound annual growth rate (CAGR) of 26% (based on IDC 2018 figures; see *http://s-prs.co/v532612*).

Given this exponential amount of data growth, enterprises need to work smart to harness this deluge of information being collected and assessed. According to IDC, 60% of organizations will have established a dedicated data ops team by 2023 in an effort to improve the quality of the data being harvested and leveraged by the typical organization. Also, according to IDC, 50% will adopt a data supermarket approach for unifying data storage, access, and governance. Finally, IDC says 70% of enterprises will have integrated their data sources in order to deliver real-time action capabilities (based on IoT, analytics, and machine learning) by 2024 (*http://s-prs.co/v532613*).

Given this change in approach, it's important to look at the tools available to take a building block-based approach. SAP Data Intelligence provides building blocks for accessing and connecting with data, governance and analytics, preparation and labeling, model development, and deployment with common scenarios. This end-to-end approach eliminates redundancies, provides enterprise-wide visibility, and ultimately increases the productivity and efficiency of the subject matter experts through operations cockpits, exploration tools, and modeling.

Figure 1.10 shows how SAP Data Intelligence fits with the rest of SAP Integration Suite. So far, we have discussed how the API Management capability and SAP Event Mesh form major pillars of the suite's capabilities. As SAP Integration Solution Advisor Methodologies describes, there are different integration capabilities available for different use cases. Now let's discuss the role of data intelligence.

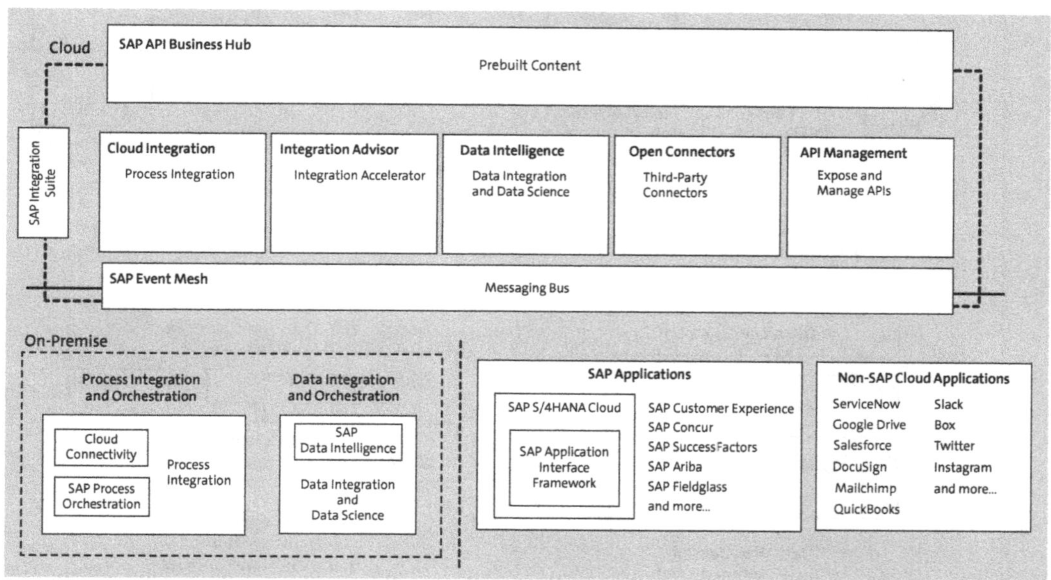

Figure 1.10 SAP Data Intelligence within SAP Integration Suite

Figure 1.11 shows the four main use cases for SAP Data Intelligence:

- **IoT ingestion and orchestration**
  For IoT ingestion and orchestration, SAP Data Intelligence enables you to tackle the challenges of integrating and analyzing vast quantities of raw sensor data and events from disparate semistructured sources, with low-level semantics and no business context.
- **Data science and machine learning**
  For data science and machine learning, SAP Data Intelligence provides a unified tool for data ingestion and processing, leveraging machine learning and advanced analytics algorithms on any mix of engines, both SAP (SAP HANA Predictive Analysis Library, SAP Data Intelligence functional services, etc.) and non-SAP (Python, R, Spark, TensorFlow, etc.), using machine learning models and prebuilt functionality.

- **Data warehousing**
  For data warehousing, SAP Data Intelligence enables the acquisition of new data sources combined with previously siloed data from traditional data warehouses, data marts, enterprise applications, and big data stores.
- **Business application transformation**
  For business application transformation, SAP Data Intelligence streamlines innovation initiatives around business applications by operationalizing machine learning algorithms into enterprise processes, democratizing their usage beyond the data science team.

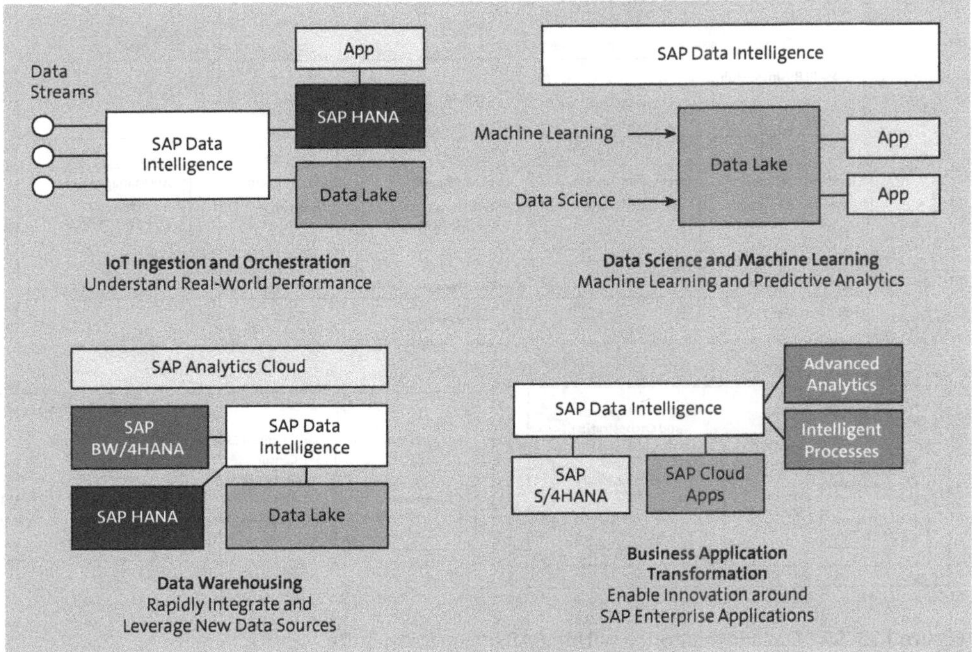

Figure 1.11 SAP Data Intelligence Use Cases

1.6.4 Process Integration

The core functional element of SAP Integration Suite is process-based integration. Process-based integration manages the transfer of information from one application to another. Within an organization there are an increasing number of applications being leveraged. Currently the average organization utilizes 450 applications, with annual growth in the 20% range.

Note

You can see a breakdown of applications used by different company sizes at *http://s-prs.co/v532614*.

As a result, the need for prepackaged integrations between applications, with their plug-and-play capabilities, is only increasing. The underlying reasons for keeping applications in sync with each other can vary. One common use case is obtaining information from a system of record. For example, many companies leverage both ERP applications and customer relationship management (CRM) applications to manage their business. The CRM system may be the system of record for customer and prospect data (addresses, quotes, customer interactions, etc.), whereas an ERP system will record customer transactions, billing, inventory availability, and the like. It's therefore common to exchange information between ERP and CRM applications.

Other application examples can be LoB-focused. For example, a human resources (HR)-focused application can be responsible for organizational reporting structures. When the ERP finance function is remitting funds to cover employee travel and expenses, the ERP system will often check HR reporting data to ensure the travel and expenses get charged to the right business unit or department.

Gartner identifies two other types of applications leveraged within an organization. In addition to systems of record, Gartner identifies systems of differentiation and systems of innovation in its Pace model (*http://s-prs.co/v532615*). *Systems of differentiation* help companies establish their value proposition for customers and partners by providing unique capabilities. *Systems of innovation* allow an organization to test new capabilities and functionality, working fast on proofs of concepts before moving them into the systems of differentiation.

In recent times, mobility has been a big area of enterprise innovation. Subsequently, in systems of differentiation, companies have built mobile-friendly applications that provide users with functionality that was previously only available on a desktop or central system. More recent innovations that are underway include the adoption of technologies such as RPA so that attended and unattended bots can help accelerate and digitalize processes that were previously analog or paper-based. Pick any industry, and you will find most companies fully implementing mobile capabilities and moving forward with RPA and AI-based initiatives to bring better customer experiences and improved productivity for their employees to the market.

As a result, the demands on the underlying systems of record and differentiation are witnessing unprecedented mobile and user demand. Leveraging prepackaged process integration capabilities can help ensure the resulting application communications are seamless, scalable, and efficient. A lot of organizations are leveraging process integration in combination with an API management layer that abstracts direct application calls, thus implementing a layer of policy and governance that helps mitigate potentially disruptive spikes in traffic requests and ensures a consistent approach from the ecosystems of application developers, who need timely access to information stored in the systems of record and differentiation.

An interesting area of application development at SAP is the adoption of the SAP One Domain Model. Over the past two decades, SAP has acquired a lot of LoB-specific applications that were developed in isolation. SAP is currently seeking to enhance these applications so that they form a seamless intelligent suite. Consequently, SAP will be working to implement reference architectures for end-to-end processes that span multiple applications. Related areas such as security and identity management and domain models will be aligned. As Rui Nogueira described it on his blog (*http://s-prs.co/v532616*):

> *"SAP One Domain Model will become the lingua franca used by the business applications of the integrated intelligent suite to semantically understand business objects in a uniform way."*

In other words, common concepts and terms will be handled in a consistent manner across applications. A customer will be named a business partner by all SAP applications, and the same approach will be taken for all other terms such as workforce person, cost center and more. The goal of SAP One Domain Model therefore will be to simplify process integrations between applications. This will consequently lead to simpler master processes across SAP applications such as lead-to-cash, hire-to-retire, source-to-pay, and design-to-operate.

SAP One Domain Model will not eliminate the need for process integration tooling, but it will effectively make plug-and-play integration even easier. Process integration will still be needed by large companies that typically leverage custom fields, terms, or processes, which need to be accommodated when sharing information across applications. Another area where process integration will be needed is integrating an SAP One Domain Model application with a third-party application. As we discussed previously, enterprise application utilization is growing around 20% per year, which is leading to an unprecedented amount of landscape heterogeneity as companies pursue innovation, differentiation, and efficient record keeping.

1.7 Related and Complementary Solutions

So far, we have been reviewing core integration suite technologies. Now it's time to look at adjacent technology that holds interest for integration experts. We'll discuss IoT and last-mile integration in the next sections.

1.7.1 Internet of Things

Gartner reports that IoT integration in general within the enterprise space is well on its way to being broadly implemented, with a whole host of specialist IoT technologies coming down the pipe: connectivity services, IoT platforms, IoT edge architecture, and event stream processing, to name a few (*http://s-prs.co/v532617*).

IoT is formed by connected machinery or objects. These items can be stationary or mobile; they can be part of a connected production process or isolated. They can be located in office locations or in remote areas. The IoT device may be constantly connected, giving continual updates, or may only communicate sporadically. Ultimately, the IoT device communicates important information about its operation so that the business has visibility into and can respond according to the priority/status afforded to those updates.

As IoT devices become more prolific and common in the enterprise landscape, the need for IoT integration with other enterprise systems and processes becomes greater. For example, for an oil and gas company, it is extremely important to keep field production systems active. Any outage or stoppage can cost the company a lot of money for every minute that associated production is stopped.

Therefore, in this situation, IoT data can be leveraged in a predictive manner to identify production-affecting issues before they occur. The IoT feed by itself does not complete the business process; the IoT data needs to be fed into other systems and applications to ensure that the operations team gets alerted. That team in turn will leverage application data centered on resourcing, scheduling, matching resources to expected work, and more. Take any enterprise working in different industries, and you will find use cases in which IoT device feeds are being increasingly leveraged to digitize and automate processes that were previously manual and analog (in the oil and gas industry, for example, previously a periodic site visit would have been needed to assess the field equipment status).

Figure 1.12 shows how IoT use cases and benefits can be grouped around improved business performance, greater efficiencies, and improved customer service. IoT data feeds are able to power business processes and capabilities that simply haven't been possible before.

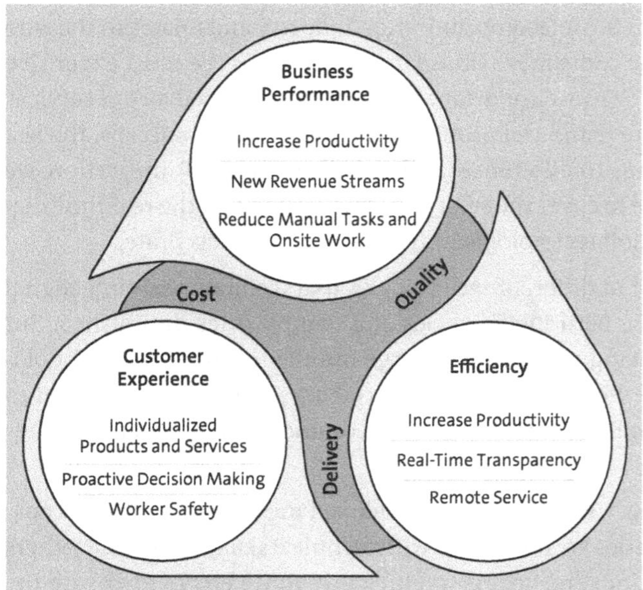

Figure 1.12 IoT Market Drivers and Dynamics

For example, employees can be empowered with contextual information, customers can receive personalized experiences, and intelligence can be added to assets, products, and factory equipment.

The SAP Internet of Things (SAP IoT) offering provides enterprises with capabilities that enable users to build, customize, and operate IoT-themed applications that support industrial IoT use cases. As the IoT layer of SAP BTP, it enables users to reimagine their business processes and models with embedded IoT services and data. In turn, SAP IoT data services enable customers to leverage IoT-enriched business insights that can be embedded in and extend business processes that deliver transformative outcomes.

SAP IoT includes reference architecture that offers business services and specific capabilities that relate to manufacturing, supply chain, logistics, assets, products, and service management industries. In addition, SAP IoT features built-in big data and analytics capabilities, as well as the ability to enrich IoT data with business context, rules, and an event-driven integration framework.

SAP IoT connects IoT devices to the broader SAP BTP to provide scalable ingestion of IoT data and device management. The service provides a secure connection to remote devices using a broad variety of IoT protocols and manages the device lifecycle from onboarding to decommissioning. SAP IoT collects and processes sensor data at scale at the edge or in the cloud and stores it in SAP BTP for use by other applications. Moreover, SAP IoT provides a multitenant architecture, allowing role-based access to device data through easy-to-use APIs.

1.7.2 Last-Mile Integration

The term *last mile* comes from the telecommunication industry and relates to the wire from the telephone poll to the consumer's house. This last mile is the most expensive part of the business but also the most important as it facilitates the delivery of services and therefore the earning of revenue and profits. In terms of the IT landscape, the last mile can mean different things to different stakeholders. In terms of integration, we will be using the term *last mile* to cover the safety net of ensuring that the real-time digital transactions that integration technologies are empowering are accurate.

So far, we have discussed a lot of different technologies that facilitate real-time digital transactions in large volumes, with inefficiencies and complexities diminished. But with all these transactions taking place across a large number of users and technologies, how do you ensure that the digital transactions taking place are accurate and in line with service-level agreements (SLAs), all while accommodating frequent changes and updates?

Vendors in the integration space have therefore developed message-monitoring capabilities. Some of these capabilities are developed with technical skillsets in mind, others with business personas in mind. The goal is to efficiently route errors to ensure the underlying business is not disrupted. In addition, automation capabilities are built in based on rules to proactively address common errors to reduce the burden on staff.

In the case of SAP, a complementary technology related to our integration offering is SAP Application Interface Framework. SAP Application Interface Framework is a means of catching errors and automating the change management process with role-based alerts and audit capabilities. This is an important tool for ensuring that the great work implemented via a seamless integration strategy is not undermined by human or business error. Figure 1.13 shows how SAP Application Interface Framework works with SAP Integration Suite at a high level, with SAP Integration Suite providing information from cloud and on-premise applications. SAP Application Interface Framework sits on top of the integration layer, providing a user experience for both integration developers and business users. The user experience provides the business personas with the ability to monitor and address errors while also providing developer capabilities for customizing interfaces. Once the interface issues have been addressed, the data can be exchanged with the underlying SAP S/4HANA or SAP Business Suite application.

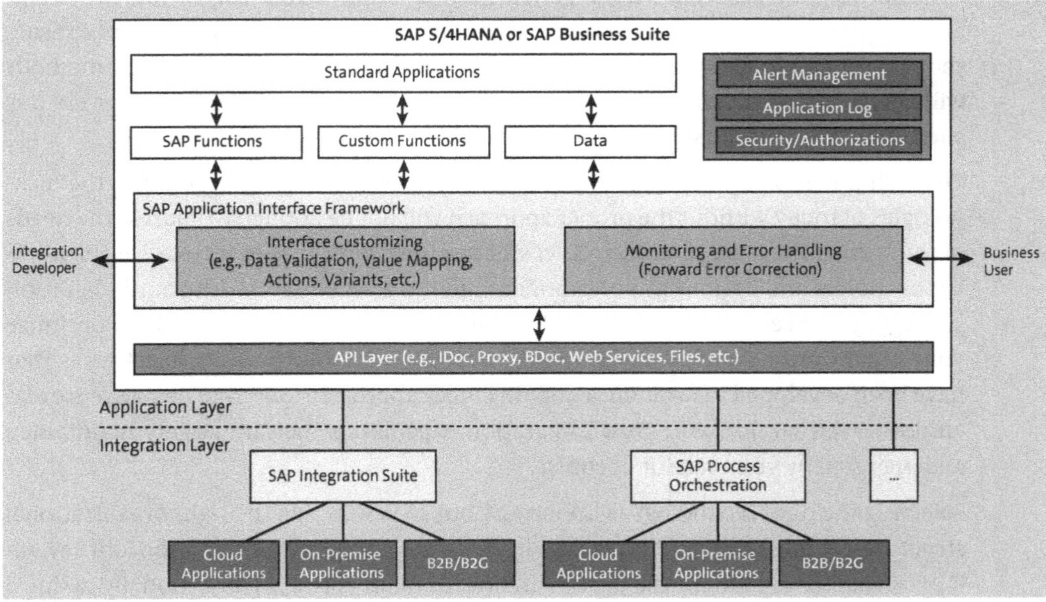

Figure 1.13 SAP Application Interface Framework

1.8 Summary

In this chapter, we discussed that it's important to recognize where your organization currently sits with regard to integration and where you should be. Once you have established the gap between where you are and where you want to be, you should determine the most effective way to get there. In response to conversations with its users, SAP has developed and made available the SAP Integration Solution Advisory Methodology and its associated community to help you gain insights, see best practices, and explore how the methodology can be applied to your organization.

The desired path for many organizations is to be sustainable, resilient, and profitable in the long term. To achieve these objectives, a company needs to be efficient, agile, scalable, secure, and innovative. SAP has outlined how companies can achieve these objectives through the concept of an intelligent enterprise. By embracing SAP software, companies can achieve best-run integrated business processes. SAP Integration Suite can help ensure that you achieve the benefits of an intelligent enterprise across legacy and new areas of your business, across SAP and third-party applications, and across all your internal and external stakeholders.

The business world is constantly changing, and the resulting business models and support mechanisms are more heterogenous than ever before. It is anticipated that the enterprise IT landscape will continue to diversify, with implications for improved company performance and associated complexity. Cloud computing is here to stay and will continue to grow; associated intelligent technologies will also become more prevalent; and the pressure to automate and operate in real time will be unrelenting. As a result, the case for a strategic approach to integration over traditional point-to-point methods will only grow bigger, with the consequences for delaying growing in impact every day such an initiative is pursued.

The technologies of yesterday are not enough to address the needs of today; the technologies of today without the proper approach will not be enough to address the needs of tomorrow. The need to work and act smarter only continues to increase as the number of choices and options grows exponentially. To be successful with your integration strategy, you need to select and implement capable technologies that help optimize your organization's efficiency and related capabilities. You need to select tools that have been developed to facilitate a building block approach, tools that enable reuse and empower self-service; you need integration capabilities that are largely maintained and operated by your vendor of choice.

Selecting the right technology is important, but so is selecting the right organizational structure. Establishing an ICoE can be a key step in a company's integration journey. An ICoE is able to implement the desires and needs of an entire organization. By taking a shared services approach, the ICoE can achieve the balance needed between an integration platform for providing frictionless data access and use and helping stakeholders achieve a self-service standing with supporting guidance regarding best practices and standards to help achieve optimal results.

With these considerations in mind, we'll now shift our focus to SAP Integration Suite in the next chapter.

Chapter 2
SAP Integration Suite at a Glance

SAP Integration Suite is SAP's enterprise integration platform as a service, offering capabilities to connect applications, processes, people, and devices. In this chapter, we'll begin with an overview of the suite and the steps to get started.

As the integration layer of SAP Business Technology Platform (SAP BTP), SAP Integration Suite simplifies integration and harmonizes processes to provide an enterprise-grade integration experience. With it, an organization can integrate business processes across SAP and third-party applications, connecting multiple landscapes (cloud, on-premise, private cloud, etc.), creating new business models, and executing industry transformations.

Integration is the foundation of an intelligent enterprise, and companies need comprehensive, flexible, reliable, scalable, and intelligent integration platforms. SAP Integration Suite follows four core principles laid out in SAP's integration strategy:

- **Out-of-the-box integration**
  Companies are looking for integration platforms that can help them to innovate quickly and integrate business processes in order to realize the real value of cloud deployments. SAP Integration Suite provides 2,000-plus out-of-the-box integration scenarios across business processes like hire-to-retire, procure-to-pay, order-to-invoice, and so on. Prepackaged content includes the integration business logic that is required to communicate between SAP and third-party applications. As actual business processes may be very complex, extension points are provided to allow companies to perform their specific business configurations. Companies have to provide the technical parameters (like URL, hostname, security credentials, etc.) and optional extension configurations for applications to be connected and deploy this content. This saves lot of implementation time and efforts for companies and helps them go live with their productive scenarios and cash in on the benefits of the cloud. Prepackaged integration content reduces the total cost of ownership (TCO) for developing and maintaining integrations as regular updates and upgrades are provided by SAP.
- **Open integration**
  Companies that are starting their cloud journeys have multiyear roadmaps and assorted application landscapes that include SAP S/4HANA, SAP ERP, and many third-party applications deployed on premise, in the cloud, in private data centers,

or hosted on different infrastructure providers. SAP provides an open integration strategy with application programming interfaces (APIs) from different line of business (LoB) solutions published in SAP API Business Hub, creating harmonized access to various business functions. SAP Integration Suite provides companies with a comprehensive, enterprise-grade integration solution that can easily connect all SAP applications and more than 160 third-party applications.

- **Holistic integration**
  Integration has many variants within a company landscape. Companies need to connect to their employees (within their application landscape), to business partners like customers, partners, and logistic providers, and to government agencies to meet compliance and regulations requirements. This requires different technical protocols and security requirements, plus strong governance. Companies want a central repository of APIs that can bring business data from various SAP and third-party applications together and offer an open and central API environment to end users for internal and external traffic. SAP Integration Suite provides a flexible and cloud-based platform to execute and automate these diverse integrations in a secure, reliable, and scalable environment.
- **Enterprise-grade integration**
  Companies need a reliable platform to exchange business messages in a secure way. In addition, for enterprise-wide use, they are looking to reduce the overall cost of development and operations. SAP Integration Suite provides a reliable, secure, and scalable platform to run these enterprise-wide integrations. It provides a low-code/no-code-based configuration approach, machine learning–based integration flow suggestions during design time, and automatic business mappings with the Integration Advisor improving overall developer productivity. Transportation lifecycle management, automated test samples during design time, APIs to realize a continuous integration/continuous delivery (CI/CD) pipeline, and integrations with various analytical tools and identity providers ensures enterprise-grade operations

In this chapter, we'll introduce SAP Integration Suite, describe its key principles, capabilities, and supported integration patterns, and discuss various aspects of operations. In addition, we'll also walk through the step-by-step process to activate a trial landscape in SAP Integration Suite.

2.1 Supported Integration Patterns

Let's look at the key integration patterns supported by SAP Integration Suite.

To begin, *application-to-application* (A2A) integration is generally used when companies need communication between various business applications within their landscapes (see Figure 2.1) in order to achieve business integration—for example, design-to-operate, lead-to-cash, and hire-to-retire. It includes integration within SAP applications or third-party applications across on-premise, cloud, and hybrid landscapes.

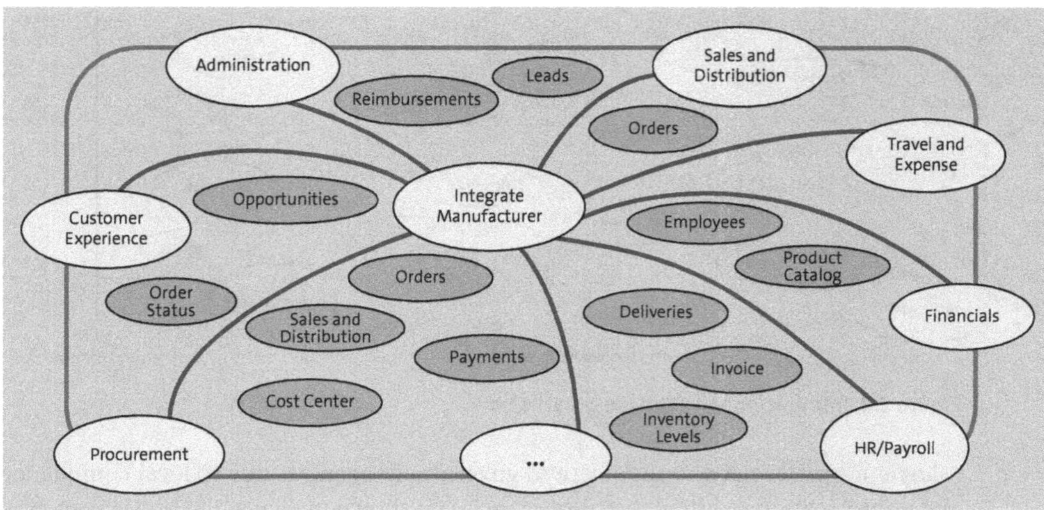

Figure 2.1 Organizational Internal Landscape

Next, bulk business transactions across organizational boundaries are defined as *business-to-business* (B2B) transactions, as shown in Figure 2.2. Organizations need to exchange transactions (e.g., orders, deliveries, shipments, invoices, etc.) with business partners (e.g., customers, vendors, logistics providers, warehouses, etc.) with predefined industry or region-specific guidelines. This incorporates various Electronic Data Interchange (EDI) formats (see Chapter 6) and a standard framework to acknowledge the messages. Procure-to-pay and order-to-cash are widely used business transactions for trading partners, customs, and shipment information with the logistics provider. This is a chain of processes across industry verticals—for example, manufacturing, retail, telecom, and so on. From a technology perspective, this requires support for multiple B2B EDI formats (multiple versions), a framework to automatically generate acknowledgments (e.g., EDI 997 and EDIFACT CONTRL messages), and adherence to SLAs signed between companies and their trading partners. In addition, a record of messages should be available for audit and compliance reasons for many years to provide legitimate proof of business transaction exchanges with trading partners. With the new modernizations, B2B integration is also shifting toward new components like APIs to share and exchange information in real time (e.g., inventory data, order status, etc.).

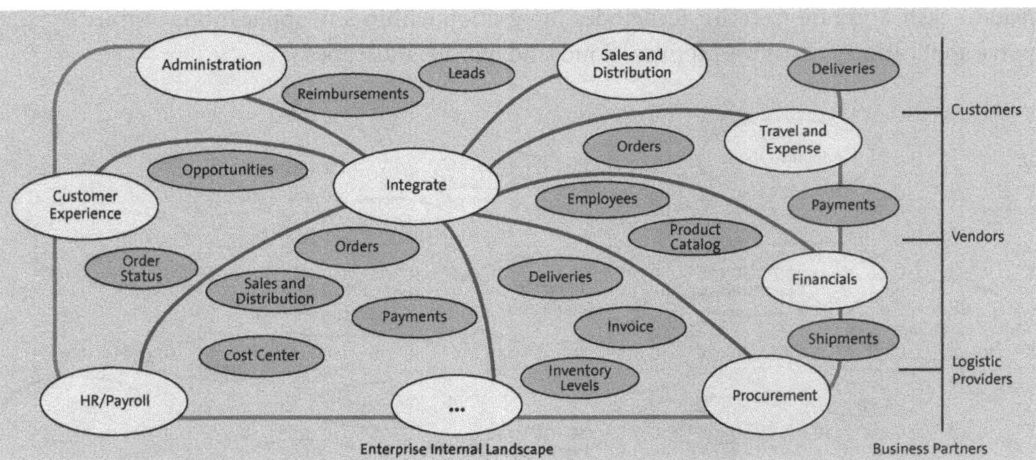

Figure 2.2 Integration with Business Partners

Many enterprises need to send data to government agencies due to local compliance and e-information required by those agencies, as shown in Figure 2.3. This is known as *business-to-government* (B2G) integration. It includes e-invoicing, payroll, tax, wages, and the like across different LoBs, like logistics, financials, human capital management (HCM), and others in different countries. Required information and security requirements vary across regions due to different governing laws. Organizations need to report the data in a timely and reliable way to meet regulatory guidelines. SAP Integration Suite provides a lot of prepackaged integration content for various country-specific e-government requirements, including goods and service tax (GST) for India, Single Touch Payroll (STP) for Australia, Pan-European Public Procurement Online (PEPPOL) for Europe, e-invoicing/value-added tax for Spain, Hungary, Chile, Italy, Portugal, and many more.

Citizen integration, meanwhile, empowers nontechnical business users to increase productivity and operate integrations seamlessly without a technology understanding. The tools are flexible enough to provide business users (with business domain and data understanding) with a required framework to design, view, and monitor business requirements. The IT department offers the solution to business users as a self-service that requires no intervention from IT teams. For example, SAP Integration Suite is used as a technology backbone by SAP Ariba Cloud Integration Gateway, which provides easy, nontechnical onboarding for trading partners (buyers and suppliers) via its partner directory. The directory defines partner identities, business transactions, and required message transformations. Although end users use a business-specific interface to onboard partners technically on the Ariba Network, all integration scenario configurations are automatically deployed in the backend without end users needing to understand the technology behind it. SAP Integration Suite is also utilized by other similar solutions offered by SAP, such as Peppol eDelivery Network and SAP Information

Collaboration Hub for Life Sciences. Another example for citizen integration is the Integration Advisor, which helps business experts develop message definitions for exchange business data, which we'll discuss further in Chapter 7.

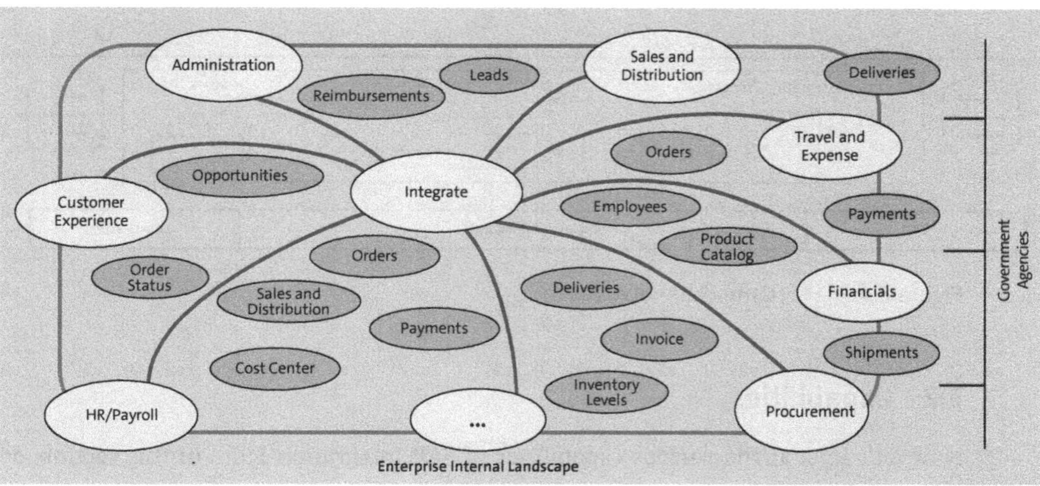

Figure 2.3 Integration with Government Agencies

Finally, companies are using APIs to abstract the complexity of heterogenous landscapes in order to provide a simplified API fabric for *omnichannel access* (mobile apps, web apps) and bimodal agility, as well as create new channels and revenue streams (through business model innovations and pay-per-use models), as shown in Figure 2.4. Increasingly, companies are building digital apps (web apps, mobile apps) for customers, employees, and partners. All modern apps use APIs for backend integration, as well as third-party apps for data and processes. With increasing demand for digital transformation and to be competitive in the market, organizations want to provide an enriched customer experience and real-time information access to business partners. There is a need for an IT department within an organization to reduce costs and to meet dynamic demand from business users for the right information and data access. IT departments have a huge responsibility to plan and renovate their landscapes with the available budgets. The information required is generally scattered across multiple systems and applications and present in different landscapes (e.g., on-premise, cloud, etc.). This information needs to be derived and filtered from various systems. Access needs to be provided to users with strong governance and security. SAP Integration Suite, along with connectivity and the API Management capability, helps set up this architecture with easy-to-design tooling and a reliable and scalable runtime. Moreover, the same access is used to create new revenue and monetization models to charge business partners for information access or for IT departments to implement charges back to business units based on usage identified by API calls.

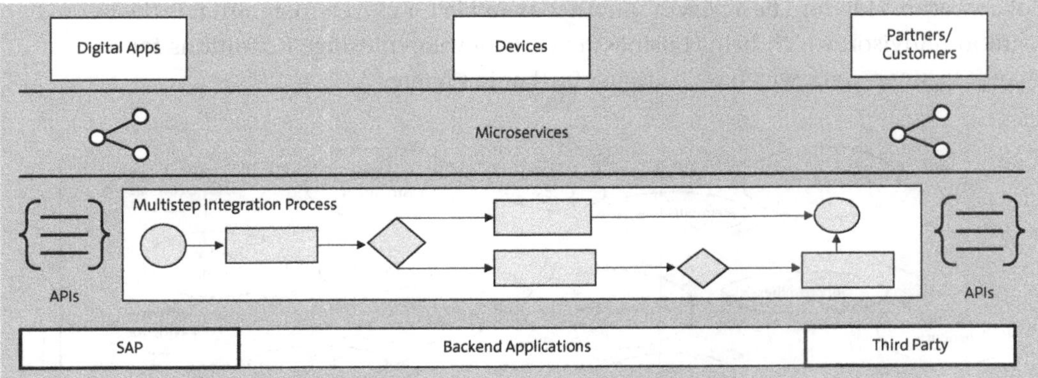

Figure 2.4 Digital Omnichannel Access

2.2 Capabilities

Now let's look at the various capabilities of SAP Integration Suite in this section, as shown in Figure 2.5.

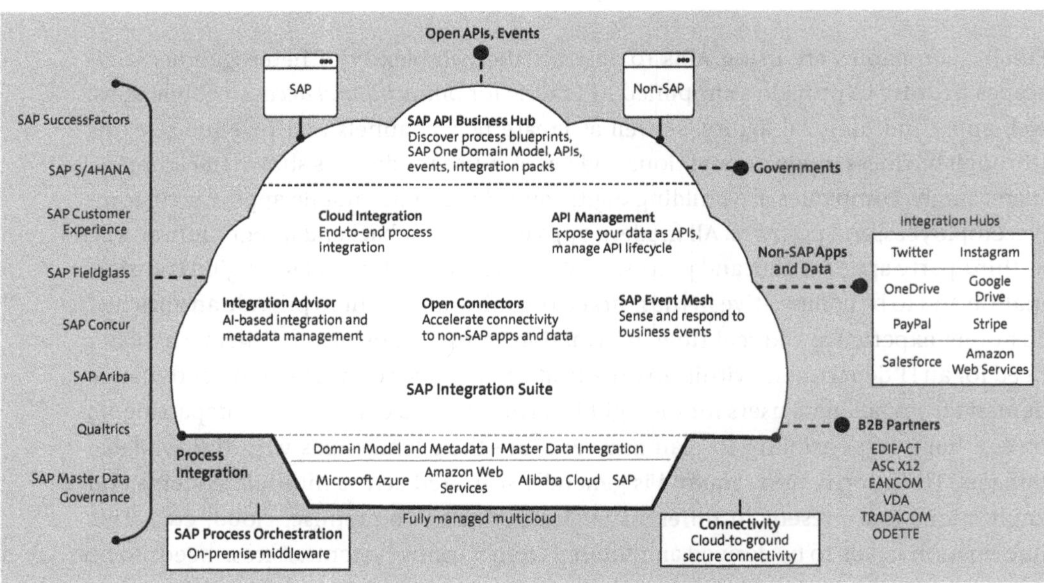

Figure 2.5 SAP Integration Suite

The key capabilities shown are as follows:

- **Cloud Integration**
  Companies are transitioning to the cloud. There will be an era of hybrid landscapes (a combination of business functionalities in the cloud and on premise) that will last for

many years. Companies need to integrate these different systems and applications in hybrid landscapes developed and maintained on different technology stacks, following different security standards, and with different business interface specifications.

The Cloud Integration capability of SAP Integration Suite enables enterprises to connect these heterogenous applications quickly in a seamless manner irrespective of landscape. It offers a web-based development environment to design, configure, and monitor integration processes. The design environment provides a flexible framework to create complex integration processes from the beginning using process steps. These process steps include security, message routing, message transformations, storage, queuing, and more. The Cloud Integration capability includes a lot of SAP and non-SAP connectivity adapters, such as JDBC, SFTP, OData, web services, email, SAP Ariba, SAP SuccessFactors, Salesforce, Microsoft Dynamics, social media, and so on. From the same web interface, integration scenarios and message processing can be easily monitored with advanced debugging options. Monitoring can also be integrated with other lifecycle management applications such as SAP Solution Manager, SAP Analytics Cloud, and SAP Cloud ALM, or custom dashboards can be created.

As a significant differentiator, SAP offers 2,000-plus out-of-the-box integration scenarios (as of the time of writing) with preconfigured business logic (integration flows) for various business processes, such as hire-to-retire, lead-to-cash, and more, across SAP and third-party applications. Prepackaged, extensible integration content helps you connect everything quickly, without the need for custom integration code. It helps realize the value of the cloud with a fast go-live and reduced effort for users. We'll discuss the Cloud Integration capability further in Chapter 5.

- **API Management**
  Enterprises are looking to transform, innovate, optimize, and scale. They want to enrich the customer experience and increase stakeholder satisfaction by automating business processes and executing new business models quickly. One of the key requirements in this transformation journey is to expose and consume business data via APIs in a governed and secure manner to provide omnichannel access to external and internal business users. The API Management capability of SAP Integration Suite provides a central layer to govern, manage, and meter APIs. It provides a code-free web-based framework to design new APIs, manage existing APIs, enrich APIs with 40-plus prepackaged security and access policies, make logical groupings using product catalogs, and offer APIs via a developer portal to the community. You can also associate rate plans with APIs to monetize API access. An API-based integration approach is about providing a plug point that can be used to support the creation of applications and extensions, as well as the integration of businesses. Companies can not only create an API fabric in their organizations but also have central access control for both internal stakeholders and external business partners. API analytics provides real-time monitoring of API performance, usage, and issues and

offers the flexibility to create custom reports. We'll discuss the API Management capability further in Chapter 4.

- **Open Connectors**
  Companies have data maintained in distributed applications throughout their landscapes. They may be using many satellite applications with their core ERP business systems for specific LoB solutions (e.g., Salesforce, ServiceNow, Workday, Microsoft Dynamics, Sage, Oracle, etc.) or may need to connect to social media, collaboration, and other open platforms. When realizing business process integrations, all these applications need to be connected, whether they're from SAP or third-party applications. Companies expect 360-degree integration to be achieved with their central integration platform as a service (iPaaS) solution.

  The Open Connectors capability of SAP Integration Suite provides 160-plus third-party connectors via a unified REST API layer, normalized authentication, error handling, and connectivity mechanisms. This ensures that developers can focus on building business integrations rather than learning about the technology used to integrate with third-party systems. The prebuilt connectors are grouped or categorized as hubs. Hubs provide a consistent resource URIs and are grouped into categories such as marketing, social, finance, field service, help desk, cloud storage, customer relationship management (CRM), human capital, payments, ERP, accounting, e-commerce, and more. The common resources functionality of Open Connectors enables companies to define a harmonized API, chaining multiple third-party adapters to reduce the cost of third-party integrations. This harmonized API can be consumed via the Cloud Integration capability and then mapped to and transformed into an SAP data model to orchestrate, mediate, transform, and route data from SAP to third-party applications. It also supports standardized events to simplify support notifications via polling and webhooks. We'll discuss Open Connectors further in Chapter 7.

- **Integration Advisor**
  Enterprises have a strong need to integrate with business partners. The majority of the integration effort in any implementation project is related to mapping business interfaces, which is executed by the functional consultant and technical developer. This is not only time-consuming but increases project costs as well. The Integration Advisor capability of SAP Integration Suite helps you accelerate development of your business interfaces and mappings. The Integration Advisor capability includes out-of-the-box content for SAP S/4HANA and various EDI industry standards. Message implementation guidelines (MIGs) help create interface specifications according to different types of systems (e.g., constants, occurrences, code lists, etc.). Mapping guidelines (MAGs) provide field-level mappings between two MIGs. Although you can create your own MAGs, proposals provided by Integration Advisor reduce your efforts a lot. Once the mappings are done, the system automatically generates the technical runtime artifacts for Cloud Integration capability. These include various pre- and postprocessing steps. We'll discuss the Integration Advisor further in Chapter 6.

- **SAP API Business Hub**
  SAP API Business Hub is a public catalog and central repository of all SAP and partner APIs, out-of-the-box integration content, events, API policies, core data services (CDS) views, and more. This helps integration and application developers to discover, experience, and consume the required assets for their integration projects. Business API definitions are available for SAP Business Suite, SAP S/4HANA, various LoB solutions such as SAP SuccessFactors, globalization, machine learning, and so on. Documentation for these APIs is provided in the OpenAPI Specification. Application developers can try out these APIs using sandbox environments without any need to subscribe to the applications. SAP API Business Hub also hosts 2,000-plus out-of-the-box integration packages that include predefined business logic to integrate many SAP and third-party business applications. Integration developers can discover and consume prepackaged accelerators and APIs via various capabilities in SAP Integration Suite, such as Cloud Integration and API Management, in an easy way. SAP API Business Hub also supports the end-to-end lifecycle of the delivered content. A central repository of integration assets helps enterprise architects create an integration architecture for their organizational landscape. We'll discuss SAP API Business Hub further in Chapter 3.

2.3 Operations

Operations are a key aspect for any solution. In this section, we'll look at operational concepts such as landscape and lifecycle management, monitoring, security, and provisioning. We'll also go over the steps to set up your SAP Integration Suite trial.

2.3.1 Landscape

First, let's explore the important aspects of running an SAP Integration Suite landscape. SAP Integration Suite is available to companies as a software as a service (SaaS) offering. Once a company subscribes to SAP Integration Suite, they can directly focus on designing their integration APIs and scenarios because the infrastructure (hardware) and platform components are operated and managed by SAP. SAP provides monthly upgrades and weekly maintenance cycles for SAP Integration Suite without disrupting a company's scenarios. The monthly updates offer new enhancements, shortening the time window for new releases so that companies don't have to wait for new innovations via long delivery cycles. Because SAP has data centers across the globe, SAP has created a time chart for the update and upgrade schedules based on various time zones. SLA information is given at *https://support.sap.com*.

In the following sections, we'll discuss the landscape setup, strategies, and lifecycle management.

Landscape Components

Let's look at some of the important concepts for setting up the SAP Integration Suite landscape. Figure 2.6 shows the landscape elements, including SAP Integration Suite tenants and their relationships. The relevant components are as follows:

- **Global account**
  Once you get into an agreement with SAP for using platform services, a unique global account number (unique identifier) is assigned to your contract. The global account includes entitlements, quotas, and resources for various services that you get under your contract. A global account administrator can create further subaccounts for actual services consumption and assigns the relevant service metrics quotas to respective subaccounts. A global account is used to manage subaccounts, members, entitlements, and quotas.

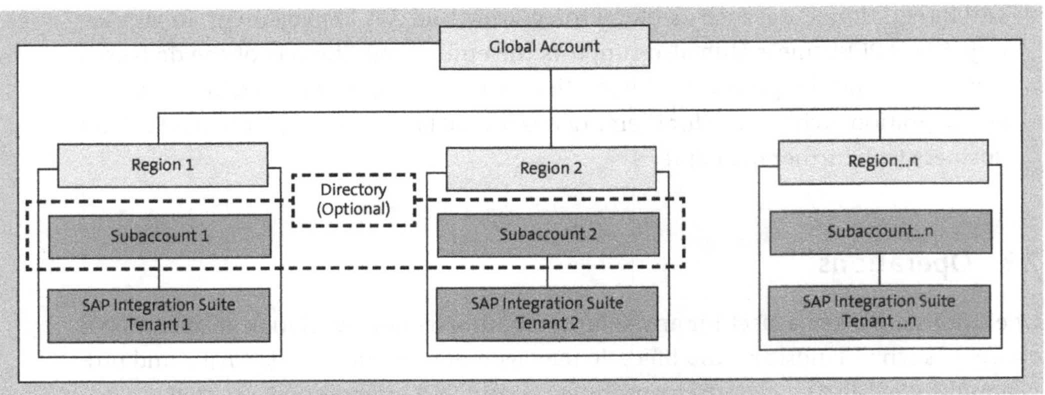

Figure 2.6 Relationships among SAP BTP Components and SAP Integration Suite Tenants

- **Region**
  The region corresponds to SAP platform data centers (SAP, Amazon Web Services [AWS], Microsoft Azure, etc.) available in different geographical areas. It denotes a physical location where your project scenarios and applications are hosted. You can select any region to provision your SAP Integration Suite tenants but cannot exceed the allocated quotas to various subaccounts. Technically you can enable various instances (as per your quota entitlements) in multiple data centers, but companies typically choose one data center to provision their entire landscape of different services, generally the one nearest to their location. We'll look at landscape examples in the next section.

- **Subaccount**
  One global account can have multiple subaccounts. Subaccounts are created in accordance with your project architecture and landscape requirements. Each subaccount is an independent entity with respect to members, entitlements, and security. Each unique subaccount can be created under one region only according to

the availability of an SAP platform. Each subaccount can contain multiple different SAP platform services per entitlement quotas assigned by the global account administrator.

- **Tenants**
A tenant is one instance of SAP Integration Suite. There's a one-to-one relationship between the subaccount and tenant. Each tenant includes all capabilities (depending on the licensed edition) of SAP Integration Suite, which can be used by companies. Companies can have multiple tenant entitlements based on their contractual agreement and landscape requirements. SAP Integration Suite is available as a self-service. Once entitlements are assigned to subaccounts, subaccount administrators can enable them via the SAP BTP cockpit. We'll look at the step-by-step procedure in Section 2.3.3.

- **Directory**
The directory is an optional governance and management layer between the global account and subaccount. A directory can contain one or more subaccounts but cannot contain other directories. Global account administrators can assign quotas to directories instead of directly to subaccounts. For example, if you have various subaccounts in your landscape, you can group relevant subaccounts in different directories according to various projects running in your organization.

Landscape Strategies

SAP Integration Suite is an enterprise iPaaS that can be used to execute different integration patterns and high-volume transactions in a reliable and scalable manner. While many companies have a central landscape, others have fragmented landscapes to meet their project requirements. Let's look at some examples:

- **Central landscape**
We have seen companies with anywhere between two and six tiers of landscapes in their organizations as part of a central, enterprise-wide integration strategy. This is required to meet their central governance policies and matching tenant requirements with their business application landscape. For example, if a company is using a three-tier (development, test, and production) ERP landscape, it will use a corresponding three-tier SAP Integration Suite landscape as well. Other tier examples include a separate sandbox environment to develop and run proofs of concept, provide trainings for their internal teams, and so on. Some companies also have a separate preproduction environment (between quality and production) as a safety layer to validate the deployments once again before transporting to production. The central landscape is mainly governed by IT departments irrespective of the applications getting integrated.

- **Project-based setup**
Companies have various implementation/transformation projects running within their organizations, and different budgets are allocated to various departments and

timelines established. These teams develop applications and solutions independent of each other. In this case, each have an independent SAP Integration Suite landscape for their respective project executions.

- **Business-critical applications**
  Some companies want to create separate landscapes for their business-critical integration scenarios. Examples include integration with business partners, government agencies, and financial transactions. As it may directly impact their business and compliance, companies define separate reliability, governance, and security procedures. Other noncritical scenarios run on a separate landscape.
- **Regional fragmented setup**
  Many companies with a global presence (businesses and subsidiaries in many regions/countries) have fragmented landscapes for their business applications, including multiple ERP deployments for their multiple regional landscapes. In these cases, companies need SAP Integration Suite tenants in the same regions to ensure close interactions with business applications, meet regional/local compliance requirements, and avoid latency issues.

Lifecycle Management

Because companies have multiple tenants in their organization for various technical, business, and regional requirements, there is a need to transport content between these tenants for content governance and lifecycles, as shown in Figure 2.7. There are multiple options available to transport content between different SAP Integration Suite tenants based on the different capabilities being used. You may need particular roles and authorizations set, depending on the content to be transported for each capability.

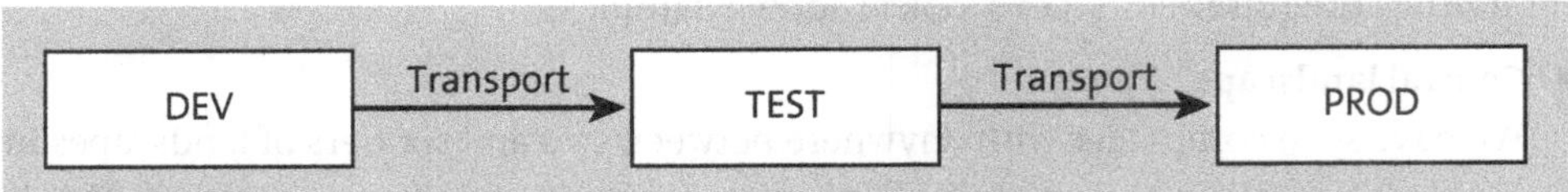

Figure 2.7 Lifecycle Management of Integration Content

Let's explore the available options:

- **CTS+**
  The enhanced Change and Transport System (CTS+) is the transport capability of the SAP NetWeaver Application Server for ABAP. This capability was enhanced (hence the +) for transport of non-ABAP objects, and it's supported by SAP Integration Suite's Cloud Integration capability. Companies running their ERP landscapes with CTS+ as the central transport management layer to transport on-premise objects want to use the same solution for transporting cloud objects as well. You can configure CTS+ in a direct mode to initiate the transfer between various tenants or download integration content in multitarget application archives (MTAR) format to provide it to CTS+.

- **SAP Cloud Transport Management**
  SAP Cloud Transport Management runs on SAP BTP and helps you transport content between various SAP BTP subaccounts, such as from developer to test and production accounts. Transport content is supported in the form of MTAR files and application-specific content, such as integration flow packages for the Cloud Integration capability. SAP Cloud Transport Management is a single application archive comprising multiple dependent and interdependent modules from various technology components that need to be together technically with the same lifecycle requirements.
- **Manual export and import**
  If your organization doesn't have CTS+ or SAP Cloud Transport Management, you can simply download the integration objects (such as API proxies via the API portal or integration flows in the design workspace) via a simple export/import option. Files are downloaded in ZIP format on your desktop and you can manually import them into the target tenant.

2.3.2 Security

Secure access to and transmission of business data is a must for all enterprises. This is one of the main criteria for choosing any enterprise software. Companies' business data is confidential, and they want strong governance procedures in place for information access via end users and automated transactions and have to follow security guidelines required to integrate with business partners and government agencies. There are various layers of security assessment provided by SAP Integration Suite:

- **Data centers**
  Data centers are a central structure used to host an organization's systems, connected with a strong network. SAP's own and its partners' data centers are geographically distributed to optimize traffic and reach throughout the world. Because they host multiple systems, telecommunication networks, and servers, security (physical access, data operations) is a key requirement. All data centers are well equipped with all key elements, such as physical access mechanisms, network, power supply, backup, building, and fire safety. Regular internal and external audits are conducted to ensure data center operations and safety. SAP data centers are International Organization for Standardization (ISO)/BS certified and conform to System and Organization Controls (SOC) 1, 2, and 3. These reports and certifications assure that an organization follows the right processes to develop, secure, and deliver operating services.
- **Secure infrastructure and access**
  SAP Integration Suite runs on SAP BTP and provides a multitenant environment, ensuring scalability and robustness. When a company subscribes to SAP Integration Suite, isolation is maintained on tenant and underlying infrastructure, such as virtual machines, memory, database schema, and so on. A tenant's runtime and data

access are controlled by infrastructure isolation, access policies, roles, and authorizations. This ensures that all data flowing through and stored in a company's tenant is completely secure. Tenant provisioning is supported by strong governance and access mechanisms (user provisioning, roles, and authorizations) that can be configured via the **Security** tab of the SAP BTP cockpit, as shown in Figure 2.8.

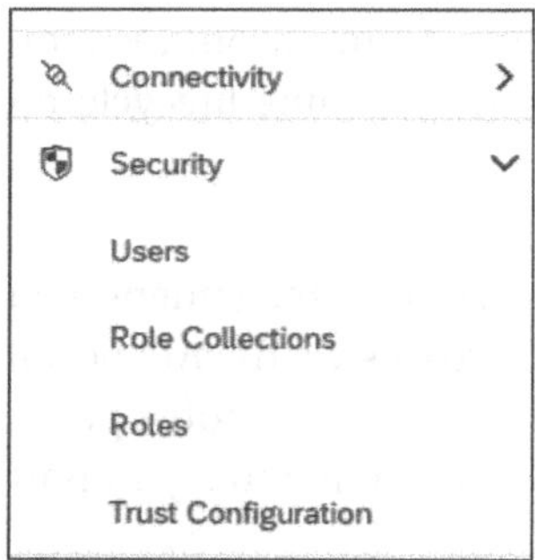

Figure 2.8 Security View of SAP BTP Cockpit

- **Users and identity provider**
  As a global account or subaccount administrator, you can create users for the various capabilities of the solution. You can create these users in the SAP ID service, which is provided as the default identity provider with all SAP BTP services. Any user with a valid P-user ID or S-user ID can be added to SAP ID service. You can get a P-user ID by creating an account on *www.sap.com*, and S-user IDs are provided to companies with valid contracts. S-user IDs are used to download software, create support tickets, do administrative tasks, and so on. S-user admins can create further S-user IDs based on various authorizations and access required. Instead of the default SAP ID service, if you want to use advanced identity federation capabilities using a separate tenant for SAP Cloud Identity Services or want to use any other identity provider, you can create trust using Security Assertion Markup Language (SAML) 2.0 with a different identity provider using **Trust Configuration**.
- **Roles and role collections**
  Roles include fine granular attributes and authorizations, created by an application to authorize certain functionality or part of that functionality. Roles are grouped together as a meaningful entity with **Role Collections**, which can be assigned to end users based on various responsibilities such as developer, administrator, business expert, and so on. When using the default SAP ID service, **Role Collections** can be applied to users. With the standalone SAP identity provider tenant, it provides more flexible options such as creating user groups and assigning role collections to user groups for easy maintenance during operations. SAP Integration Suite provides various predefined role collections for performing different activities across capabilities. Only those users with the right roles can access certain functionalities of SAP Integration Suite.

- **Secure runtime scenarios**
  Integration solutions must provide a secure means of exchanging messages between applications. There are many open specification standards that different applications have implemented to securely exchange information:
  - **Transport-level security**
    Different applications integrate using diverse technical protocols and follow different specifications. SAP Integration Suite ensures transport-level security with different protocols and adapters (e.g., OData, web services, email, SFTP server, etc.), based on HTTPS, SSH, SMTPS, and so on, followed by different authentication mechanisms (basic authentication, certificate-based client authentication, OAuth authentication, SAML).

Certificate Handling

If companies want to enable mutual client-based certificate authentication (the recommended approach from a security guidelines perspective), certificates need to be exchanged by both SAP Integration Suite and connecting applications based on inbound or outbound traffic. One side of a certificate authority (CA)-signed certificate is provided by SAP while provisioning SAP Integration Suite, but another certificate needs to be arranged by the company and needs to be signed by certificate authorities. This process takes anywhere from one to three or more weeks depending on various factors (process, country-specific compliance, etc.). Companies first need to generate a public-private KeyStore, followed by a certificate signing request (CSR), and give it to a CA for signing. Once it's signed, the public key needs to be uploaded in SAP Integration Suite.

  - **Principal propagation**
    SAP Integration Suite supports SAML 2.0-based principal propagation. SAML 2.0 is an open standard that allows identity and security credentials to be shared across applications for achieving single sign-on. This is useful if companies do not want to forward the identity of a user to any mediated layer, but rather want to send security credentials directly to integrating applications to have the application user in the source system or application be validated directly from the backend or target application.
  - **Message-level security**
    Once the transport-level secure connection is established, the second level of security can be applied at the message layer. There are many scenarios in which organizations require an actual message on the secure wire to be further encrypted and digitally signed (e.g., when exchanging financial documents with banks or business partner). The Cloud Integration capability of SAP Integration Suite supports Pretty Good Privacy (PGP), PKCS #7, XML digital signatures, WS-Security, and other supported techniques.

- **API policies**
  The API Management capability provides groups of policies that can be used to configure security-related aspects of configured APIs. For example, the access control policy allows you to limit access to APIs to only a specific IP or group of IPs. The spike arrest policy allows you to limit the number of concurrent calls to avoid denial of service attacks. The verify API key policy provides verification of the application key to access certain APIs. Other advanced-level security policies include JSON threat protection and regular expression protection. In addition, there is strong governance available for API access.
- **Secure on-premise connectivity**
  The company landscape is becoming more and more heterogenous and has data maintained in various distributed systems. One of the abilities of SAP Integration Suite is to provide seamless integration to multiple business applications, including SAP and third-party applications in the cloud or on-premise. Companies have concerns related to on-premise connectivity with their business applications and have strong security guidelines. For this, SAP provides the cloud connector, an on-premise agent that needs to be installed in the company's landscape. It acts as a reverse proxy and creates a secure tunnel with SAP BTP, ensuring that data entering organizational boundaries is received from components/applications deployed on SAP BTP–subscribed account(s) only. You can develop applications on SAP BTP and can connect to on-premise systems and applications. The cloud connector provides support for RFC, SMTP, HTTP, and so on. It's just a routing and reverse proxy mechanism; it doesn't provide the capabilities of an iPaaS or of middleware (e.g., connectivity protocols, message transformations, encryption, signing, integration processes, etc.).

  Companies can also use other reverse proxy software, such as SAP Web Dispatcher—but it takes a lot of time to get security approval to open the respective ports/IP addresses in their firewalls. The cloud connector eliminates the need for this complex process because it gets installed in the demilitarized zone (DMZ) and initiates connectivity from an on-premise agent to SAP BTP in the cloud and creates a secure tunnel. Once the secure tunnel is established, data transfer happens in this secure layer. The cloud connector administrator has full control over which on-premise systems are exposed to the cloud. A single cloud connector instance can be connected to multiple subaccounts on SAP BTP.

2.3.3 Provisioning

Companies can get SAP Integration Suite instances via following channels:

- **Subscription**
  This is a licensing model by which a company can subscribe up front to the required SAP Integration Suite instances (tenants) or any other SAP BTP service. Once the

contract is signed, subscribed entitlements are available in the global account. For example, if a company subscribes to four instances of SAP Integration Suite standard edition, four tenants will be available in the global account. The global account administrator can allocate a quota to different subaccounts and can then create various subaccount administrators.

- **Cloud Platform Enterprise Agreement/pay as you go**
  In this licensing model, companies get flexibility to consume cloud credits. In the Cloud Platform Enterprise Agreement (CPEA), a company gets upfront cloud credits; in pay as you go (PAYG), a company can consume cloud credits without an upfront commitment. Companies can activate any service on SAP BTP and get monthly credits and an invoice based on the consumed credits. The global account administrator is in charge of this account and can enable entitlements of various subaccounts.
- **Trial**
  SAP offers a trial of SAP BTP's SAP Integration Suite. Companies can activate and use the services and try out various scenarios. Once a company activates the trial, the end user is provided a subaccount directly with one tenant quota. The company can start configuring the next steps of provisioning SAP Integration Suite.

Let's look at the step-by-step approach to get your SAP Integration Suite instance running.

Registering

You first need to register at SAP's website and activate your user ID. This email address/user ID will be used to log into your SAP BTP trial account cockpit. This is a mandatory step to use the trial environment. Follow these steps:

1. Go to *www.sap.com* and click the **Log on** icon, as shown in Figure 2.9.

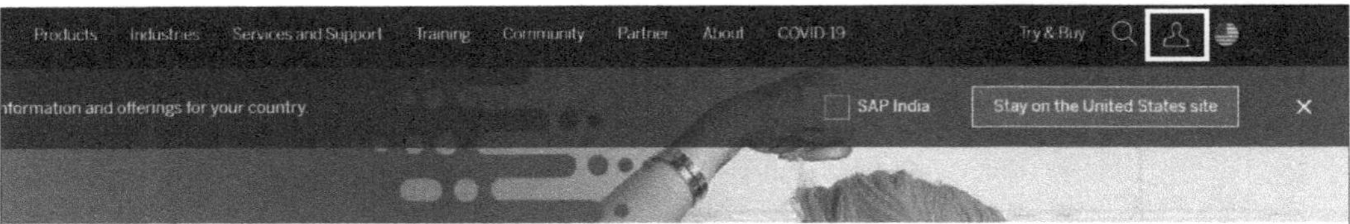

Figure 2.9 Logging in to SAP Website

2. If you already have an account on the SAP website, you can click **Log In** as shown in Figure 2.10 ❶ and go directly to the next section. If you don't have an existing SAP account, you can register by providing the required details and clicking the **Submit** button ❷.

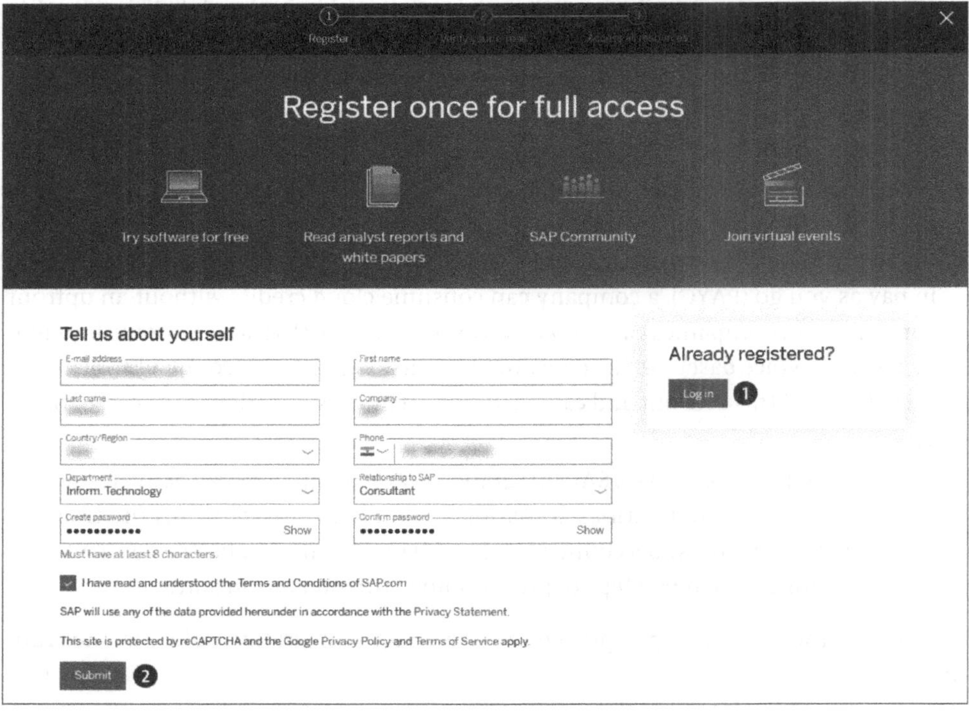

Figure 2.10 Registering on SAP Website

3. An email will be sent with your account activation link, as shown in Figure 2.11.

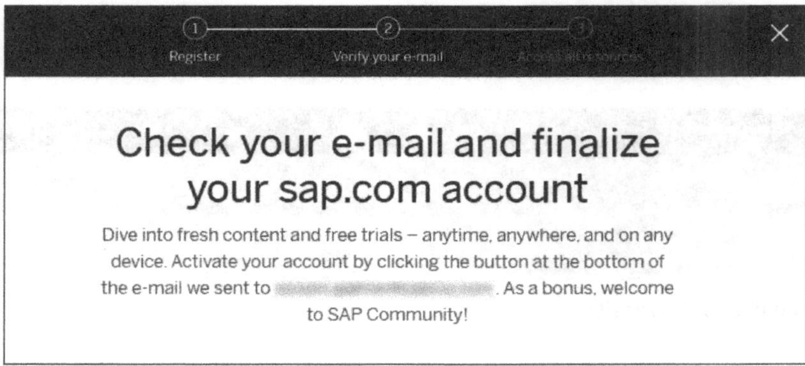

Figure 2.11 Registration Process Completion

4. Once you click the activation link, it will activate your account and will take you to SAP logon page. You can also see a P-user ID generated by the system, as shown in Figure 2.12.
5. You can use your user ID or email address to log into the system. Once logged in, click the **Try & Buy** button, as shown in Figure 2.13.

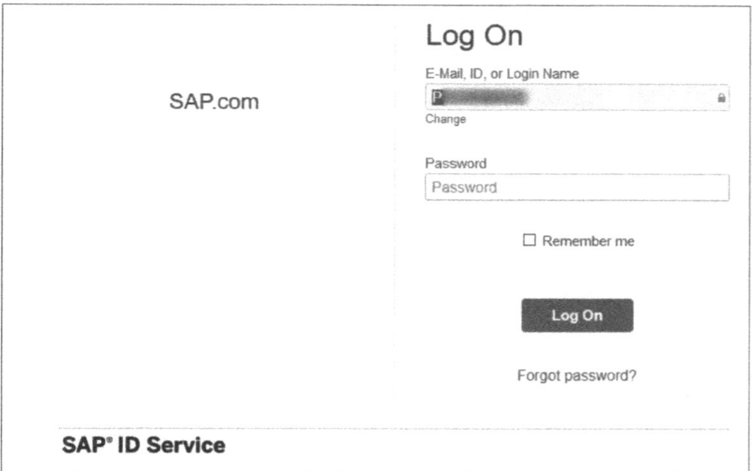

Figure 2.12 Account Activation

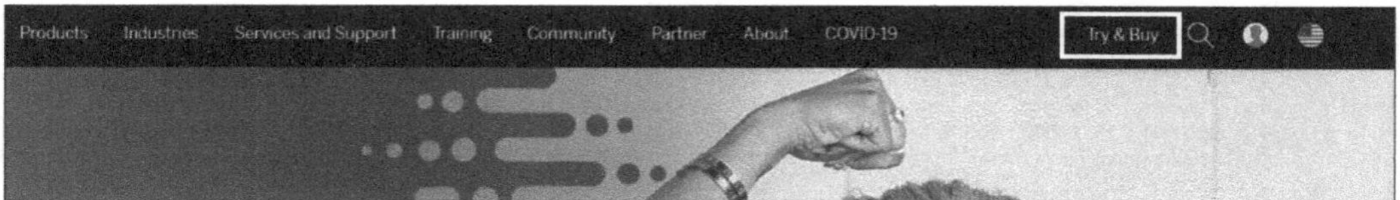

Figure 2.13 Getting into Trial Page

Activating the Trial

After registration is done, you can activate the SAP Integration Suite trial. Follow these steps:

1. There are multiple SAP applications and services listed under **Free product software trials**, as shown in Figure 2.14. Select **SAP Business Technology Platform**.

Free product software trials

ERP and Finance
SAP S/4HANA Cloud
SAP S/4HANA
SAP Business ByDesign

CRM and Customer Experience
SAP Consumer Sales Intelligence
SAP Event Ticketing
SAP Marketing Cloud

Supply Chain Management
SAP Innovation Management
SAP Enterprise Product Development

Business Technology Platform
SAP Business Technology Platform
SAP HANA Cloud powering SAP BTP
SAP Analytics Cloud powering SAP BTP

Analytics and Data Warehousing
SAP Analytics Cloud
SAP Crystal Reports
SAP Crystal Server
SAP Data Warehouse Cloud
SAP Lumira, discovery edition

Artificial Intelligence
SAP Conversational AI
SAP Intelligent Robotic Process Automation

Figure 2.14 Getting into the Trial Site

2. As you've already registered and received a user ID/password, you can proceed to the trial activation by clicking the **Enter Your Trial Account** button, as shown in Figure 2.15.

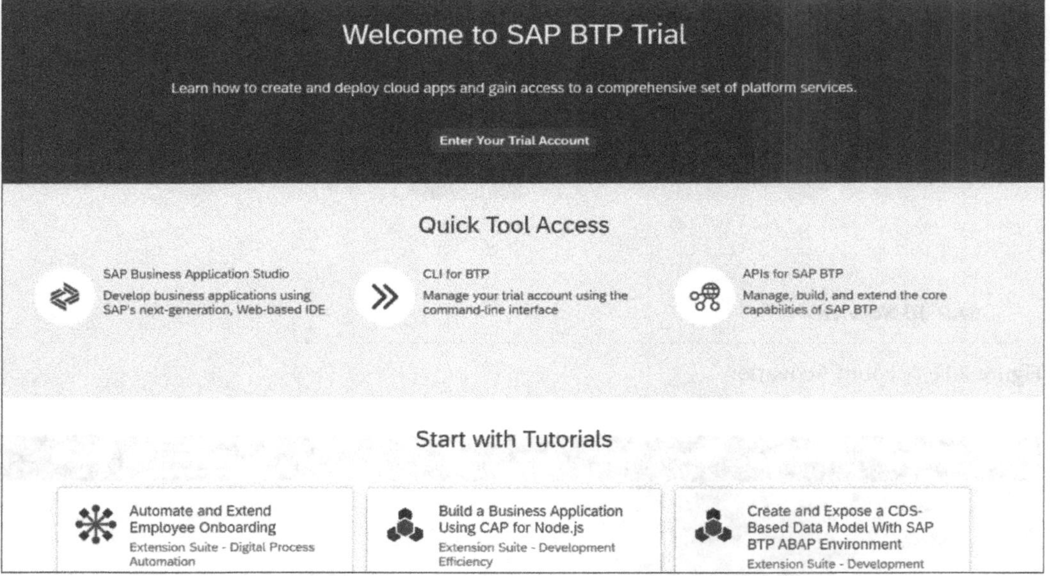

Figure 2.15 Selecting the Trial for SAP BTP

3. Choose the region that is closest to you and click **Create Account** to create a subaccount, as shown in Figure 2.16. Click **Continue** on the next screen.

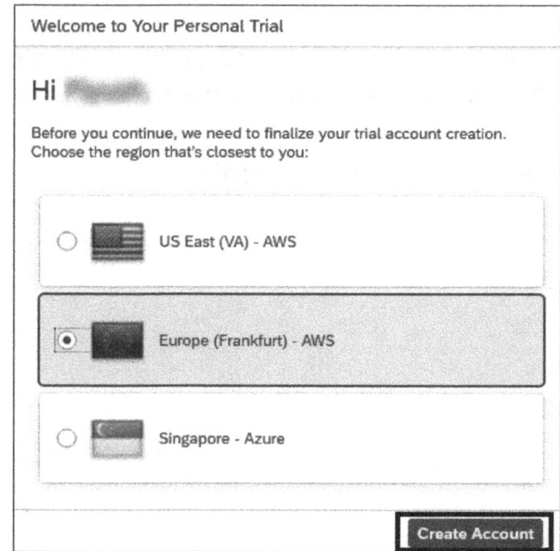

Figure 2.16 Select Region

4. Click on the **trial** tile, as shown in Figure 2.17.

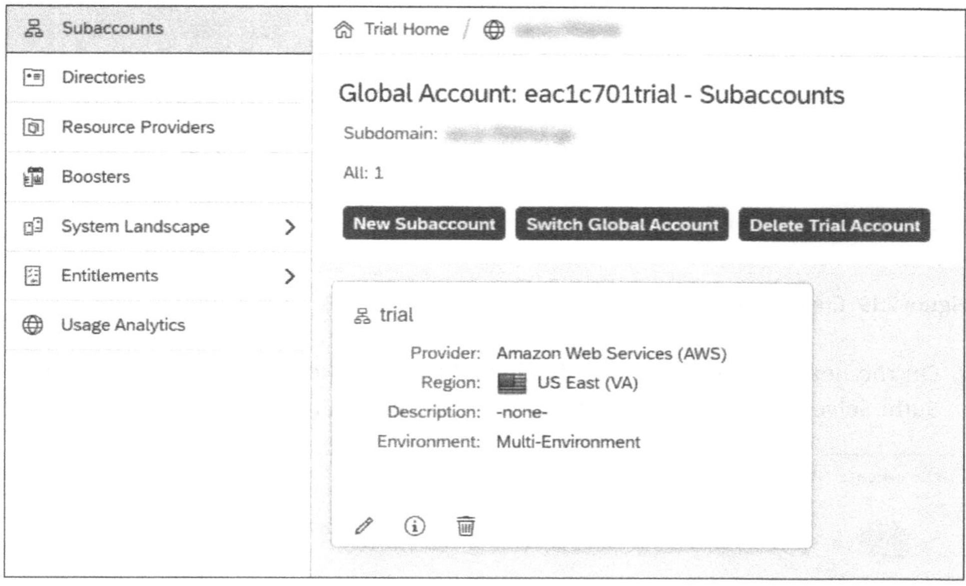

Figure 2.17 Trial Tile

5. Navigate to the **Service Marketplace** tab and search for "integration suite," as shown in Figure 2.18. Once you get the search results, click the **Integration Suite** tile.

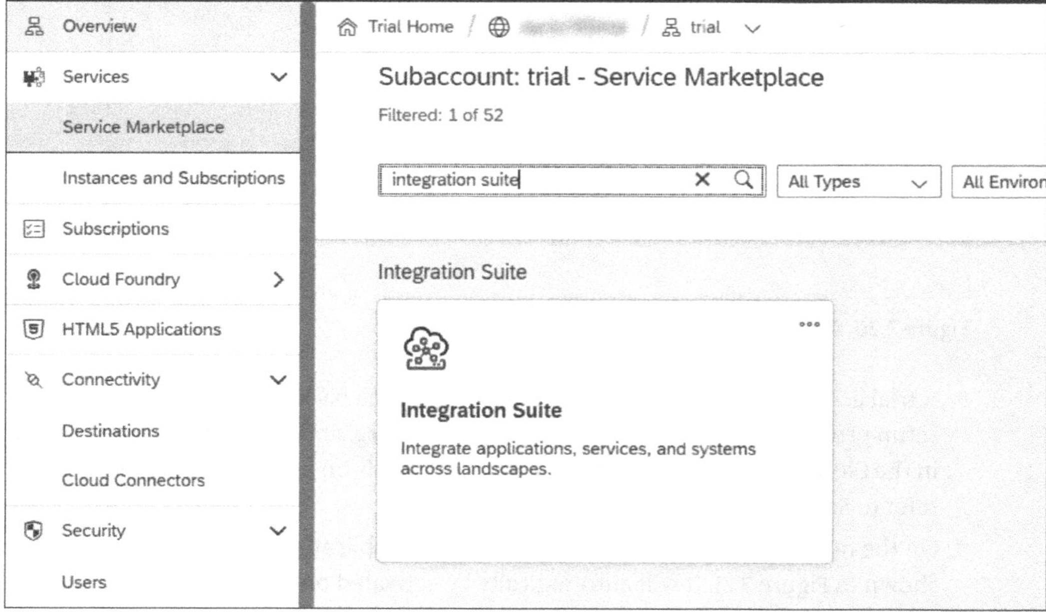

Figure 2.18 Service Marketplace

6. Click the **Create** button, as shown in Figure 2.19, to create a new subscription.

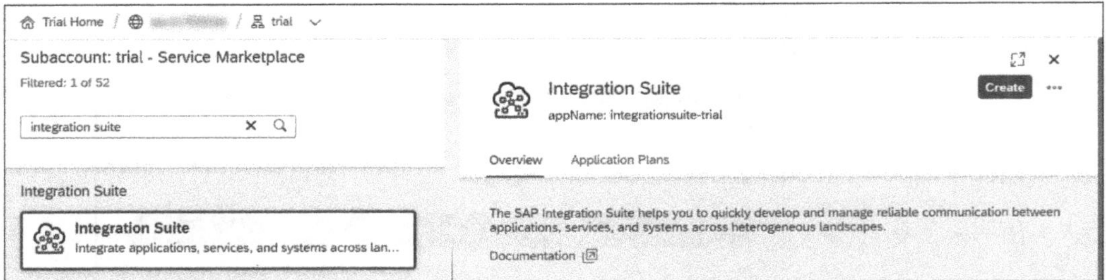

Figure 2.19 Create a New Subscription

7. On the next screen, the **Service** field will be automatically filled with **Integration Suite**. Select **trial** in the **Plan** field and click the **Create** button as shown in Figure 2.20.

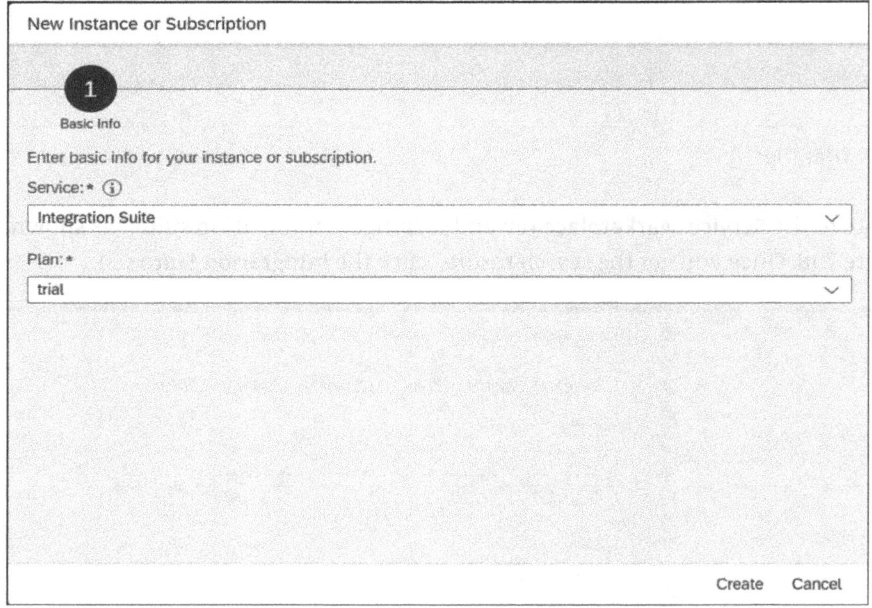

Figure 2.20 Account Setting Window for Personal Trial

8. A trial account for you is automatically set up in the background. Once the account setup process is complete, a unique subaccount, org, and space are assigned to you in the Cloud Foundry environment. For more details on understanding these terms, refer to Section 2.3.1.
9. On the next screen, you'll notice that the **Go to Application** link is now available, as shown in Figure 2.21. It will automatically be activated once SAP Integration Suite is subscribed successfully in your subaccount.

Figure 2.21 SAP Integration Suite Subscription

10. Click **Go to Application** to access the SAP Integration Suite landing page.

Accessing the SAP Integration Suite Tenant

In the previous steps, you registered on the SAP website and activated a trial from the SAP BTP cockpit. Now let's look at preparing the SAP Integration Suite tenant for first use. Follow these steps:

1. Click **Go to Application** to see the SAP Integration Suite landing page, with no capabilities listed. Click **Add Capabilities**, as shown in Figure 2.22, to enable the desired functionality in your SAP Integration Suite tenant or subaccount.

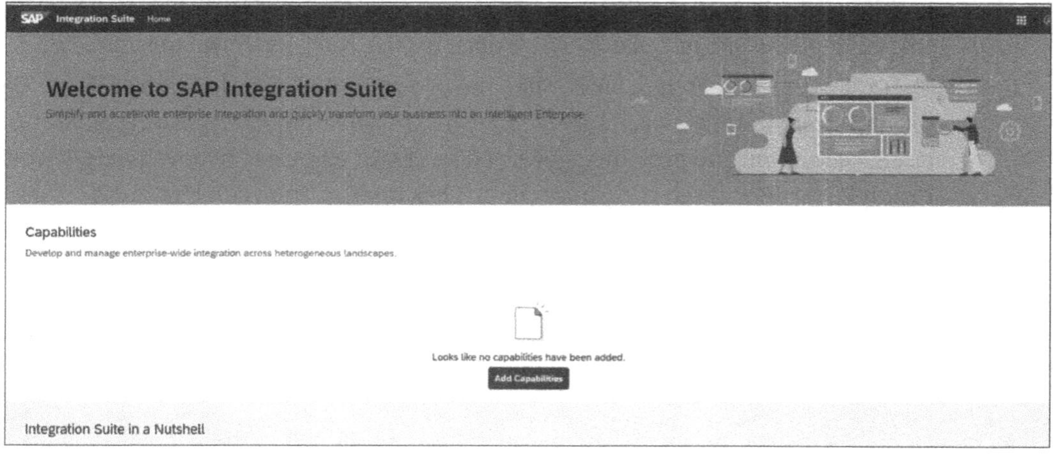

Figure 2.22 SAP Integration Suite Landing Page

2. You can select one or more capabilities of SAP Integration Suite based on your business use case requirements (refer to Section 2.2 for the details of each capability). For this example, select all capabilities of SAP Integration Suite. Click **Next** after you've selected capabilities, as shown in Figure 2.23.

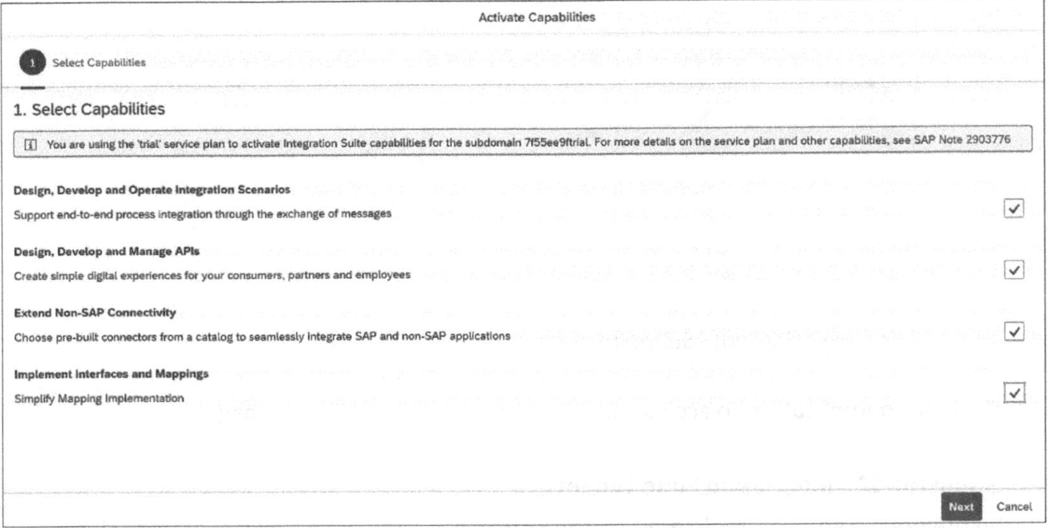

Figure 2.23 Select Capabilities of SAP Integration Suite

3. The Cloud Integration capability offers a **Message Queues** functionality to process messages in a reliable manner. It can be activated for use cases in which messages need to be reliably delivered to the target system. Optionally, messages can be stored in queues and can be processed one by one for next steps or sending to the target system. In case of failures, messages aren't lost and are available for retries. In this example, we are not selecting this option. Click **Next**, as shown in Figure 2.24.
4. The API Management capability offers developer portal functionality that can be used by organizations to create an API Business Hub Enterprise by listing all APIs in a central location and providing governed access for end users or developers. In this example, we are not selecting this option. Click **Next**, as shown in Figure 2.25.

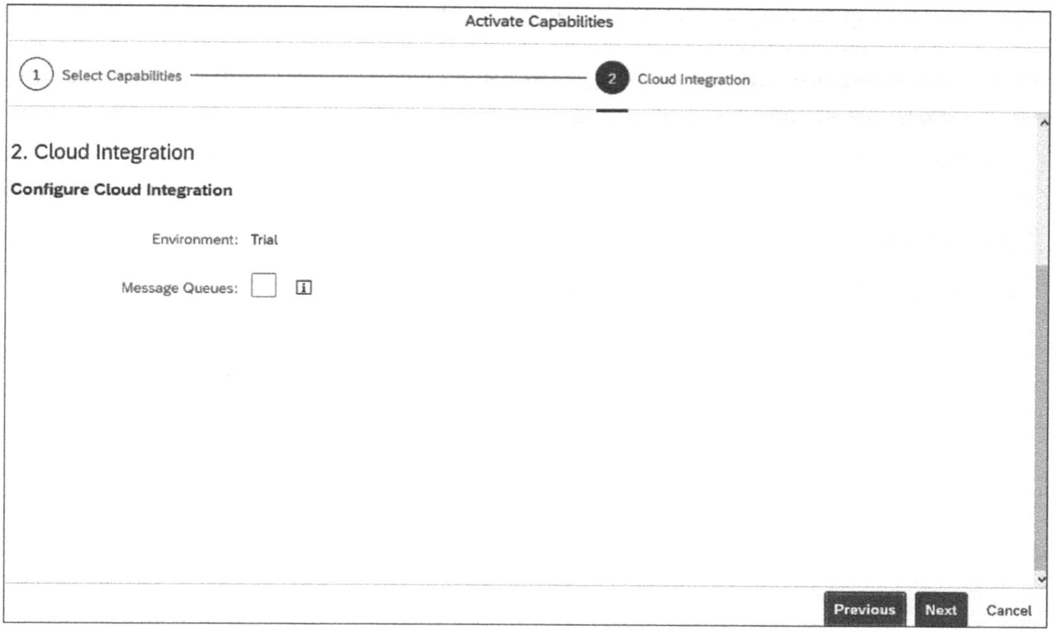

Figure 2.24 Select Optional Functionalities for Cloud Integration

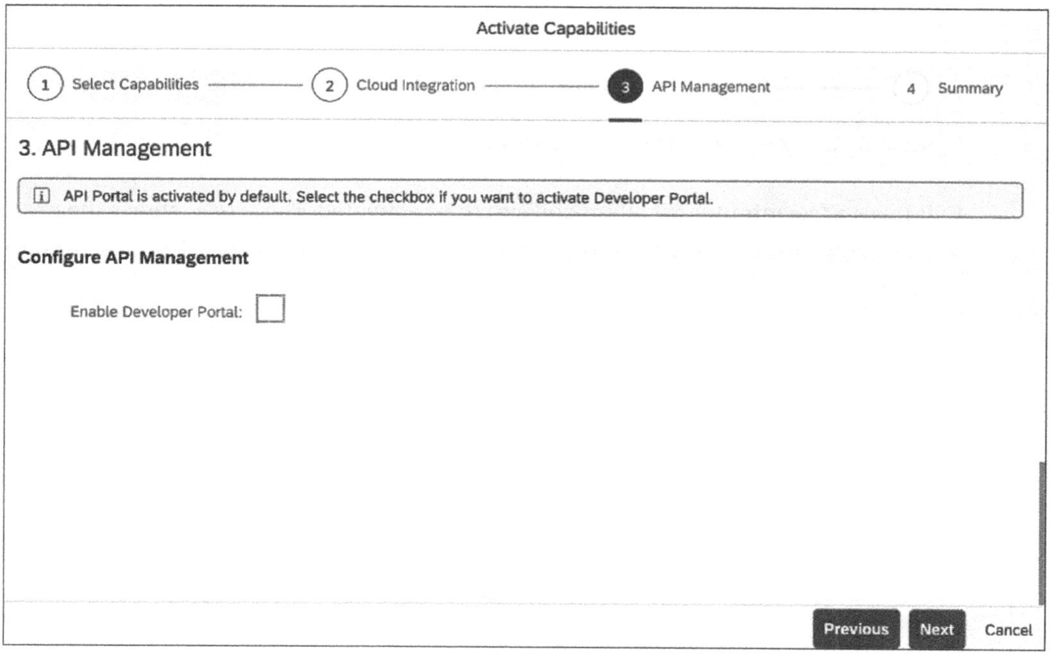

Figure 2.25 Select Optional Functionalities of API Management

5. Take a quick look at the selected capabilities in the summary view. If any changes are required, you can click **Previous** and reconfigure your selections. Once you're done, click the **Activate** button to activate the SAP Integration Suite tenant with the selected capabilities, as shown in Figure 2.26.

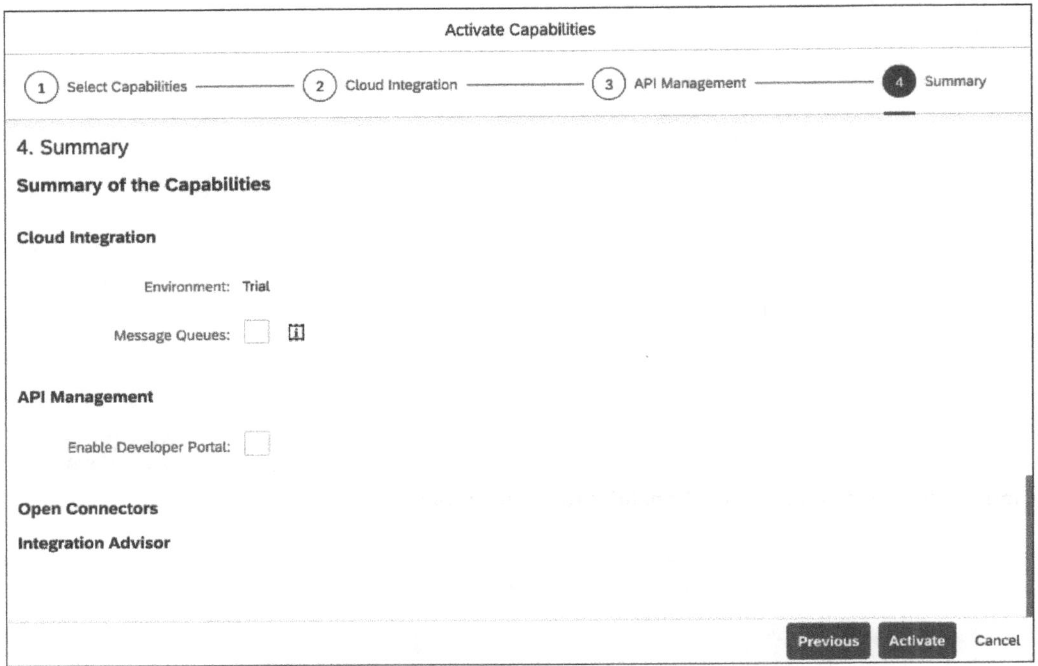

Figure 2.26 Quick Check of Selected Capabilities

6. It takes a few minutes for all capabilities to be activated. The screen shows the progress of each capability, and at the end you'll get a confirmation for all activated capabilities, as shown in Figure 2.27.

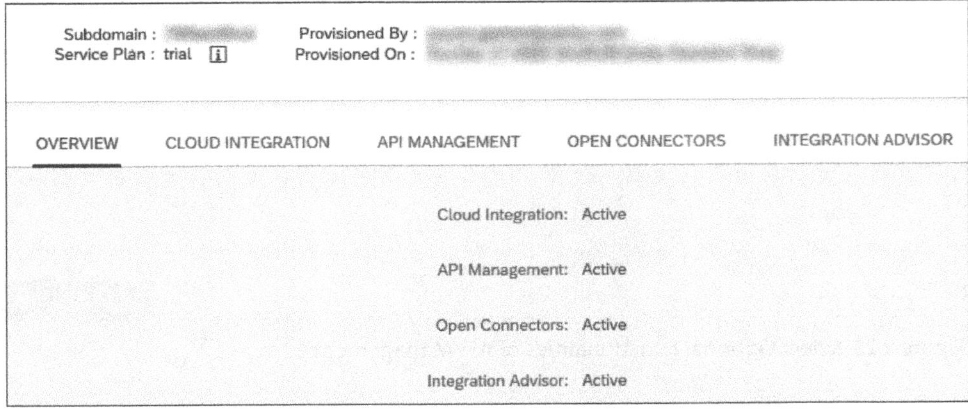

Figure 2.27 Summary of Activated Capabilities, Plus Account Details

7. When you're all set up, you'll get a view of all activated capabilities on the SAP Integration Suite landing page, as shown in Figure 2.28.

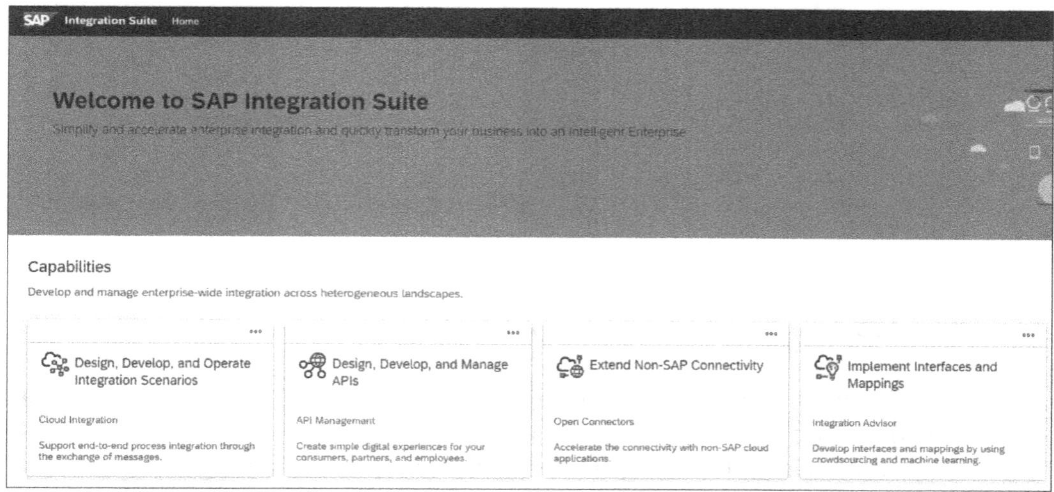

Figure 2.28 SAP Integration Suite Landing Page, Showing Activated Capabilities

Adding Roles and Authorizations

After all the required capabilities are activated, the next step is assigning the right authorizations to end users based on their roles and responsibilities. Follow these steps:

1. Go back to the SAP BTP cockpit and click **Users**, as shown in Figure 2.29. Your user ID is displayed. You can add more users as well. We'll focus here on adding roles and authorizations to existing available user IDs. Click your user ID under **User Name** and a new window will appear to assign a role collection.

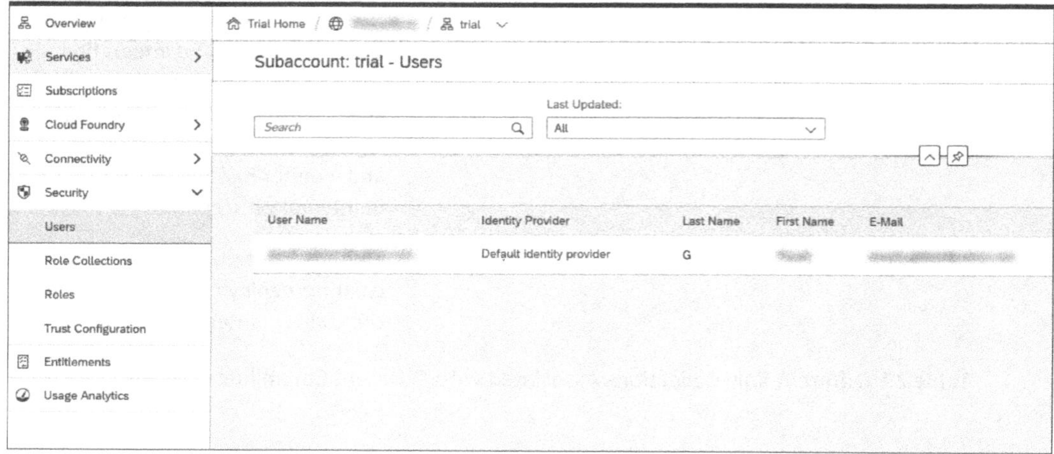

Figure 2.29 Security Configuration View in SAP BTP Cockpit

2. Click **Assign Role Collection**. A new window will open (as shown in Figure 2.30) with all available **Role Collections**.

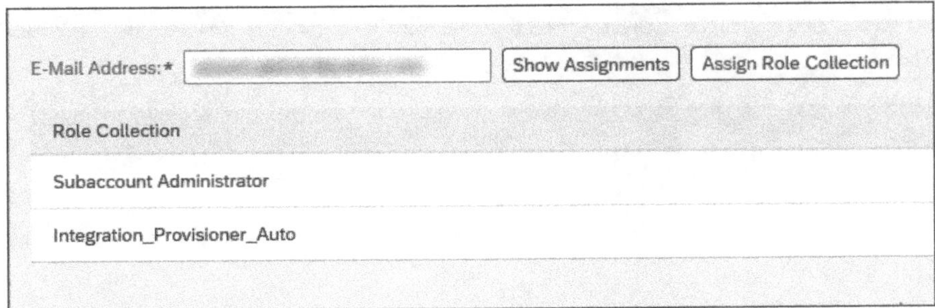

Figure 2.30 Add Role Collection to Existing User

3. There are many different roles available for each capability. The personas who will be working with SAP Integration Suite as developers or end users determine how these role collections will be assigned. For trial use cases, you can assign the role collections as shown in Table 2.1.

Capability	Role Collection	Tasks
Cloud Integration	`PI_Business_Expert`	This grants required authorizations to a nontechnical user who wants to check the status of deployed integration scenarios and monitor the integration messages. If a payload is attached to an actual integration flow, it also provides the right to read messages and payloads.
Cloud Integration	`PI_Administrator`	This provides administrative rights to end users, such as deploying security artifacts and integration content and performing some advanced activities, like deleting messages from a transient store and monitoring/checking the status of integration scenarios.
Cloud Integration	`PI_Integration_Developer`	This provides rights to developers for creating, deploying, and monitoring the status of integration flows.

Table 2.1 Different Role Collections Associated with Different Capabilities

Capability	Role Collection	Tasks
API Management	`APIPortal.Administrator`	This provides administration rights to end users for using the API portal UI and its associated services, such as discovering, developing, and configuring APIs and associated rate plans and monitoring the traffic.
API Management	`APIManagement.Selfservice.Administrator`	This provides the roles required during the onboarding process, like setting virtual hosts.
Integration Advisor	`trial-content-developer`	This authorization provides administration rights to end users for creating and managing content in the Integration Advisor.
Open Connectors	`OpenConnectors_User`	This authorization provides access to the Open Connectors capability from the SAP Integration Suite landing page.

Table 2.1 Different Role Collections Associated with Different Capabilities (Cont.)

All steps for provisioning and role assignment are now complete. You can return to the SAP Integration Suite main landing page, as shown in Figure 2.31, and start accessing activated capabilities.

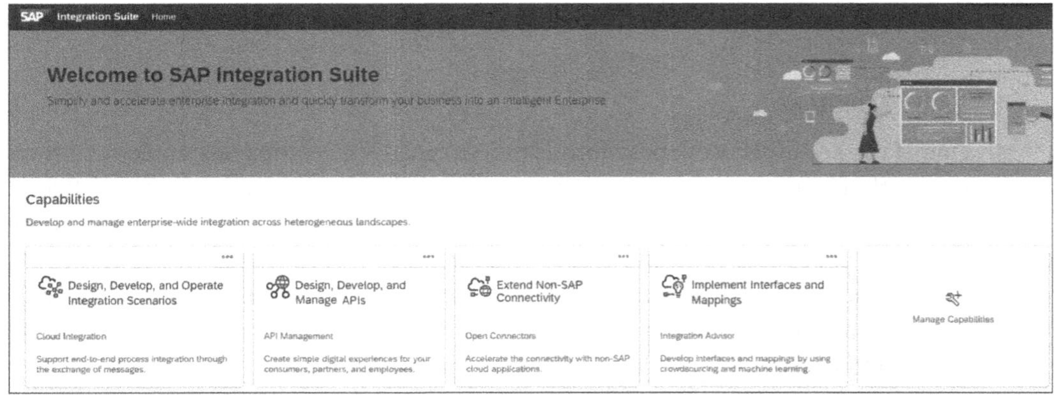

Figure 2.31 SAP Integration Suite Landing Page with Activated Capabilities

You can access any capability by clicking its respective tile or clicking **Manage Capabilities**. There you can view URL links for access, as shown in Figure 2.32.

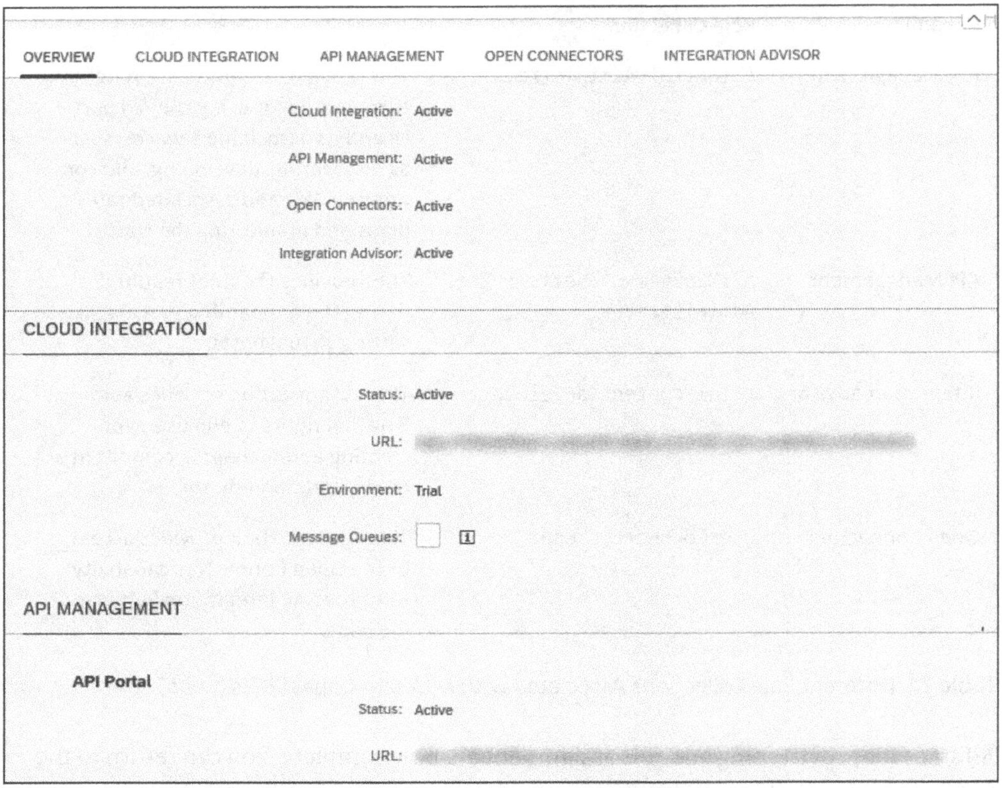

Figure 2.32 Capabilities with Access Links

This completes your trial onboarding.

2.3.4 Monitoring

One of the key aspects of operations is monitoring. Once companies implement their integration scenarios or APIs, monitoring is needed for basic health checks, configuration management, viewing the status of runtime message processing, and using advanced debugging techniques to find issues. SAP Integration Suite offers various flexible monitoring channels. We'll review the various dashboards that are available out of the box to monitor the status of integration artifacts in the following sections, as well as additional solutions that can be used for monitoring.

Cloud Integration Monitoring Dashboard

The central Cloud Integration monitoring dashboard can be accessed by clicking the **Monitor** option on the left-hand menu. It provides different views, as shown in Figure 2.33, to check the status of deployed content and the message processing status (**Failed**, **Successful**, **Retry**, **Processing**, etc.). You can also enable log levels (**None**, **Info**, **Debug**, and **Trace**). By default, the system is set to log level **Info**. Detailed investigations can be performed using log level **Debug** or **Trace**.

Note

The trace feature is very convenient as it displays the payload value after each step of an integration scenario wherever it was recorded. But it should be used only for detailed investigations because it is resource-intensive. Payload storage time is also limited.

Security tiles provide information for certificates, keystores, user credentials, JDBC drivers, and data sources. The **Connectivity Tests** tile provides a very important functionality during any integration implementation project. It takes a lot of time to troubleshoot the connection to the integrating application due to firewall and network issues. Without even starting integration scenario development, administrators can check here whether all connections are successfully implemented in a secure manner for various protocols.

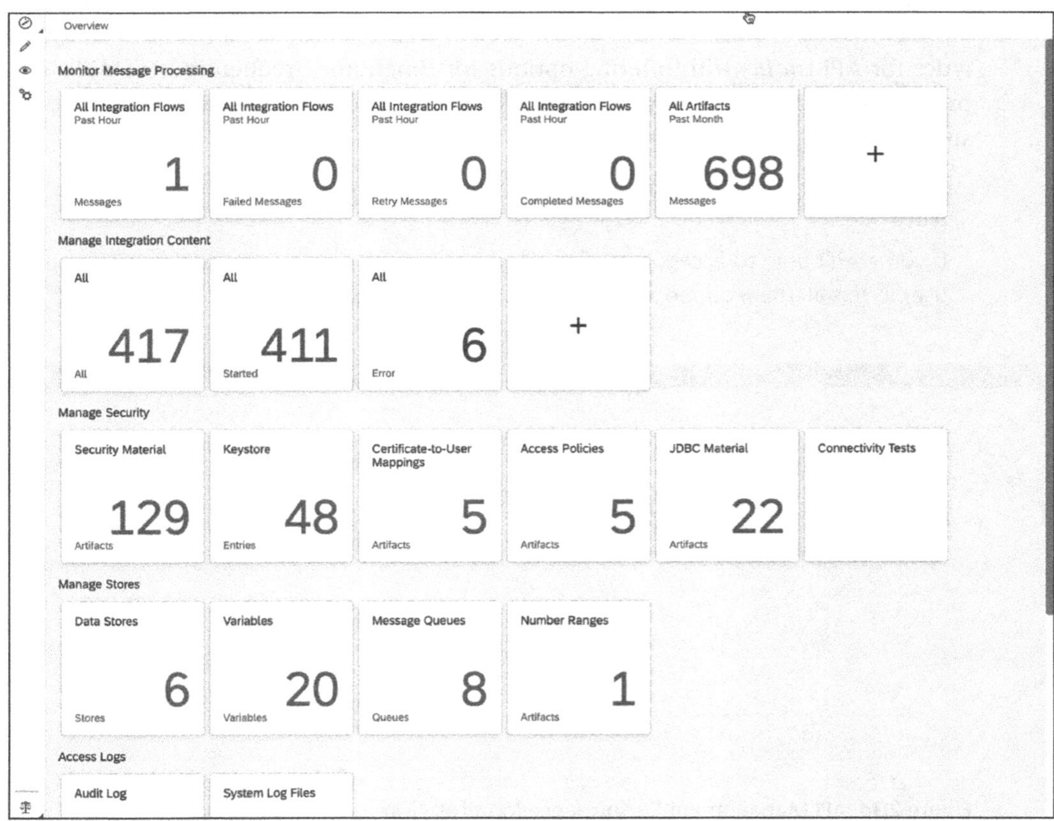

Figure 2.33 Cloud Integration Monitoring View

Message stores, available under **Manage Stores**, provide information related to the current status of configured transient data stores, variables, and number ranges in various integration scenarios.

> **Note**
>
> For important integration scenarios, you can also create a separate monitoring tile by clicking + and selecting the required integration scenario you want to monitor. This will create a new tile in the **Monitor** window.

API Monitoring Dashboard

It's important for administrators to get the usage status of API traffic moving through the systems because the API Management capability is used as the central API fabric for multiple business applications and important transactions. SAP Integration Suite has an API activity monitoring UI available via the API portal's **Analytics** option, as shown in Figure 2.34. It provides a quick summary view into API deployment status, rate plans, API errors, applications, and products in use. API Management's advanced analytics dashboards have performance and error views that provide detailed and advanced analytics for API traffic, with filtering options for timeframe, frequently used APIs, API proxies, products, and applications. It also provides information about APIs with issues such as failures and slow response times.

> **Note**
>
> If you aren't able to access any of the dashboard functionality, you can check if your user ID has all the required authorizations to view/edit various views.

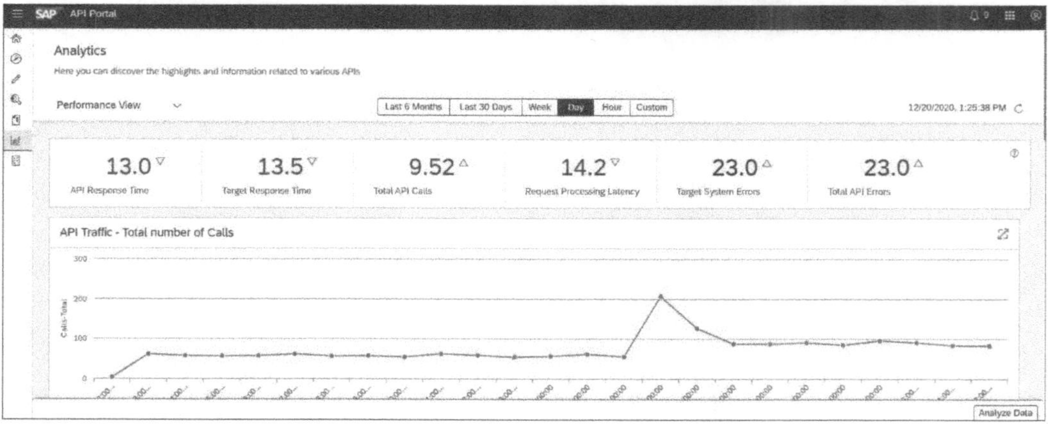

Figure 2.34 API Management Performance Monitor View

Custom Dashboards

The UI dashboard provides a majority of the needed information, but many companies need custom monitoring dashboards based on various real-time data usages. Depending on the capabilities you are using, SAP Integration Suite provides well-documented OData APIs and policies to extract the information and feed in the data to your custom dashboards. In addition, the API Management capability also provides an automatic way to create your own custom dashboard under the **Analytics** option using various parameters and dimensions for deeper issue analysis, such as response times of a specific APIs/applications, API providers, timeframe, and so on.

Additional Solutions

In addition to the dashboards we've mentioned, the following solutions can be used in tandem with SAP Integration Suite for monitoring purposes:

- **SAP Solution Manager**
  SAP Solution Manager is an on-premise application lifecycle management solution for your various enterprise applications running in heterogenous environments. SAP Solution Manager is well adopted among SAP companies running on-premise business applications and is used as a central tool for lifecycle management and deep technical monitoring. SAP Process Orchestration (SAP's on-premise middleware) provides out-of-the-box connectivity with SAP Solution Manager to view and to perform deep-dive analyses of issue resolution. The Cloud Integration capability of SAP Integration Suite also provides out-of-the-box integration with SAP Solution Manager to provide companies an integrated view of integration scenarios running in the cloud. Default templates include performance- and exception-related key performance indicators (KPIs), such as erroneous integration flows' average response time. You can create alerts and tickets in case of errors or performance problems. As a prerequisite, connectivity needs to be established between SAP Integration Suite and SAP Solution Manager before you start monitoring the content.
- **SAP Cloud ALM**
  Whereas SAP Solution Manager is an on-premise application lifecycle management tool, SAP Cloud ALM is for companies that use only cloud solutions from SAP and need an application management tool in the cloud. This solution recently launched as of the time of writing. The Cloud Integration capability of SAP Integration Suite provides monitoring content with individual message status, information about respective integration flows, and exceptions.
- **SAP Analytics Cloud**
  SAP Analytics Cloud is a SaaS solution offered by SAP that runs on SAP BTP and provides analytical capabilities such as prediction, planning, and business intelligence based on data from multiple sources, such as a CRM system, an ERP, various databases,

third-party solutions, and more. SAP Integration Suite provides SAP's business content template out of the box to enable the Cloud Integration analytical dashboard in SAP Analytics Cloud. It helps companies to use SAP Analytics Cloud as a central analytics tool to monitor business data. You can filter the values based on various dimensions, such as integration flow names, artifact types, message status, and timeframe. Refer to Figure 2.35 for an example view of message processing.

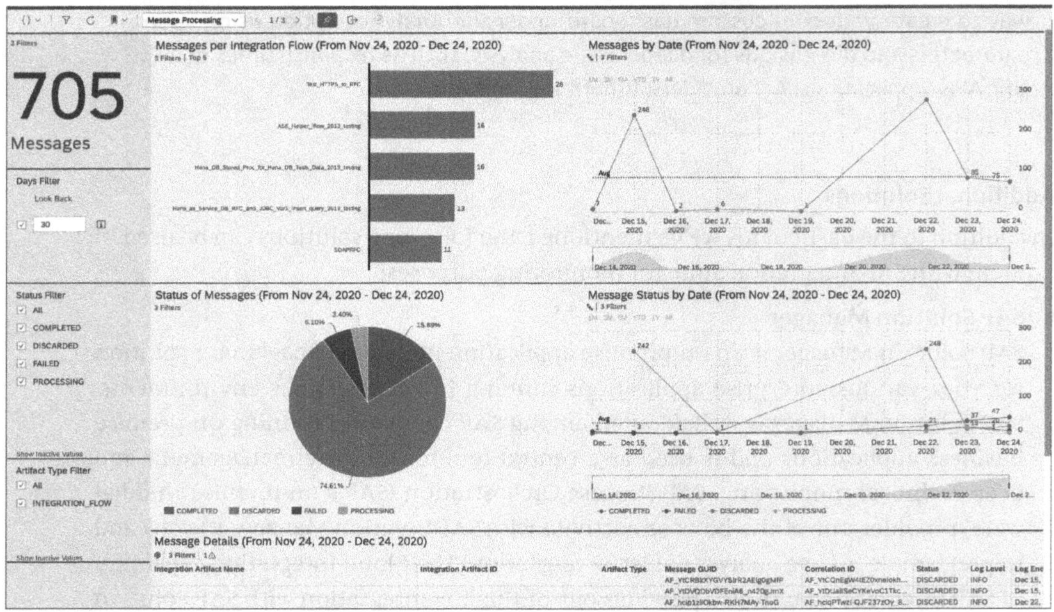

Figure 2.35 Cloud Integration Data in SAP Analytics Cloud Dashboard

- **SAP Alert Notification service for SAP BTP**
  SAP Alert Notification service for SAP BTP is included in selected editions of SAP Integration Suite. It can be used to subscribe to events from the platform and send alerts to consumers via various channels, such as email, Microsoft Teams, ServiceNow, Slack, webhooks, and so on, in a reliable manner. During integration scenarios, proactive and automated alerting is needed as the administrator is not always actively monitoring scenarios via dashboards. For example, imagine that a certificate is about to expire and the security administrator needs to be notified or a high-value sales order is not delivered due to system unavailability and the sales head needs to be informed immediately. Basic email alerting also can be utilized with Cloud Integration email adapter functionality, but companies need to configure their own email servers.

2.4 Summary

In this chapter, we introduced SAP Integration Suite and its importance for various integration scenarios such as applications, business partners, and government agencies. In addition, companies are looking at SAP Integration Suite for digital transformations and using APIs as a central key strategic layer for data access to various internal and external stakeholders. Out-of-the-box prepackaged integrations accelerate a company's integration journey.

SAP Integration Suite provides a multitier landscape to support companies' internal development and lifecycle management policies, with various options to transport content between tenants, such as SAP Solution Manager, SAP Cloud Transport Management, CTS+, and manual mode. It provides flexible monitoring modes with out-of-the-box dashboards, APIs, and policies to create custom control centers and embedded monitoring with SAP Solution Manager and SAP Analytics Cloud dashboards. Security is one of the most important components supported by SAP Integration Suite, irrespective of connectivity to cloud, on-premise, and private data centers. SAP Integration Suite supports secure API policies, transport and message-level security, secure access via roles/authorizations, and physical security of data centers via certified operations. We guided you through the steps to provision and get started with an SAP Integration Suite trial system.

In subsequent chapters, we'll go deep into individual capabilities of SAP Integration Suite and look into the detailed functionalities to design, develop, operate, and maintain integration scenarios. We'll start with prepackaged integrations and SAP API Business Hub in the next chapter.

3

Chapter 3
Prepackaged Integrations

In this chapter, we'll take our first look at SAP API Business Hub. We'll walk through the journey of discovering and consuming digital content from SAP and selected partners to accelerate your integrations.

Digital transformation (a term picking up the pace and buzzing in everyone's ears) is no longer a surprise element in modern living; the recent past paved the way for its arrival, and it has arrived naturally today with a pervasive, profound presence. More and more industries offering goods and services to consumers are becoming digital, thereby increasing online sales. E-commerce, the internet, and the use of smartphones have penetrated our lives so deeply that we cannot afford any kind of service disruption. E-commerce companies are striving hard to reduce the latency in their turnaround time and provide a seamless customer experience.

Behind the scenes, these companies are constrained by complex and hybrid IT landscapes and huge data lakes and also have to be on top of the latest and the greatest innovations. To drive business value, companies have to reimagine their landscapes, redesign their architecture to deliver personalized and intelligent experiences, accelerate/squeeze project timelines from months to days across diverse technologies, and flexibly adapt to rapidly changing business needs.

For organizations in the digital transformation journey and eager to become intelligent enterprises quickly, SAP offers integration packs. SAP API Business Hub (*https://api.sap.com*) is the central catalog for prepackaged integrations, application programming interfaces (APIs), events, workflows, and business processes to jump-start and accelerate your digital transformation journey.

The purpose of this chapter is to walk you through the central catalog of out-of-the-box content offered by SAP and selected partners. You'll see how users can search, discover, test, and consume integration and API content and engage with SAP API Business Hub. To do this, you'll start your journey by logging into SAP API Business Hub, experiencing different content types, building apps and configuring integrations, learning about the community and partner collaboration in the SAP API Business Hub, and determining best practices and reviewing guidelines for selecting a development approach.

3.1 SAP API Business Hub

To simplify and accelerate your integrations, SAP API Business Hub is available, a web-based application hosted by SAP (see Figure 3.1). As of the time of writing, it provides 1,700-plus APIs and events and 2,000-plus integrations for end-to-end scenarios such as hire-to-retire, lead-to-cash, source-to-pay, travel-to-reimburse, and more. You can discover, test, and consume these integrations and APIs to create apps and extensions.

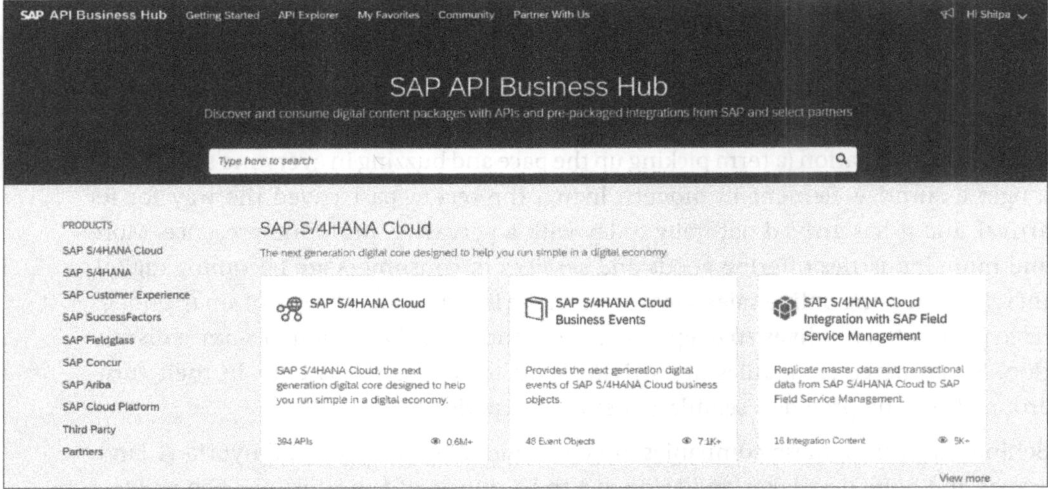

Figure 3.1 SAP API Business Hub Home Screen

SAP API Business Hub simplifies the development process and reduces the effort for the following personas:

- Application developers looking for APIs to quickly build apps and extensions, be they mobile applications or web applications
- Integration developers who are looking to develop end-to-end integrations (application-to-application [A2A] or business-to-business [B2B]) and developers who are working on event definitions that help their businesses to "run simple" in a digital world
- Process owners who want to work on the business processes and workflows in a low-code/no-code environment

In this section, we'll discuss the key features of SAP API Business Hub and take a tour of its offerings.

3.1.1 Salient Features

SAP API Business Hub provides three key features, as shown in Figure 3.2.

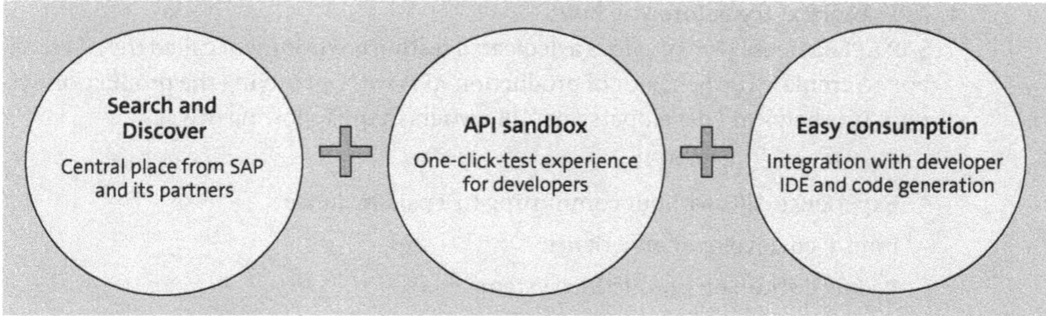

Figure 3.2 SAP API Business Hub's Salient Features

Let's take a closer look at the features provided by SAP's rich catalog to explore digital content:

- **Search and discover**
  After logging in at *https://api.sap.com*, you'll see the search bar (see Figure 3.3), in which you can type the search term or name of the content you're looking for to view and navigate to the associated digital content. To further simplify the user experience and ease your day-to-day job, you can select from the available options under **Refine By** to refine the search based on **Type**, **Products**, **Industries**, **Line of Business**, **Vendor**, and **Country/Region**.

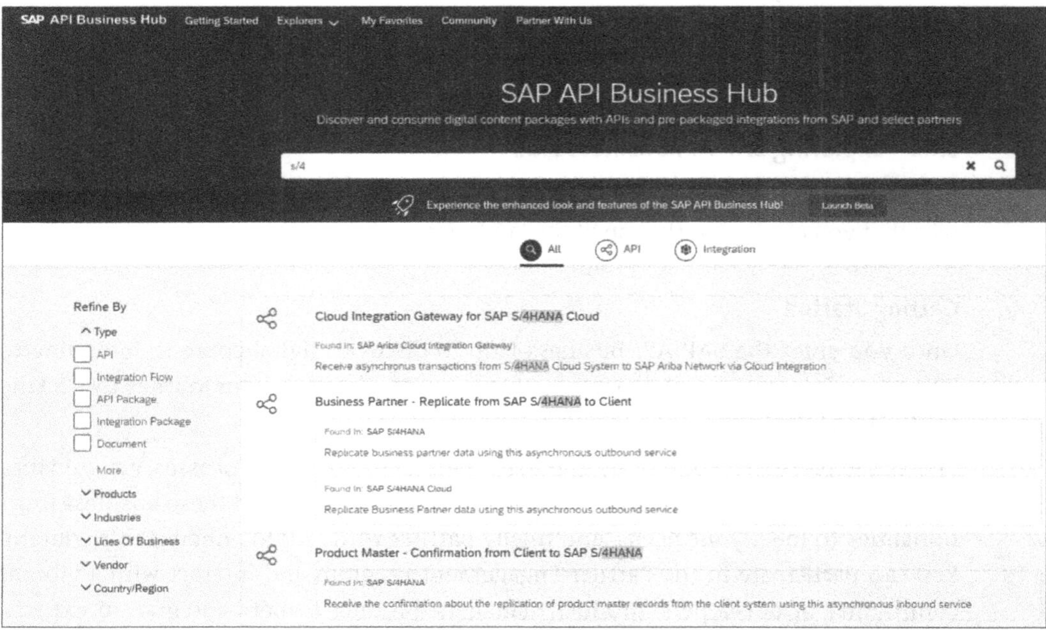

Figure 3.3 Search Options for Integrations and APIs

- **API sandbox: try before you buy**
  SAP API Business Hub provides a dedicated testing environment called the API sandbox to emulate the behavior of production APIs without buying the production system. This helps API developers and API owners in the following ways:
  - Cost-free testing of APIs
  - Experience APIs without committing to a paid instance
  - Improved developer experience
  - Reduced strain on production systems

> **Note**
> In the SAP API Business Hub, all sandbox APIs are protected by an API key, which is generated per user. The API sandbox in the SAP API Business Hub is intended solely for evaluation purposes and should not be relied on for any productive use.

- **Ease of consumption**
  The SAP API Business hub provides a seamless consumption experience for API consumers. It helps developers to consume the APIs by generating codes and providing integration with popular IDEs—for example, SAP Business Application Studio—to quickly build prototypes. It also provides tight integration with the Cloud Integration capability of SAP Integration Suite. Integration packages, which are available on SAP API Business Hub, can be consumed in the Cloud Integration capability, which we'll discuss further in Chapter 5.

3.1.2 Exploring SAP API Business Hub

In this section, we'll walk through SAP API Business Hub. We'll explore each of the key tabs that you can find at the top of the landing page.

Getting Started

Once you enter the SAP API Business Hub to discover digital content, intending to build apps, integrations, and extensions, you can quickly start your journey by clicking the **Getting Started** tab, as shown in Figure 3.4.

You can explore the world of APIs, integrations, and business processes, get guidance on working with digital content organized into packages, extend these business functionalities to meet your needs, and finally partner with SAP to enhance this content. You can participate in the Partner Engagement program and interact with a vibrant community of developers, product managers, product owners, integration experts, enterprise architects, integration consultants, and other stakeholders to learn and to get tips and tricks to improve the experience.

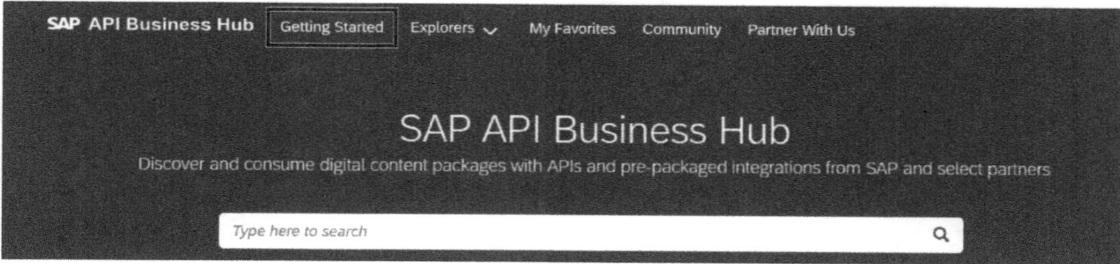

Figure 3.4 Get Started with SAP API Business Hub

The digital content on SAP API Business Hub is organized into packages. A package on the SAP API Business Hub is a logical bundling of related artifacts that enables developers to easily build applications, be they mobile, iOS SDKs, SAP Fiori, or others, and to accelerate integrations easily. We will deep dive into each type of package in Section 3.2.

Explorers

You can mark your frequently used APIs in the **Explorers** tab, as shown in Figure 3.5.

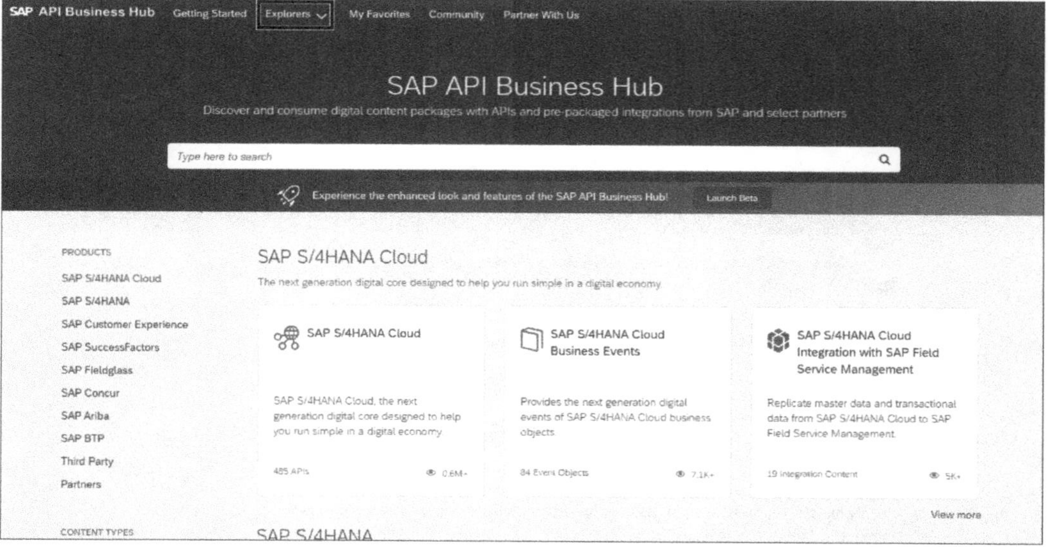

Figure 3.5 Different Explorer Options

For example, say you want to work with SAP S/4HANA Cloud APIs and would like to check the sales order response daily. To get it to your personal space via the **API Explorer** option, follow these steps:

1. Select **API Explorer** from the **Explorers** dropdown to arrive at the screen shown in Figure 3.6.
2. Click the **Add** (+) button.

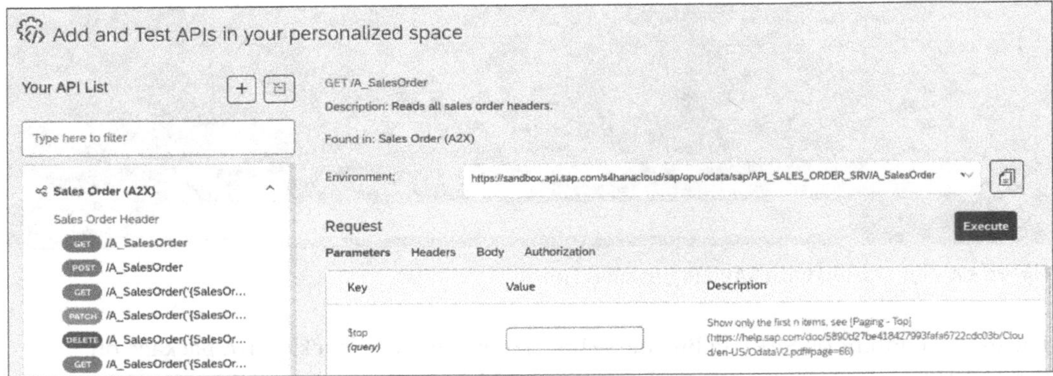

Figure 3.6 API Explorer Personalized Space

3. Search for the APIs to be tested in your personalized space—for example "Sales Order (A2X)"—check the relevant boxes, and add them by clicking the **Add** button, as shown in Figure 3.7.

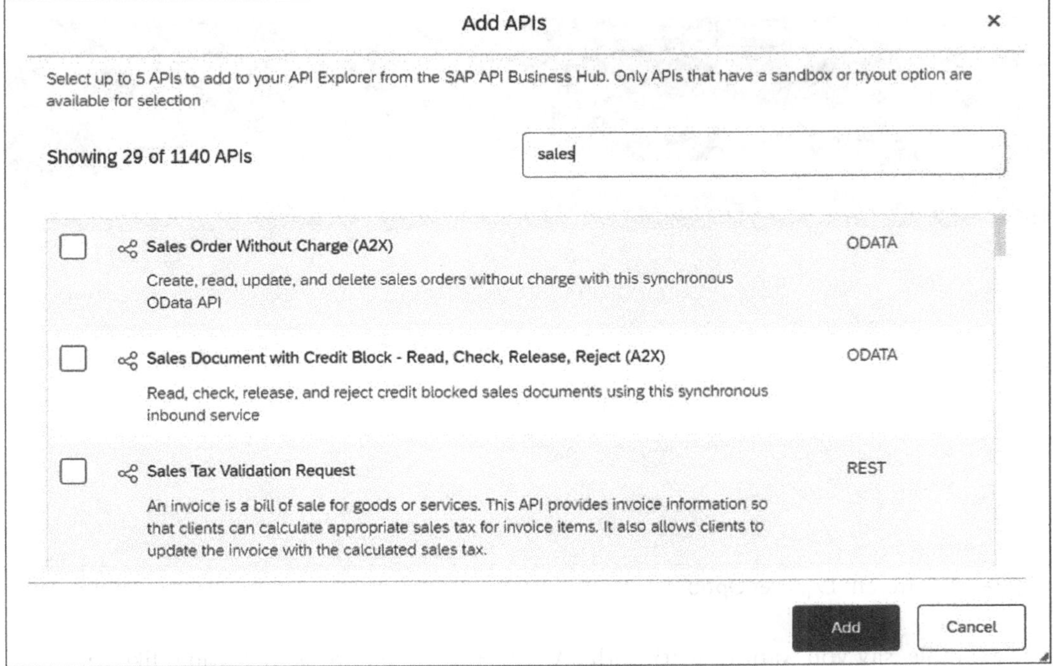

Figure 3.7 Personalized Space for Testing APIs

4. The APIs are added under **Your API List**, as shown in Figure 3.8.
5. You can execute them by clicking the **Execute** button to check the results, as shown in Figure 3.9. The result here should be the sales order coming from the sandbox system.

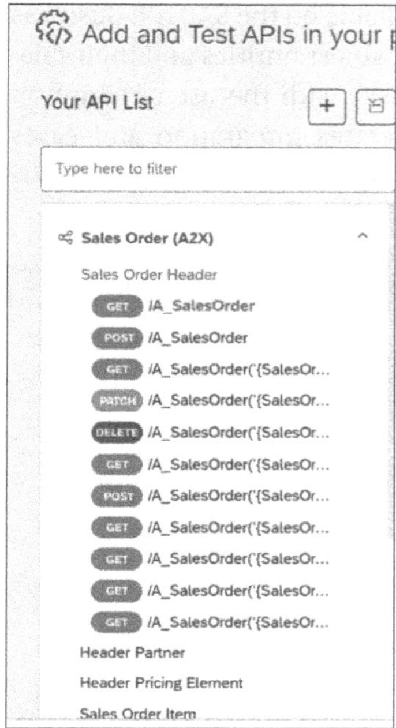

Figure 3.8 List of APIs

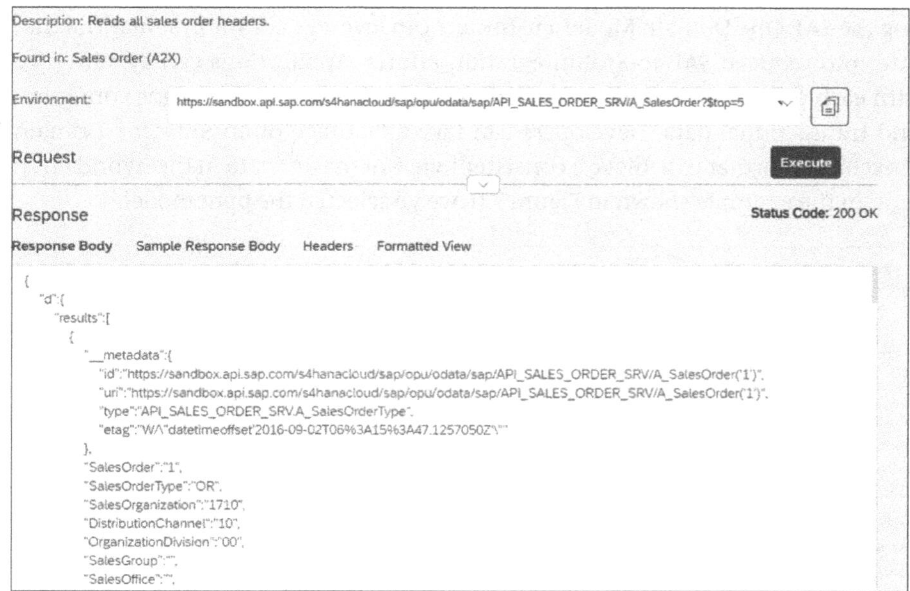

Figure 3.9 Test Results for API

The other option underneath the **Explorers** tab is **SAP One Domain Model**, as shown in Figure 3.10. SAP One Domain Model is the lingua franca of the intelligent suite. This common language is used across SAP solutions. SAP One Domain Model, a single and coherent domain model for SAP's intelligent suite, is available on the SAP API Business Hub. It focuses on providing visual documentation of business entities and their relationships to each other and synchronizes business objects with the use of common semantics for end-to-end business processes. It accelerates integration and eases extension use cases by providing consistent APIs across business processes for the intelligent enterprise.

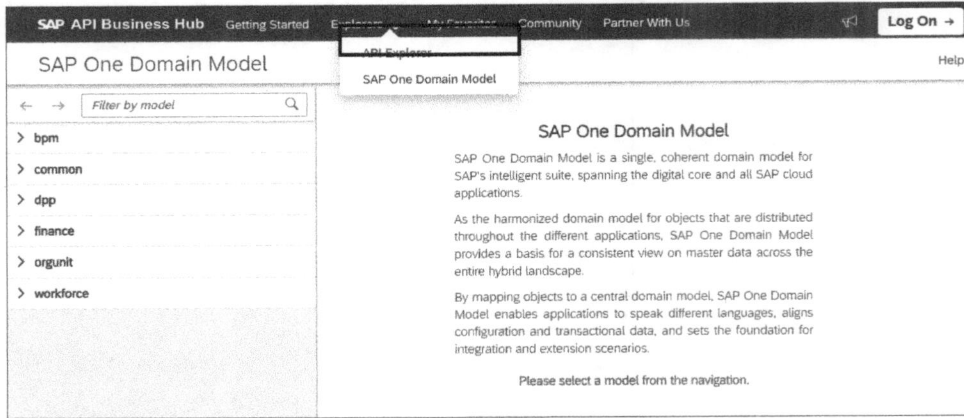

Figure 3.10 SAP One Domain Model on SAP API Business Hub

By using the SAP One Domain Model, customers can leverage common semantics that translate into reduced SAP-to-SAP integration efforts. Applications can synchronize well with each other although they speak different languages, and it aligns configuration and transactional data. Developers can take advantage of an SAP One Domain Model exchange format to achieve a consistent view of master data in the hybrid environment. In the example shown in Figure 3.11, we've selected the **bpm** model.

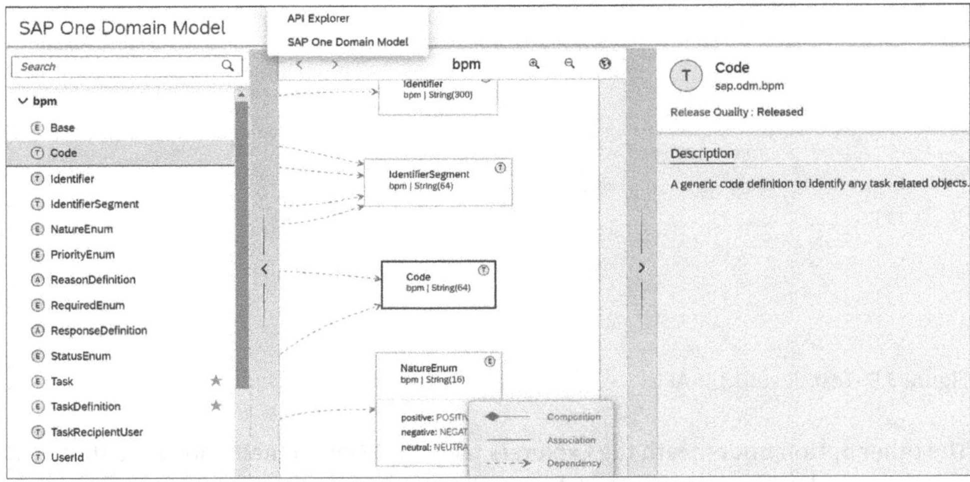

Figure 3.11 Details of SAP One Domain Model

My Favorites

When you want to access your integrations quickly with shortcuts, you can log into the SAP API Business Hub and add favorites. For instance, say you want to make a list of e-invoice packages, to keep them handy and monitor the patterns of how the government has introduced a goods and service tax (GST). To add favorites, follow these steps:

1. Log into the SAP API Business Hub.
2. Search for the integration package you want to mark as a favorite—such as "e-Invoice".
3. Click the **Mark as Favorite** button, as shown in Figure 3.12.

Figure 3.12 How to Mark Your Favorite APIs

4. Navigate to the **My Favorites** tab from the top menu bar. You can see that the package is listed, as shown in Figure 3.13.

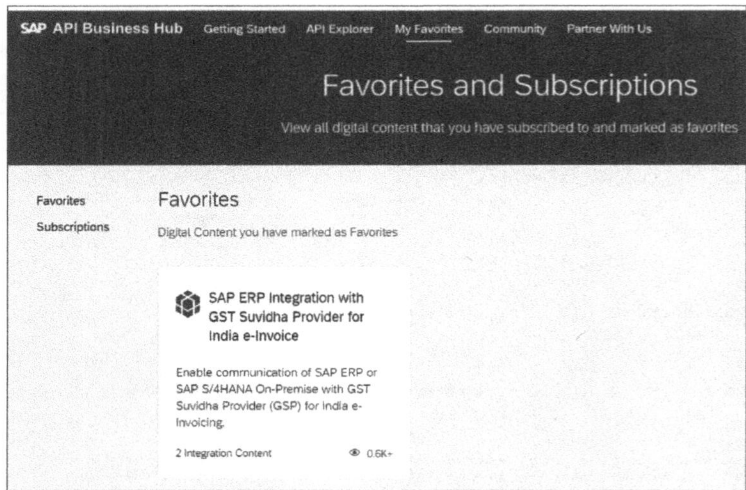

Figure 3.13 List of Your Favorite Integrations

Community

If you are in a company just starting its digital transformation journey, you might notice that digital transformation requires a drastic shift in both mindset and behavior. To help, there's a vibrant SAP Integration Suite community focusing on APIs and integrations, which you can navigate to via the **Community** tab, as shown in Figure 3.14.

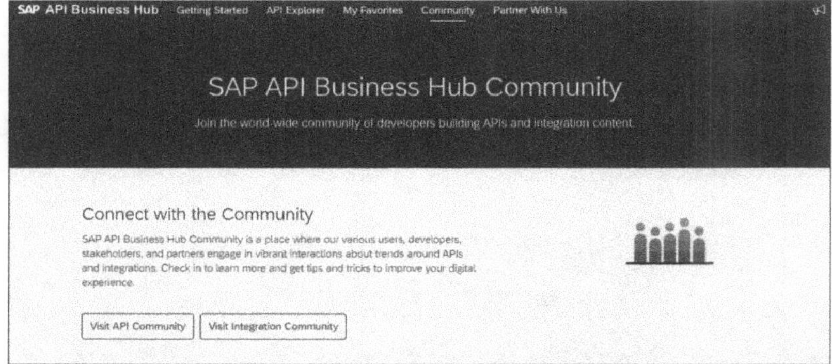

Figure 3.14 SAP API Business Hub Community

Partner with Us

SAP works closely with members and partners with aligned goals to create, codevelop, and nurture an effective network. This allows you to partner with SAP to publish your APIs and integration packs on the SAP API Business Hub. You can explore and experience the rich catalog from other partners and SAP under the **Partner With Us** tab, as shown in Figure 3.15. SAP recommends best practices and guidelines to quickly get you started with your integration journey. We will also discuss partner onboarding in detail in Section 3.4.

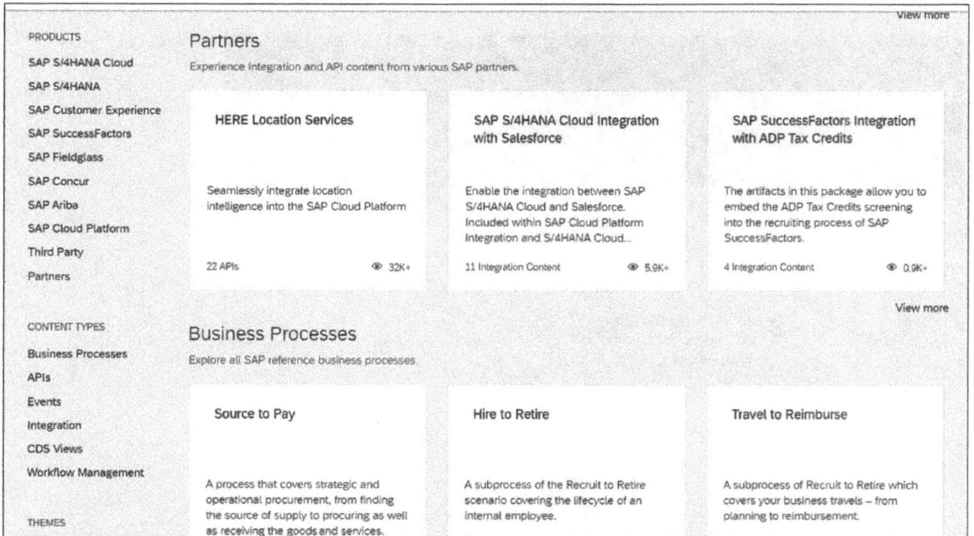

Figure 3.15 Partner Digital Content

3.2 Prepackaged Accelerators

SAP API Business Hub is the place to get started with building apps, extensions, and business networks and to integrate with applications, business partners, and government agencies. Creating a strong value proposition and new business opportunities/ revenue models, plus evolving during the digital transformation, is now made faster using prepackaged accelerators.

As of the writing this book, SAP API Business Hub has more than 3,000-plus digital artifacts—business processes, integration flows, APIs, events, adapters—across 335 packages, with millions of API calls in the API sandbox. Let's dive into prepackaged accelerators and see how each one of them can help to accelerate your integration journey. The following are the different types of packages available:

- **API packages**
  An API package contains building blocks (APIs) and the related documentation for developers to extend solutions, integrate to collaborate with the business, and perform innovations on top. They also contain preshipped templates from SAP for securing, managing, and governing the APIs so that customers foster innovation instead of reinventing the wheel.

  Customers can view the basic details of the packages, preview the package tile details, and collect information such as the type of the package, number, version number, type of digital content available, and usage of the content. Each integration and API package has a name and brief description in the header section of the package, as shown in Figure 3.16.

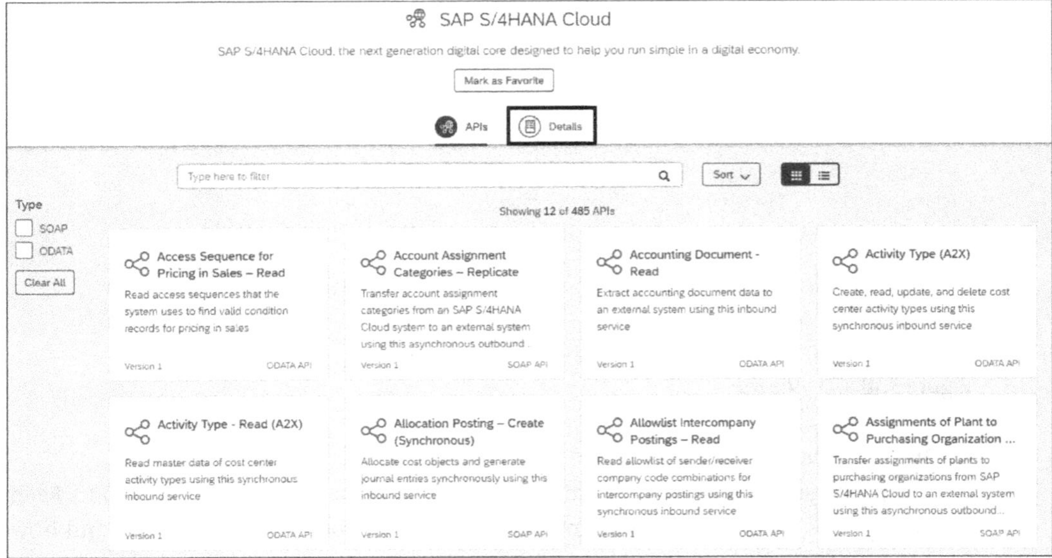

Figure 3.16 API Package

To know the details of the package, navigate to the **Details** tab, which provides complete information about the package, including business tags and documents.

- **Integration packages**
  Integration packages, as shown in Figure 3.17, contain different integration artifacts bundled together as integration flows, message mappings, and value mappings. Integration packages can be copied to your workspace and deployed to set up communication between different sender and receiver participants.

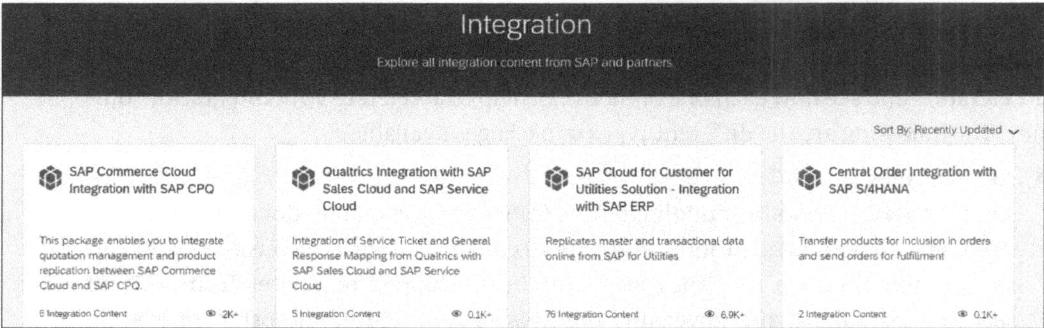

Figure 3.17 Integration Packages

- **CDS views**
  To find all the core data services (CDS) view-related information—for example, the properties of the CDS view—the metadata and data types are listed in this package (see Figure 3.18).

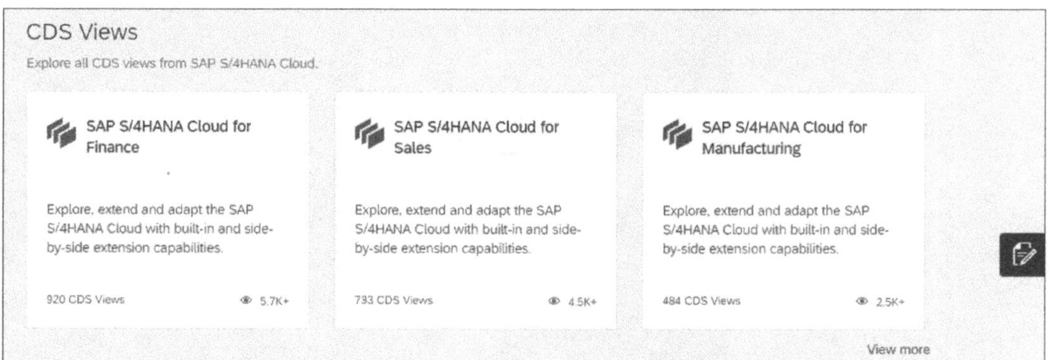

Figure 3.18 CDS Views

- **Open Connector packages**
  This package contains APIs for third-party integration, as shown in Figure 3.19. Each package has its own set of documentation to learn about the connectors and how you can use them in your integration scenarios.

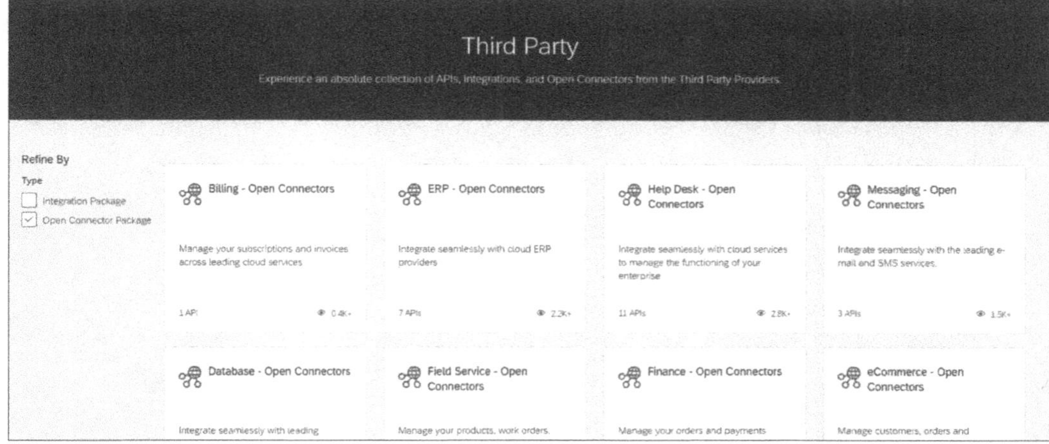

Figure 3.19 Third-Party Packages

- **Event packages**
  A digital event catalog can be found in the event packages, as shown in Figure 3.20. Underpinning agents for each event catalog are the APIs for receiving and sending notifications in case of a change of objects in the backend system, such as SAP S/4HANA.

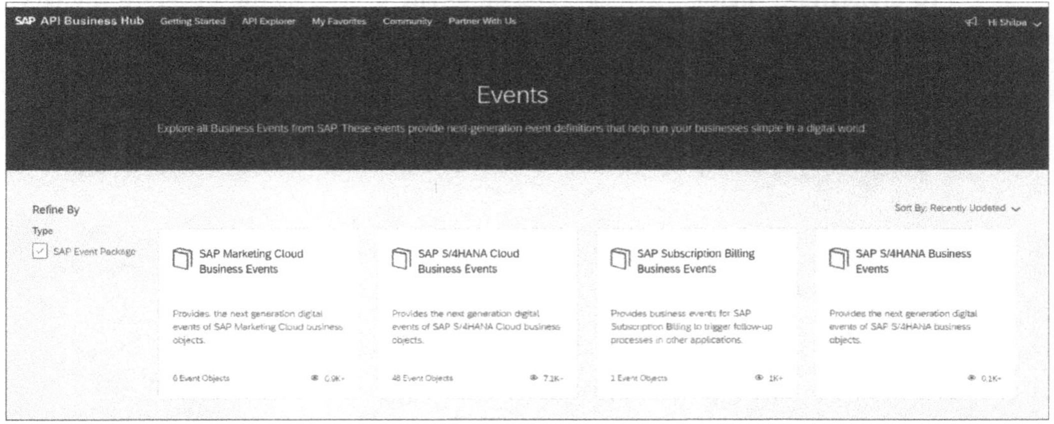

Figure 3.20 Event Packages

- **Business process packages**
  For end-to-end business interactions such as lead-to-cash, hire-to-retire, source-to-pay, and so on, packages are available that contain an end-to-end flow of activities to support each business process, as shown in Figure 3.21.

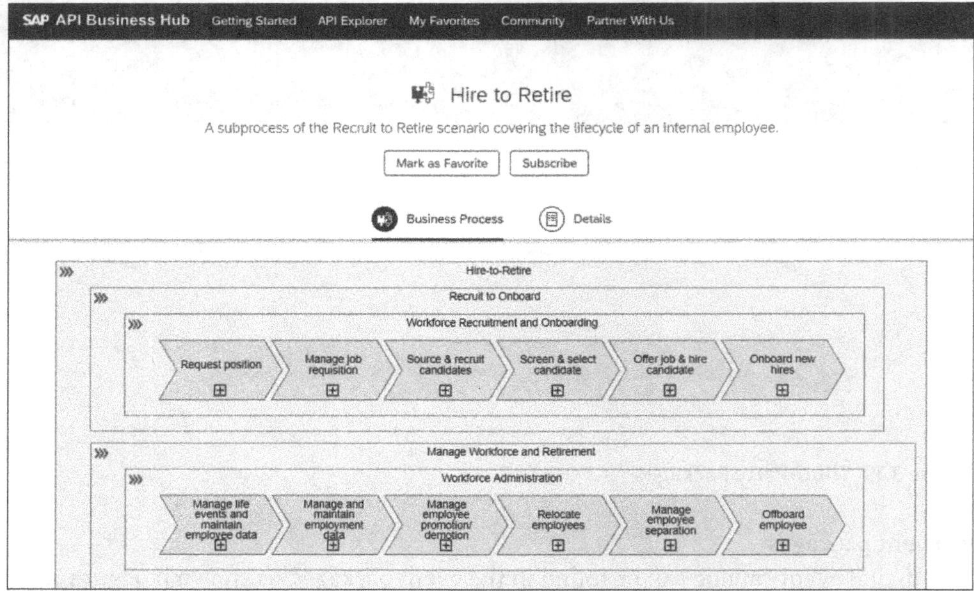

Figure 3.21 Business Process Packages

- **Workflow management packages**
  To orchestrate, digitize, manage, and automate decisions, the workflow management packages, as shown in Figure 3.22, can be leveraged to configure the workflow in a low-code approach. Users can build, run, and gain end-to-end process excellence by providing one view of the process.

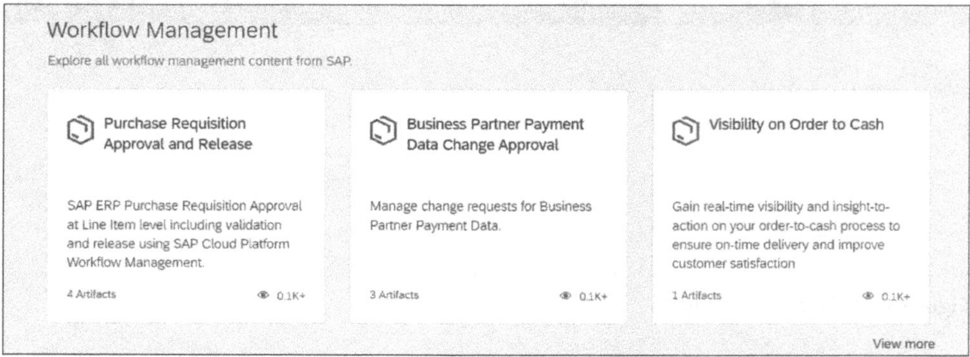

Figure 3.22 Workflow Management Packages

3.3 Building Applications, Extensions, and Integrations

To build sample applications, extensions, and integrations with SAP and third-party or non-SAP applications, SAP API Business Hub is the one-stop shop to get started. In this section, we'll walk through the key processes to discover, test, and consume APIs and other integration content.

3.3.1 Discovering APIs

How can you find the right APIs to meet your business needs? Exploring the SAP API Business Hub should be your first choice to discover the right API. SAP applications, backend systems, and business networks continue to expose their data through APIs to facilitate codevelopment, innovation, and wins in the digital economy. Let's start the journey of building an SAP Fiori application by discovering the right SAP S/4HANA APIs. Follow these steps:

1. Navigate to *https://api.sap.com*.
2. Click **APIs** under **CONTENT TYPES** to discover APIs from SAP SuccessFactors, SAP S/4HANA Cloud, Concur, SAP Ariba, SAP Business Technology Platform (SAP BTP), and 1,700-plus more APIs, as shown in Figure 3.23.

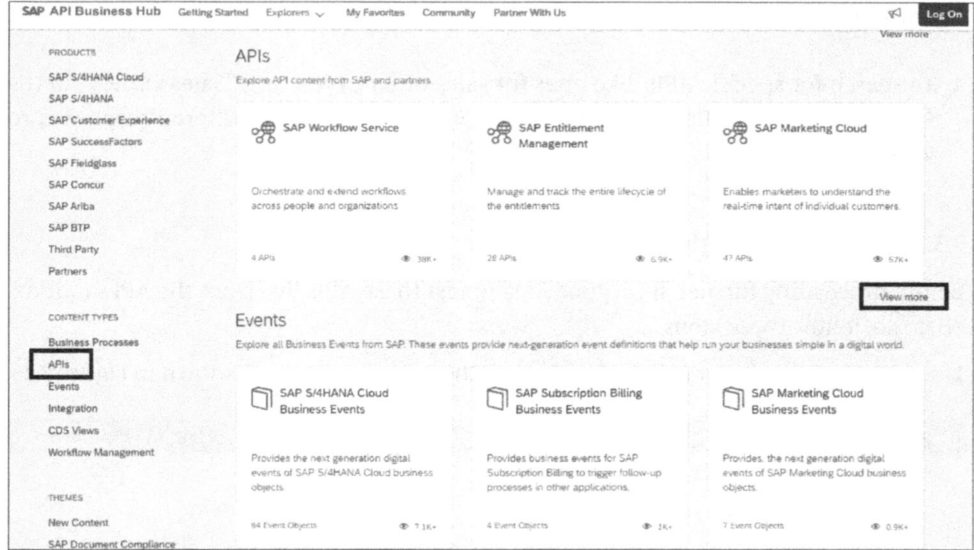

Figure 3.23 API Packages

3. Click **View More** to view all the API packages. For this example, select the **SAP S/4HANA Cloud** API package to see all the APIs specific to SAP S/4HANA Cloud, as shown in Figure 3.24.

> **Note**
>
> SAP S/4HANA Cloud contains a mix of OData and SOAP APIs. OData APIs are useful in building digital applications like SAP Fiori applications or mobile applications. The SOAP service is useful in system integration scenarios.

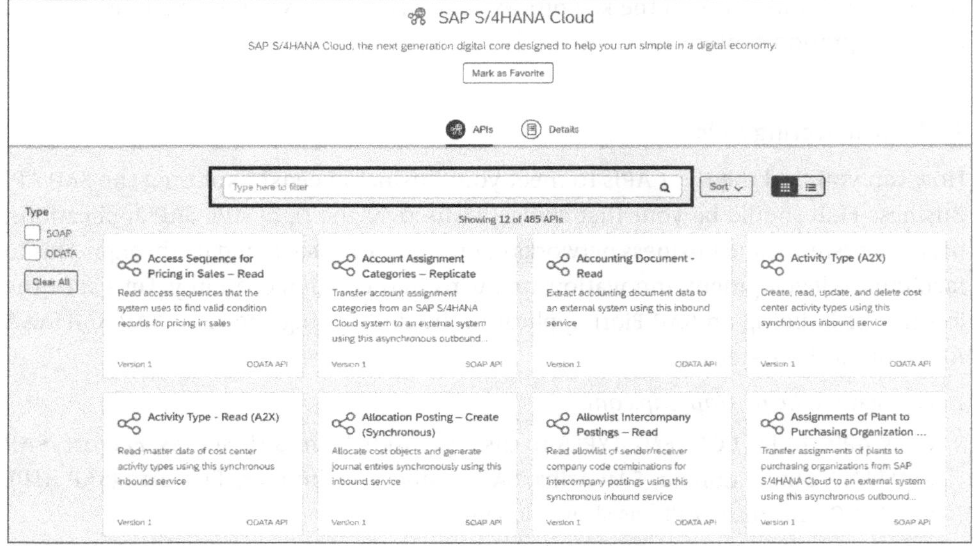

Figure 3.24 SAP S/4HANA Cloud APIs

4. To search for specific APIs, like ones for sales orders, type, say, "Sales Orders" in the **Filter Artifacts** text field. Click the Sales Orders API from the filtered responses to view the API details.

3.3.2 Testing APIs with the API Sandbox

Before proceeding further, it's a good idea to test these APIs live from the API sandbox. To do so, follow these steps:

1. Select the **Sales Order (A2X)** API and click the **Try out** button, as shown in Figure 3.25.

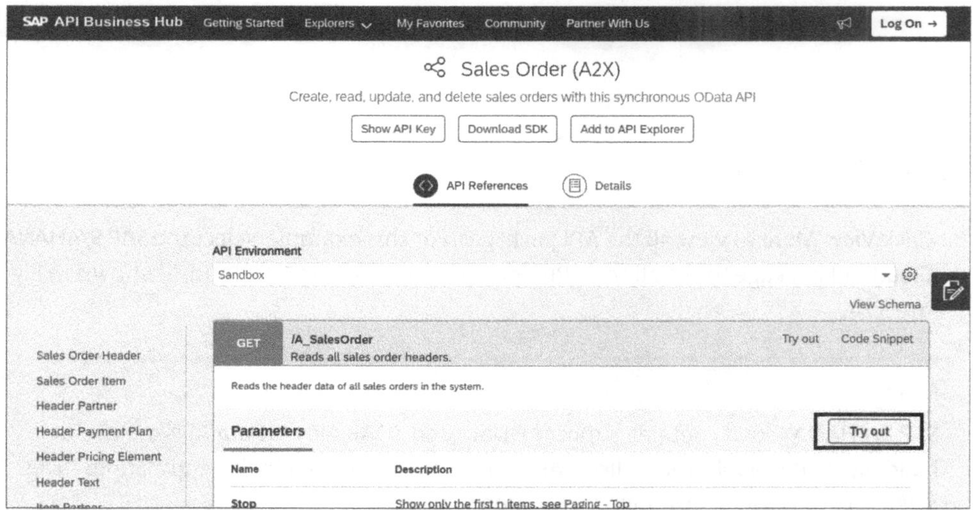

Figure 3.25 Try Out APIs

3.3.3 Consuming APIs

SAP Business Application Studio is a powerful, web-based integrated development tool to improve end-to-end application development by providing more control over the environment and code. It enables developers to quickly discover APIs from SAP API Business Hub to generate SAP Fiori applications. Let's get started and see how we can consume APIs.

Create Destination

Once the Sales Orders API has been explored and tested from the SAP API Business Hub, we can consume this API to build an SAP Fiori application in SAP Business Application Studio. Follow these steps:

1. Navigate to *https://account.hanatrial.ondemand.com/cockpit*.
2. Click **Your Trial** and navigate to your trial account.
3. Under **Connectivity**, click **Destinations**, as shown in Figure 3.26.

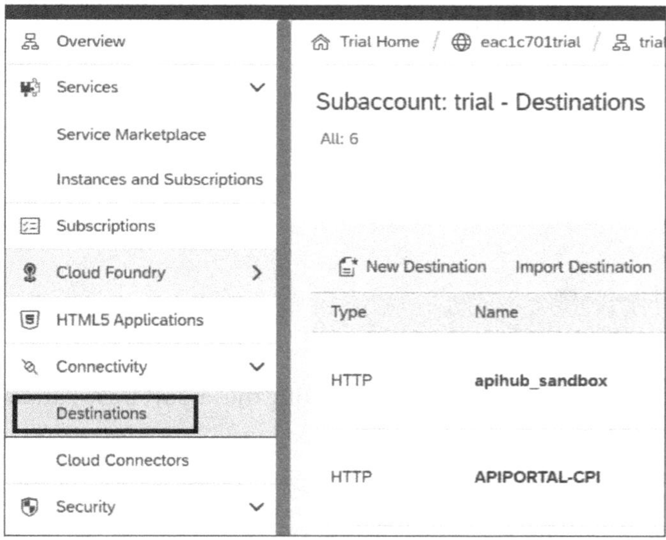

Figure 3.26 Create Destination

4. You need two destinations. The first one points to the SAP API Business Hub, and the second one points to the SAP API Business Hub sandbox, which provides service metadata and allows you to run the application. To create a destination, save the following two destinations separately in text files:
   - The first destination is shown in Listing 3.1. Its purpose is to connect to SAP API Business Hub.
   - The second destination is shown in Listing 3.2. Its purpose is to connect to the SAP API Business Hub sandbox.

```
#
#Fri Oct 30 09:47:54 UTC 2020
Description=SAP API Business Hub
Type=HTTP
TrustAll=true
HTML5.DynamicDestination=true
Authentication=NoAuthentication
WebIDEUsage=apihub_catalog
Name=SAP_API_Buisness_Hub
WebIDEEnabled=true
ProxyType=Internet
URL=https\://api.sap.com
WebIDESystem=SAP API Business Hub
```

Listing 3.1 Connect to SAP API Business Hub

```
#
#Fri Oct 30 09:47:53 UTC 2020
Description=SAP API Busniess hub sanbox
Type=HTTP
HTML5.DynamicDestination=true
Authentication=NoAuthentication
WebIDEUsage=apihub_sandbox
Name=apihub_sandbox
WebIDEEnabled=true
ProxyType=Internet
URL=https\://sandbox.api.sap.com
```

Listing 3.2 Connect to SAP API Business Hub Sandbox

5. Next, import them under the **Destination** section by clicking the **Import Destination** button, as shown in Figure 3.27.

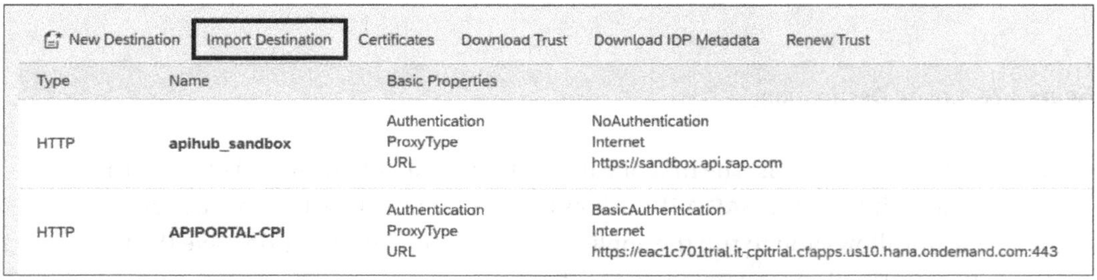

New Destination | Import Destination | Certificates | Download Trust | Download IDP Metadata | Renew Trust

Type	Name	Basic Properties	
HTTP	**apihub_sandbox**	Authentication ProxyType URL	NoAuthentication Internet https://sandbox.api.sap.com
HTTP	**APIPORTAL-CPI**	Authentication ProxyType URL	BasicAuthentication Internet https://eac1c701trial.it-cpitrial.cfapps.us10.hana.ondemand.com:443

Figure 3.27 Import Destination

6. You'll arrive at the screen shown in Figure 3.28, and all the fields should be prefilled. You can click the **Save** button.

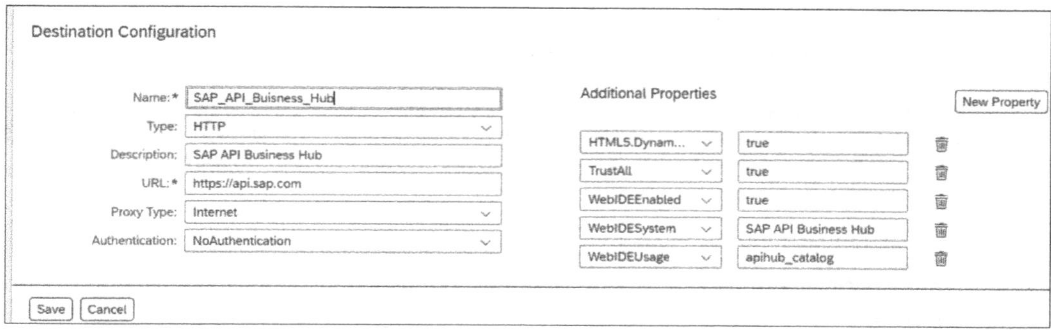

Figure 3.28 Destination Required to Establish Connection between SAP API Business Hub and SAP Business Application Studio

Creating an SAP Fiori Project

In this section, we'll walk you through creating an SAP Fiori project using SAP Business Application Studio. Follow these steps:

1. Navigate back to **Service Marketplace** on the left-hand side and select the **SAP Business Application Studio** tile.
2. Click the **Go to Application** button, as shown in Figure 3.29.

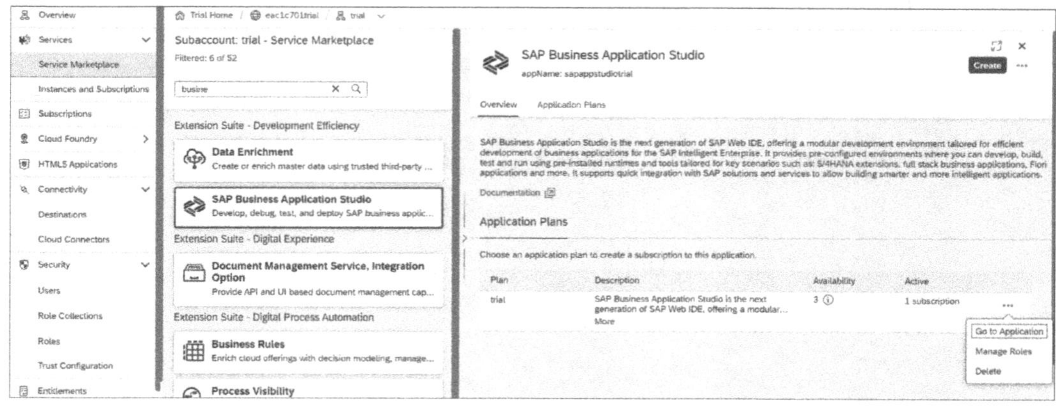

Figure 3.29 Navigate to SAP Business Application Studio

3. Create your dev space by clicking the **Create Dev Space** button, as shown in Figure 3.30. This is where you'll create your SAP Fiori project.

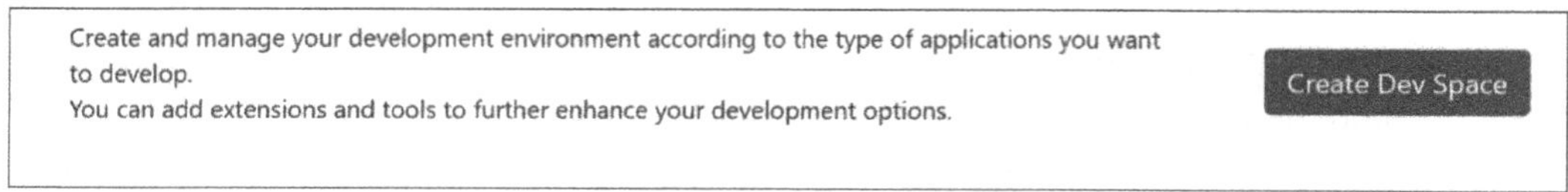

Figure 3.30 Creating Dev Space

4. Give a name to your application in the upper-left-hand corner, select **SAP Fiori**, and click **Create Dev Space**, as shown in Figure 3.31.

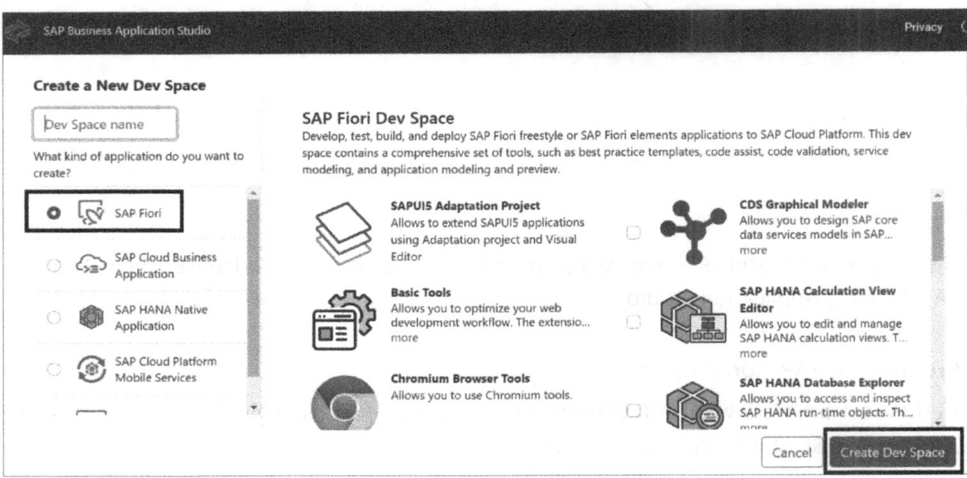

Figure 3.31 Name Dev Space

5. You can see your created space. Navigate inside your space and click **File • New Project from Template**, as shown in Figure 3.32.

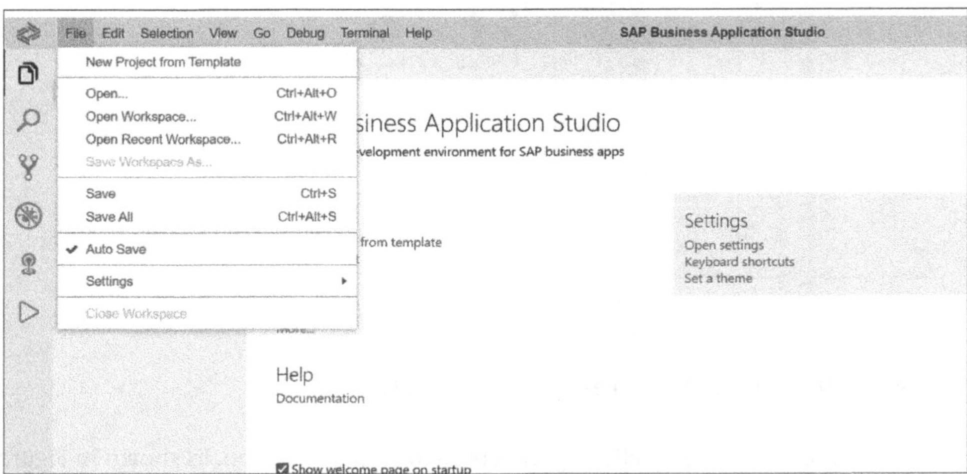

Figure 3.32 Creating Project from Template

6. Click **SAP Fiori Freestyle Project** to begin building your own SAPUI5 freestyle app, and then click **Next**, as shown in Figure 3.33.
7. Select the **Cloud Foundry** environment, select **SAPUI5 Application** as the template, and then click **Next**, as shown in Figure 3.34.
8. Enter your **Project Name**, as shown in Figure 3.35, and select **Next**.

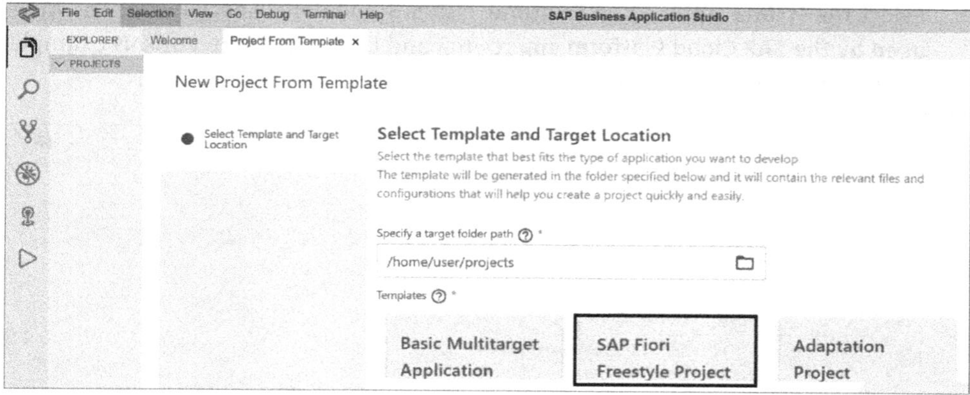

Figure 3.33 Create SAP Fiori Freestyle Project

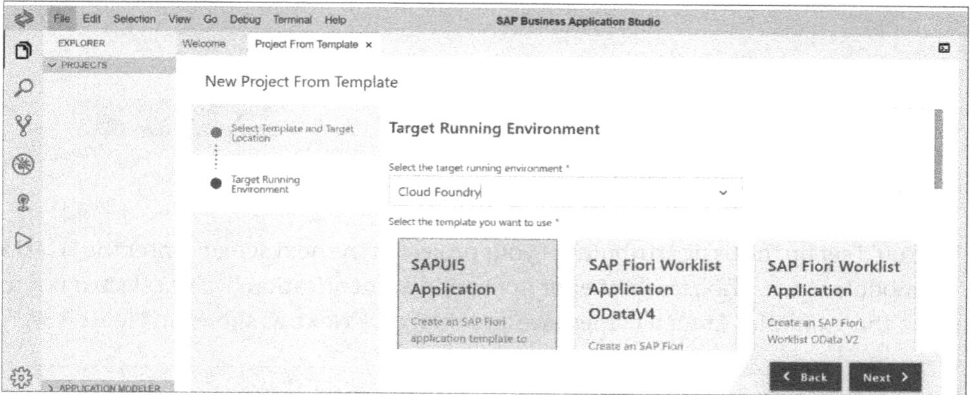

Figure 3.34 Select Environment

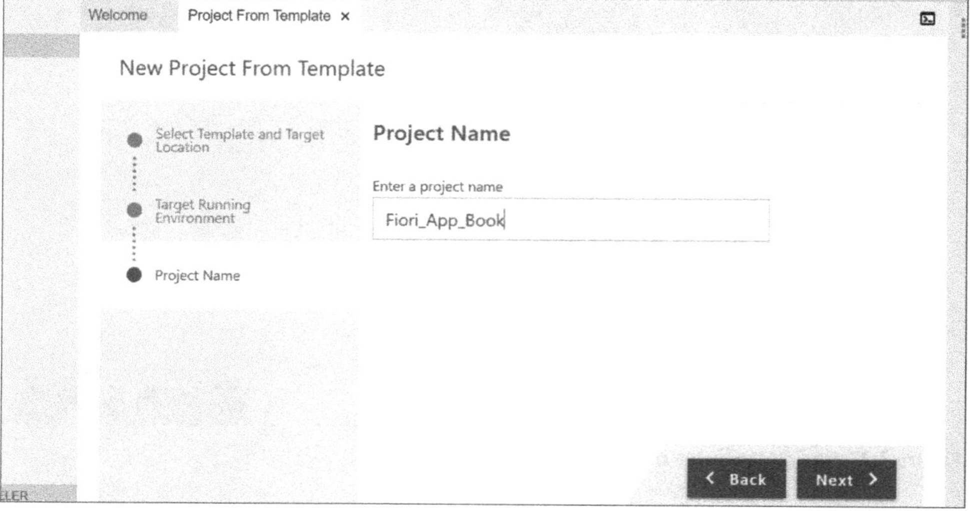

Figure 3.35 Your Project Name Screen

9. Select the **HTML5 Application Runtime**, either **Standalone Approuter** or the **Managed by the SAP Cloud Platform** app router, and then click **Next**. For this example, select **Standalone Approuter**, as shown in Figure 3.36.

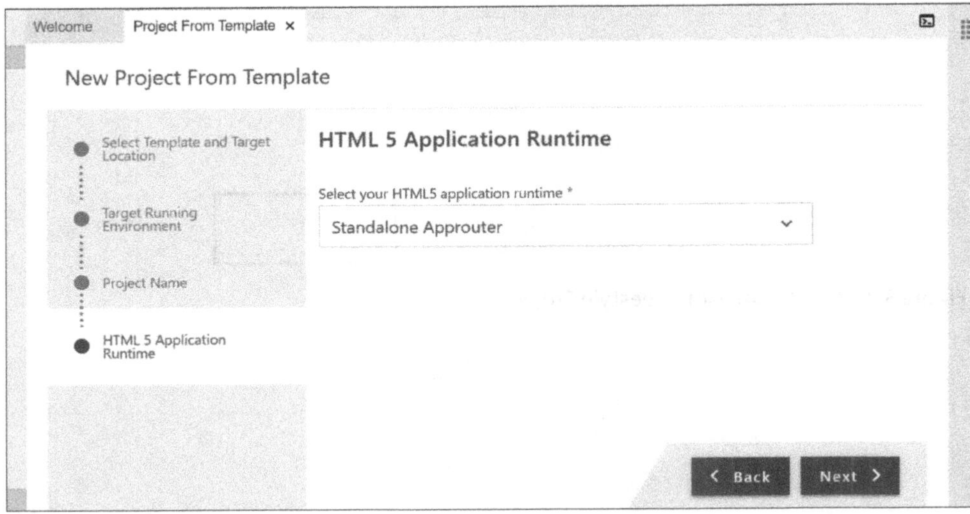

Figure 3.36 Your Application Runtime

10. You'll set up the basic attributes of your project in the next screen. Enter the HTML5 module name. Choose whether or not to add authentication (we've selected not to in this example). Enter a namespace and then click **Next**, as shown in Figure 3.37.

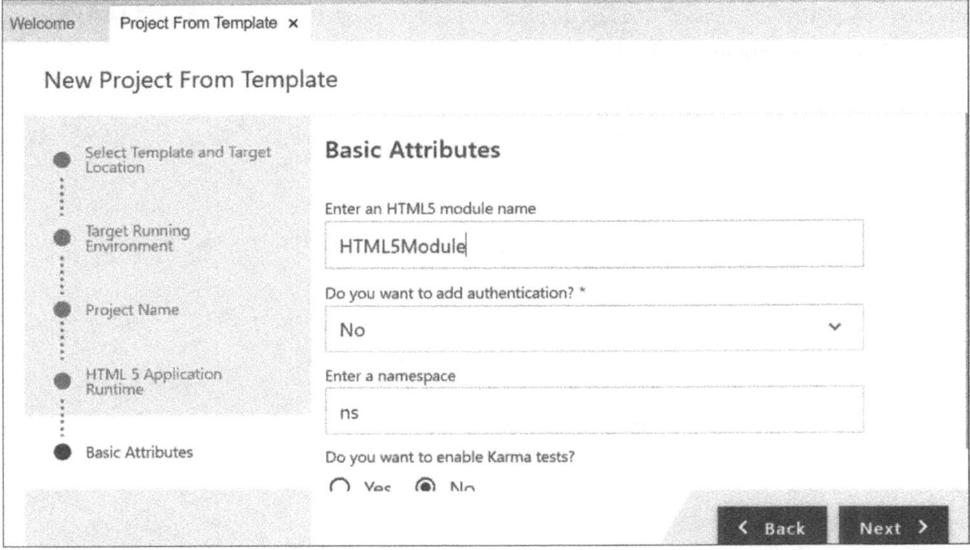

Figure 3.37 Basic Attributes of Your Project

11. Give a name to your view, select **Yes** to add a data service, and click **Next**, as shown in Figure 3.38.

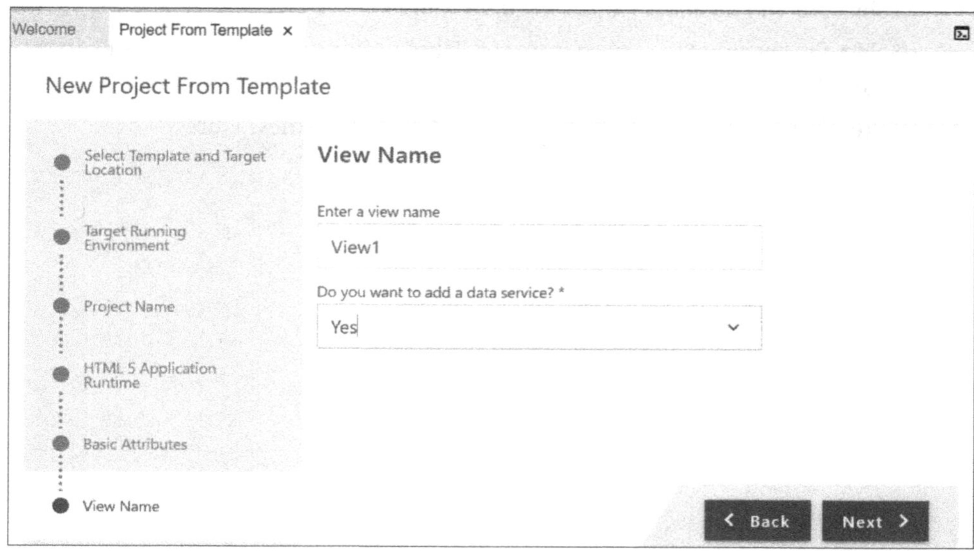

Figure 3.38 View Details

12. In the **Consume Services** step, choose the relevant data source. You can consume your service from different data source providers. Select the SAP API Business Hub in this case and click **Next**, as shown in Figure 3.39.

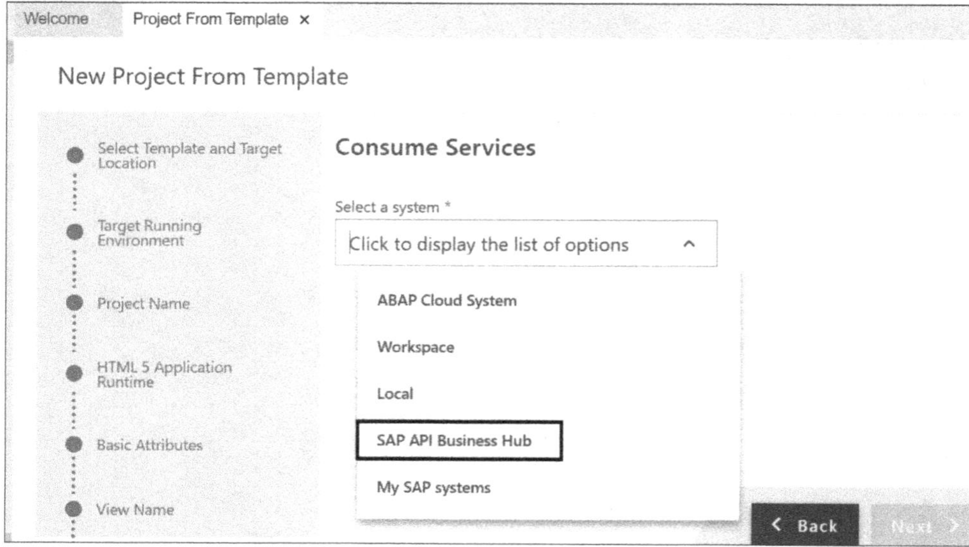

Figure 3.39 Consume Services

13. Select the following options to set up the SAP API Business Hub details, as shown in Figure 3.40:
    - Select **SAP API Business Hub** as the data source.
    - Select the specific API service from the list of options (our sales order API, in this example).
    - Enter your login credentials to access the SAP API Business Hub.

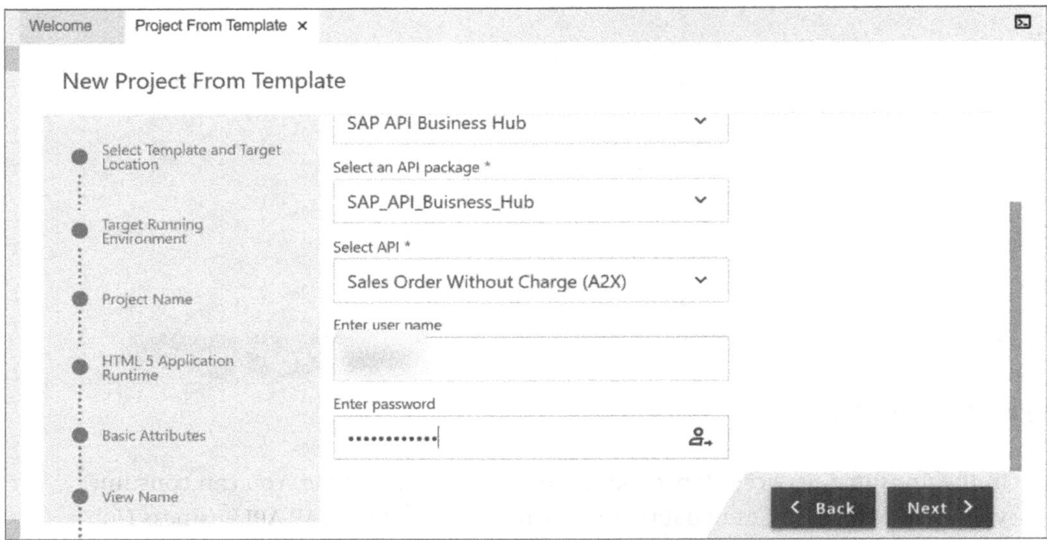

Figure 3.40 SAP API Business Hub Details

14. After clicking **Next**, the project will be created in your workspace, as shown in Figure 3.41.

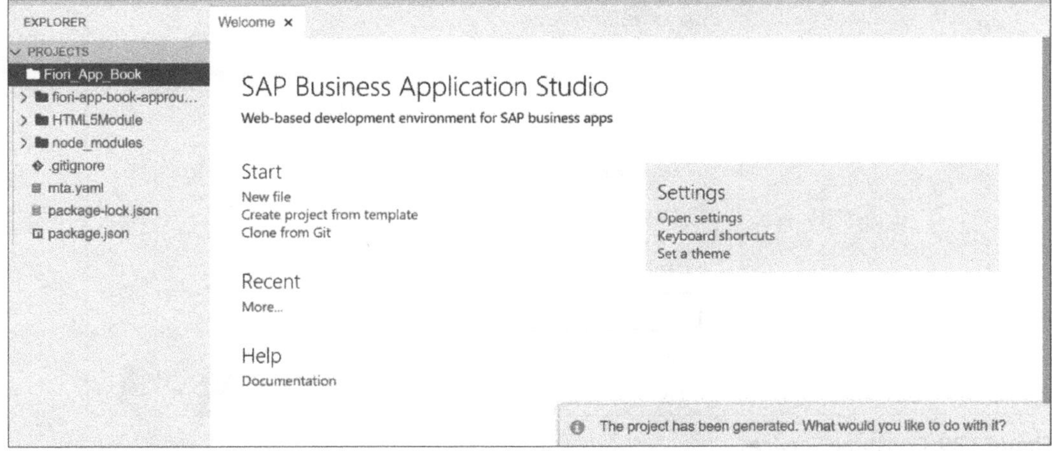

Figure 3.41 Created Project Details

15. Listing 3.3 is used to connect to the sandbox destination.

```
"dataSources": {
"mainService": {
"uri": "<sandboxdestination><servicepath><servicename>
```

Listing 3.3 SAP API Business Hub Connectivity

16. Now that the project is created, define and bind the sales order entities. Open the created project in the workspace by selecting it from the left-hand side, then open the *view.xml* file in the layout editor, as shown in Figure 3.42.

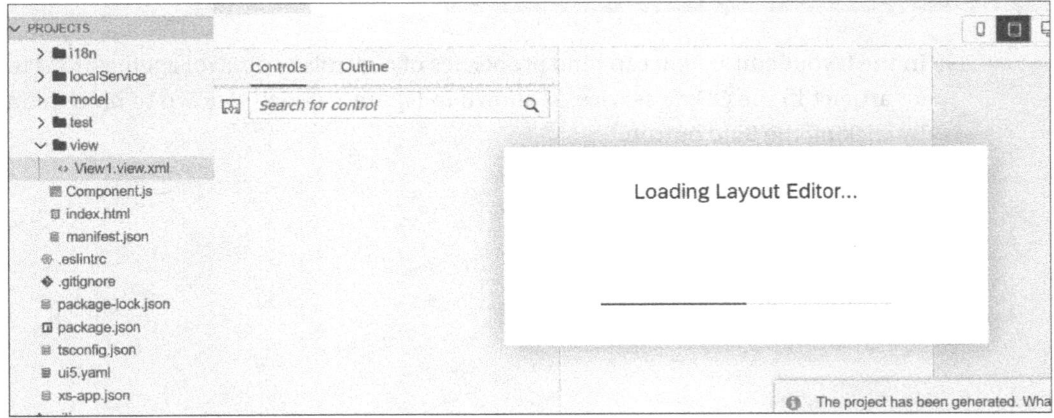

Figure 3.42 Layout Editor

17. Click **Undefined Entity Set** to define it and select the **Define entity set and set the selected control as template** option, as shown in Figure 3.43. This is selected to define the entity set as a template.

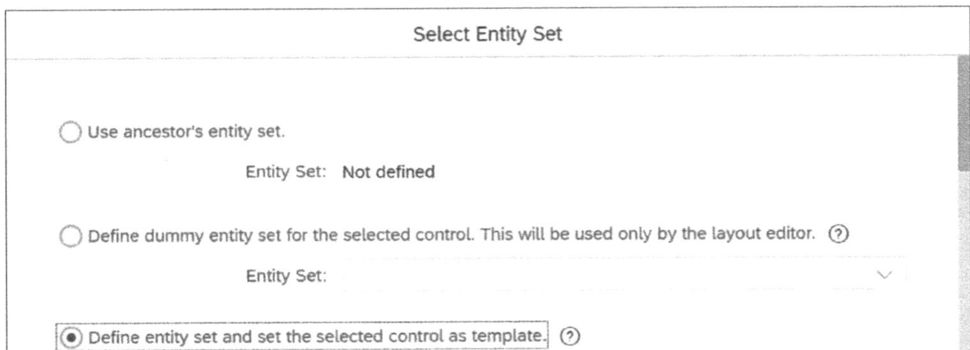

Figure 3.43 Define Entity Set

18. Select the **Entity Set**. You can choose any entity set; for this example, enter "/A_SalesOrderWithoutCharge", as shown in Figure 3.44.

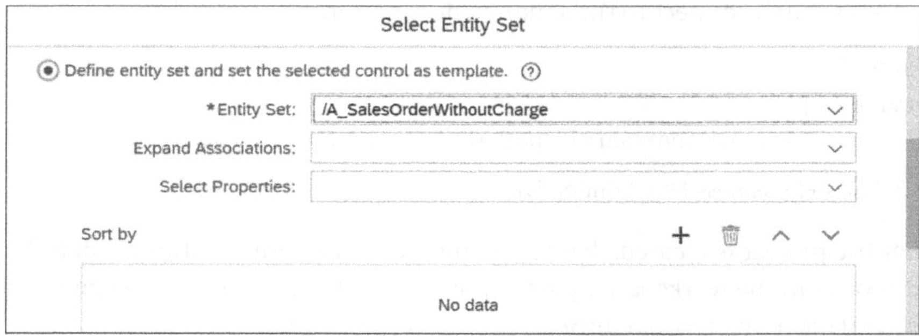

Figure 3.44 Select Entity Set

19. In the layout editor, you can bind properties of controls or control aggregations to an artifact in the OData service, as shown in Figure 3.45. Scroll down to bind them by clicking the **Bind** button.

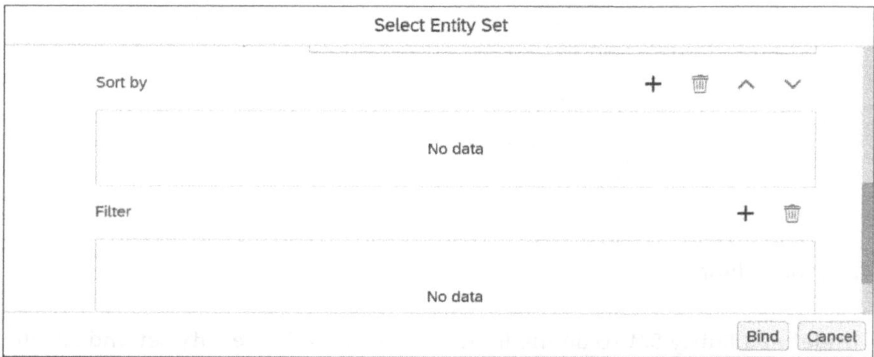

Figure 3.45 Bind Entity Set

20. Bind the data by selecting the property title from the list, double-clicking the entity, and then clicking the **Bind** button, as shown in Figure 3.46. In this example, select **CreatedByUser** by double-clicking it.

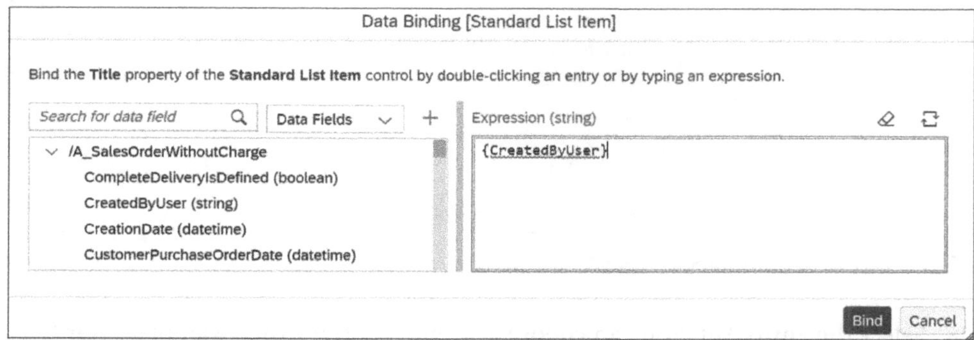

Figure 3.46 Bind Data to Entity Set

Running the Project Locally

To see the results, run the project locally. Follow these steps:

1. From the left-side menu, click **RUN CONFIGURATIONS** (the play icon), as shown in Figure 3.47, and click **+** to create a new configuration.

Figure 3.47 Option to Run Project Locally

2. Select the runnable file (HTML) and the SAPUI5 version to be used for this run configuration, as shown in Figure 3.48.

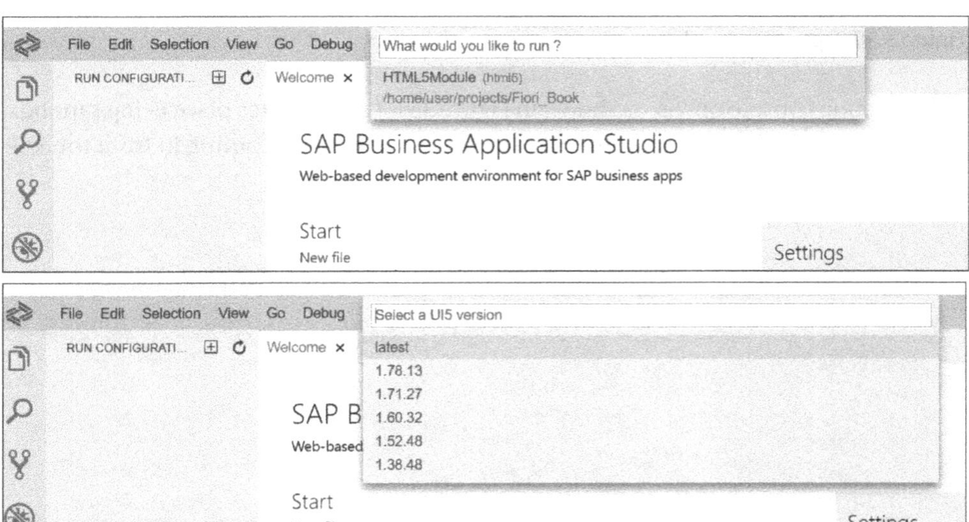

Figure 3.48 Options to Select Configuration to Run Project

3. A configuration tree appears in the **RUN CONFIGURATIONS** view containing the run configurations that were created for the runnable objects. Select the dropdown arrow to display an object's bindable services, as shown in Figure 3.49.

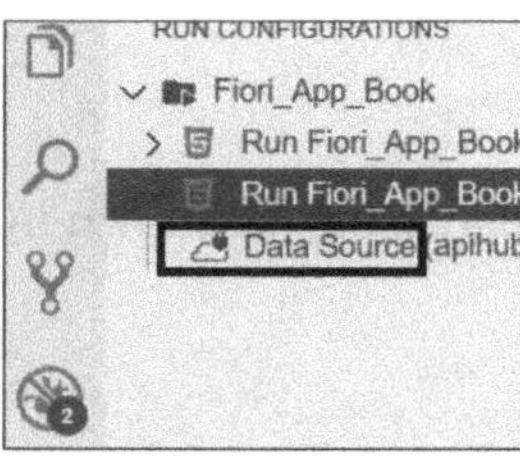

Figure 3.49 Option to Bind Object

4. Click **Bind** to bind the SAP API Business Hub data source.
5. Select the SAP API Business Hub sandbox data source, and enter your login credentials to access the SAP API Business Hub.
6. Click **Run Module** (the green play icon) to run the project, as shown in Figure 3.50.

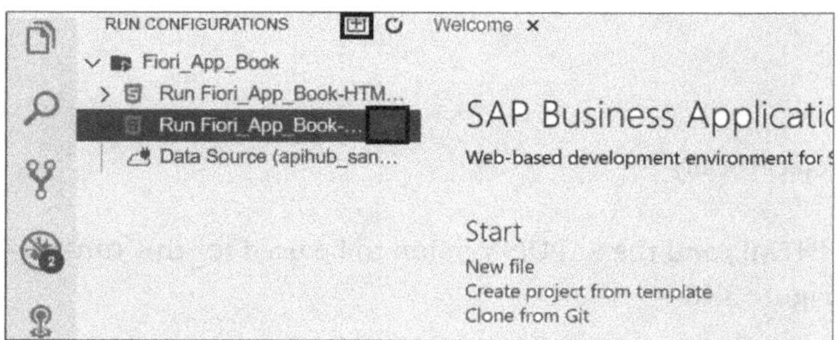

Figure 3.50 Option to Run Project

7. The debug console opens, as shown in Figure 3.51. The run takes place using sandbox data from SAP API Business Hub. These are the sales orders coming in from the SAP API Business Hub sandbox.

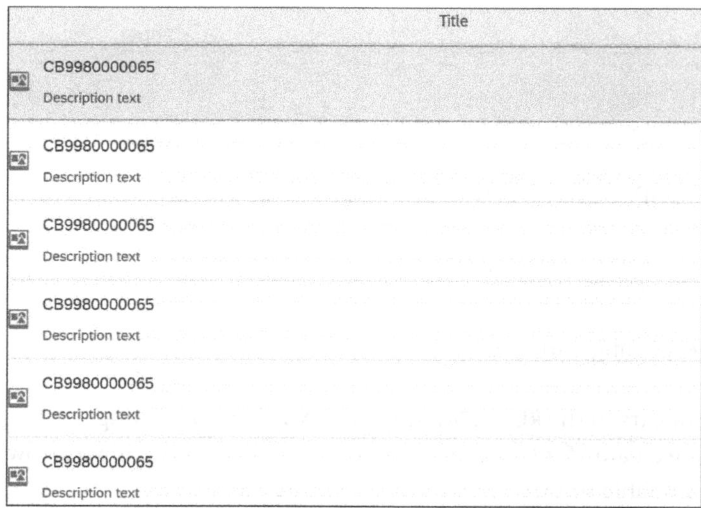

Figure 3.51 Data Coming in from SAP API Business Hub

3.3.4 Discovering Integration Packs

SAP API Business Hub is a one-stop shop for all the integrations and APIs from SAP and selected partners. Prepackaged integrations are the key differentiators to kick-start your integration journey. This out-of-the-box prepackaged integration content includes integration flows, mappings of interfaces between SAP and non-SAP applications, adapters, and value mappings and is delivered with SAP Integration Suite; therefore, SAP-to-SAP messages are free of cost with any SAP Integration Suite license if the integration flows have not been modified. Let's consider a scenario in which the currency rate replication model is maintained in the SAP S/4HANA system and this rate has to be replicated to SAP SuccessFactors because the company needs to pay its employee their monthly wages in a different currency.

For this integration scenario, we will use SAP Integration Suite as our integration platform as a service (iPaaS) and the required integration flow, which is available as preconfigured content on the SAP API Business Hub. We'll discuss how to monitor the messages that are processed across SAP Integration Suite. To discover the available integration packs, follow these steps:

1. Navigate to *https://api.sap.com* and search for "SAP S/4HANA Cloud Integration with SAP SuccessFactors Employee Central". This is prepackaged content to replicate the currency exchange rate from SAP S/4HANA to SAP SuccessFactors.
2. Now copy this content to the Cloud Integration tenant. You first need to enable the SAP Integration Suite on a trial basis (refer to Chapter 2, Section 2.3.3) if you haven't already. To do so, navigate to *https://account.hanatrial.ondemand.com/cockpit*.
3. Click **Enter Your Trial Account** and navigate to your trial. Select the **Integration Suite** tile.
4. Click the **Go to Application** link, as shown in Figure 3.52.

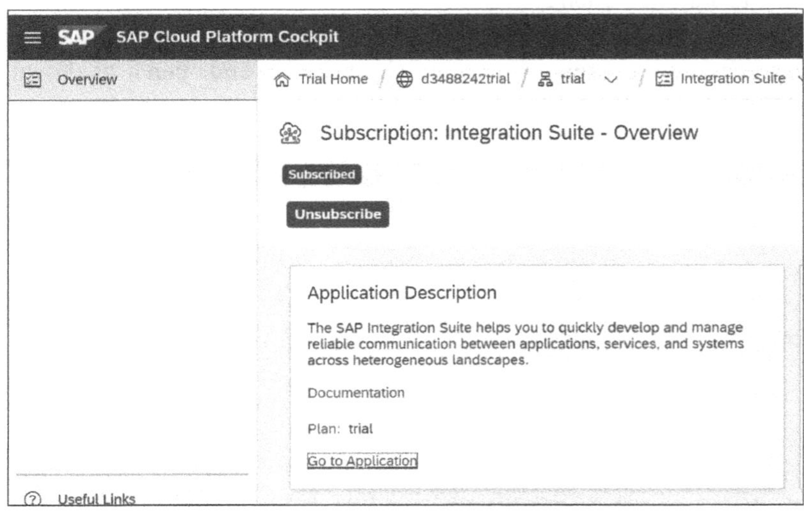

Figure 3.52 Navigate to SAP Integration Suite Application

5. Provide your username and password and click **Log On**, as shown in Figure 3.53.

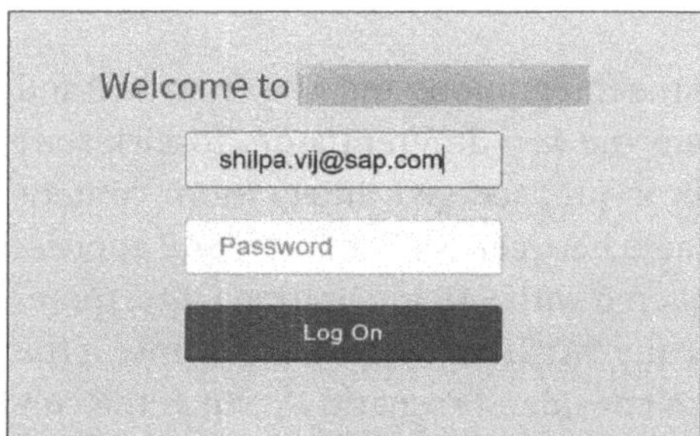

Figure 3.53 User Details Screen and Log On Option

6. Click the **Design, Develop, and Operate Integration Scenarios** tile, as shown in Figure 3.54, which takes you to your Cloud Integration tenant.

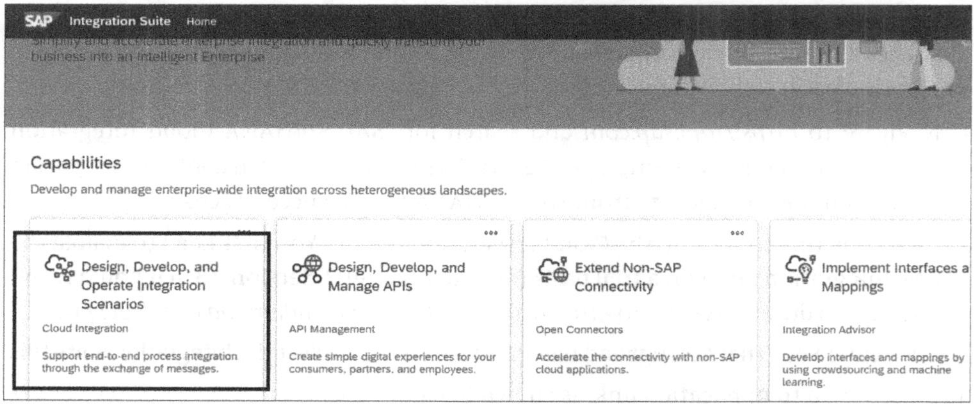

Figure 3.54 Navigate to SAP Integration Suite

7. Search for the prepackaged content, as mentioned previously, and open it, as shown in Figure 3.55.

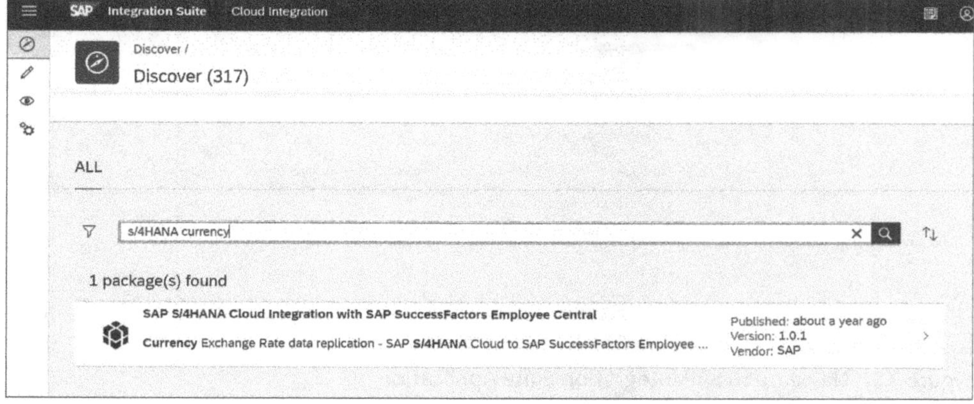

Figure 3.55 Search Options for Prepackaged Content

3.3.5 Consuming Integration Content

Prepackaged content can be copied to your workspace and configured based on your needs. Follow these steps:

1. Click the **Copy** button to copy the content to your workspace, as shown in Figure 3.56.

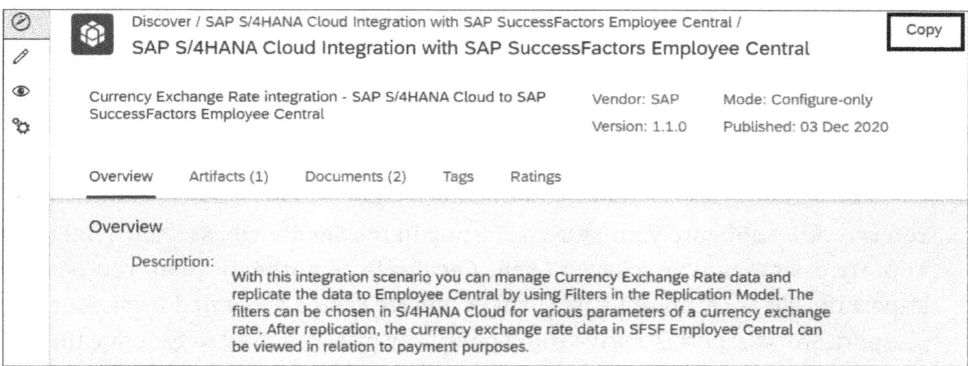

Figure 3.56 Copy Prepackaged Content to Your Tenant

2. Navigate to **Design** (the pencil icon) to configure the endpoints, as shown in Figure 3.57.

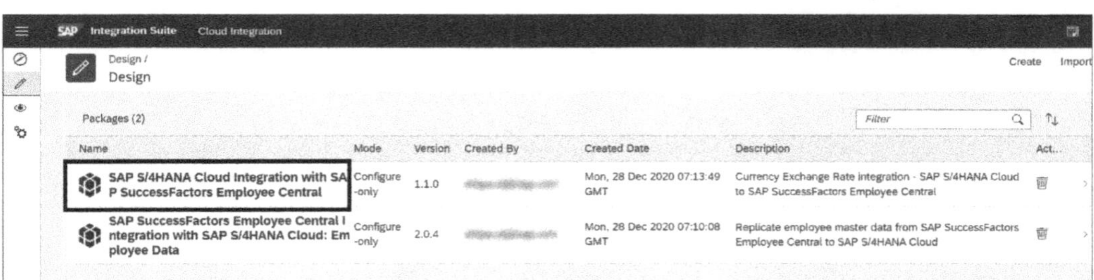

Figure 3.57 Copied Integrations in Design Tab

3. Open the package by clicking it and navigate to the **Artifacts** tab, as shown in Figure 3.58. Select your artifact's checkbox.

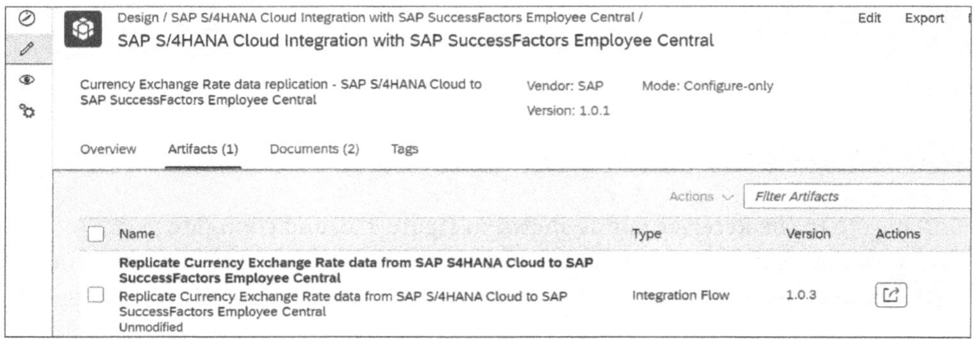

Figure 3.58 Navigate to Artifacts Tab

4. From the **Actions** menu, select **Configure**, as shown in Figure 3.59.

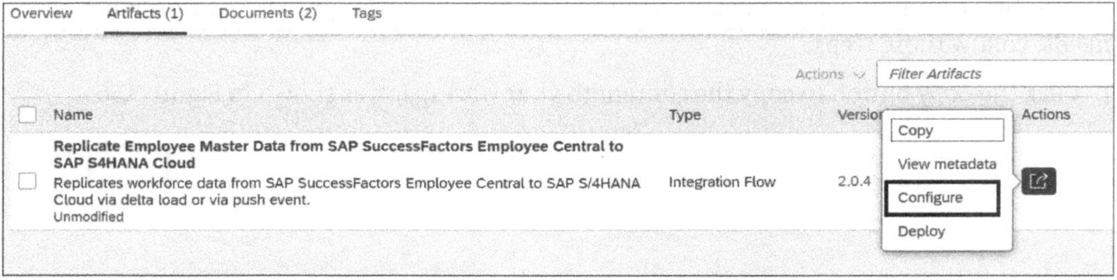

Figure 3.59 Configure Package

5. You can now configure your system starting in the **Sender** tab, as shown in Figure 3.60. Here, since you've selected **Client Certificate** as **Authentication**, you need to import the client certificate. The client certificate can be generated from your SAP passport. SAP passport is a private and public key pair. You need to generate the certificate from it. The simplest method is to import the passport as a key pair in the SAP Integration Suite KeyStore. Once it's imported, click **Download Certificate** to get client certificate. Once it's generated, you can select your client certificate, click **Select**, and browse the certificate from your local system.

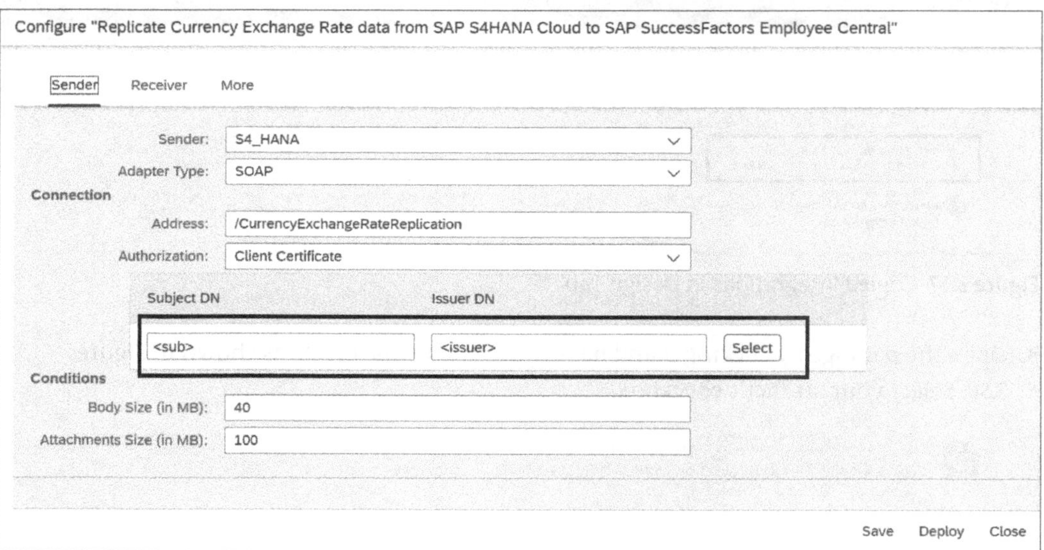

Figure 3.60 Configure Your Sender Parameters

6. Navigate to the **Receiver** tab, as shown in Figure 3.61, and configure your SAP SuccessFactors system by filling in the address of the SAP SuccessFactors system and the credentials, which are deployed on the tenant. When you're done, click **Deploy**.

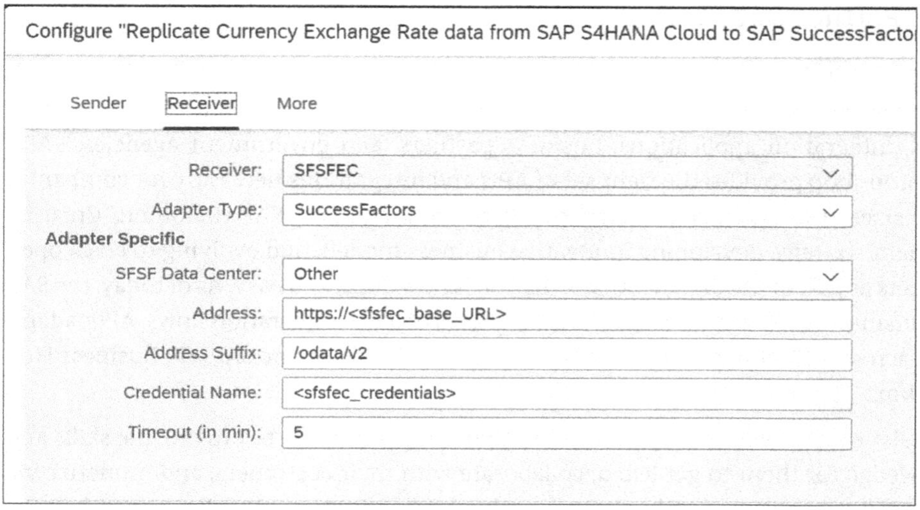

Configure "Replicate Currency Exchange Rate data from SAP S4HANA Cloud to SAP SuccessFacto

Sender Receiver More

Receiver: SFSFEC

Adapter Type: SuccessFactors

Adapter Specific

SFSF Data Center: Other

Address: https://<sfsfec_base_URL>

Address Suffix: /odata/v2

Credential Name: <sfsfec_credentials>

Timeout (in min): 5

Figure 3.61 Configure Your Receiver Parameters

7. Navigate to **Monitoring** (the eye icon), as shown in Figure 3.62, and check the status. The integration flow should be in a **Started** state.

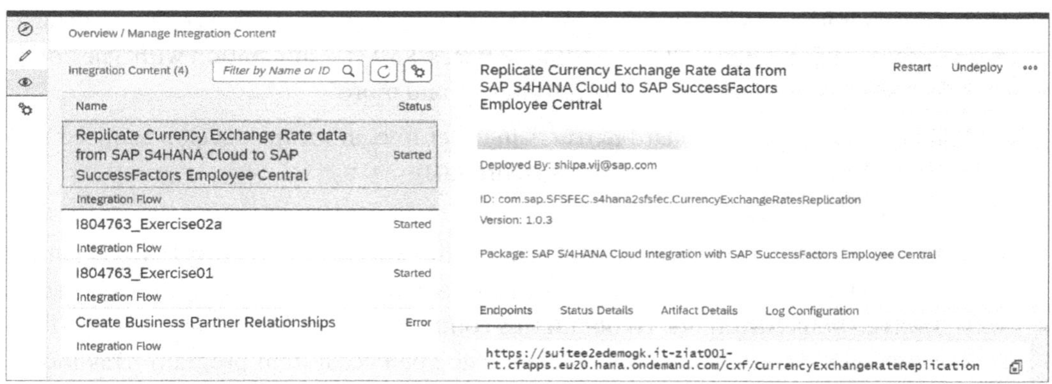

Figure 3.62 Monitoring UI to Check Integration Flow Status

Once the replication is triggered from your SAP S/4HANA system, the configured integration flow will run and replicate the message in the SAP SuccessFactors system. Also, you can see that the message in the monitoring UI will be **Completed**, as shown in Figure 3.63.

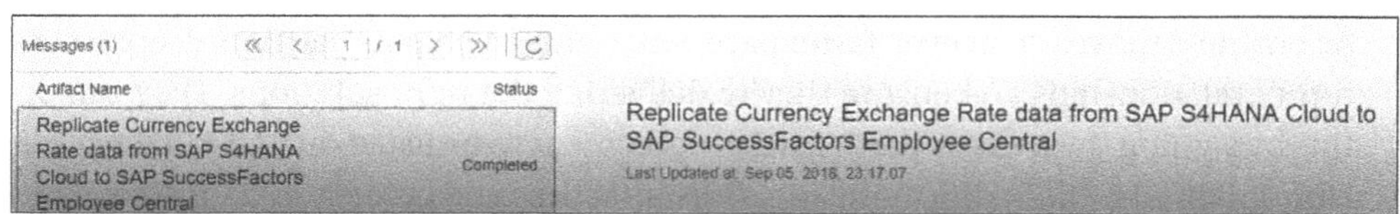

Figure 3.63 Integration Flow Status after Message Is Processed Successfully

3.4 Partner Onboarding

The SAP API Business Hub provides the foundation that enables digital businesses. From discovering the right APIs for your businesses to building applications for extensions, integration applications, business partners, and government agencies, SAP's one-stop shop provides the right set of APIs and integrations necessary for companies of all sizes. It has gained immense popularity among many SAP customers. Creating the right strategy, developing innovative business models, and evolving to crack open markets as part of the digital transformation is now quick and easy. As of today, the SAP API Business Hub has more than 3,000 digital artifacts—integration flows, APIs, adapters—across 300-plus packages, with millions of API calls in the SAP API Business Hub sandbox.

To assist with growing requests from partner ecosystems and to provide the skills and knowledge for them to get listed, collaborate with their customers, and monetize via SAP API Business Hub, SAP has been focusing on creating a solid, well-planned foundation for onboarding partners' digital content. SAP and partners can collaborate throughout the complete framework for publishing digital content on SAP API Business Hub. SAP supports partners through the complete lifecycle (onboarding, enablement, publishing, and monetizing) of their digital content.

Digital content (APIs and integrations) from selected partners is listed on SAP API Business Hub—for example, HERE Location Services, SAP ERP integration with Salesforce, Office 365 Cloud Adapter by Rojo Consultancy, and more.

Follow these steps to onboard partners in order to coinnovate, develop at speed and scale, and engage with like minds (see Figure 3.64):

1. **Join the SAP partner program**
   Developing a long-term successful relationship with partners is very critical in order to provide wider choices by building solutions with SAP or by offering SAP's solutions to your customers. To obtain this common goal and sustainable competitive advantage, SAP offers the SAP PartnerEdge open ecosystem program. This model allows SAP's partners to engage with SAP at zero cost and minimal contract obligation. Partners can register for and get access to the SAP partner portal to try out a variety of products, learn and explore trainings, and enable resources. It also provides a deeper and expanded scope through the SAP PartnerEdge open ecosystem specialized partner program. You get access to exclusive enablement offerings, and SAP provides tools that fit well with your needs.
2. **Sign the SAP API Business Hub participation agreement**
   As business moves at an ever-faster pace, SAP wants to help its partners deepen customer relationships and ensure they're delivering the right solutions. They can do this by signing the partner participation agreement to be listed on SAP API Business Hub, followed by adhering to the set of best practices and guidelines defined by SAP.

3. **Provide a sandbox for APIs and follow SAP Best Practices and guidelines**
   Partners have to provide a testing environment for each of their APIs that are planned to be listed on SAP API Business Hub. Once you develop your APIs in the OpenAPI 2.0 or OpenAPI 3.0 format and make sure the APIs' packages follow the guidelines to be published, you can now submit the packages for review and get a quality score. This score determines the extent to which your package adheres to the guidelines.
4. **Certify your content via SAP Integration and Certification Center**
   Once partners decide to transform customer businesses by providing best-quality coinnovation, certification, and technical services, they can join the SAP Integration and Certification Center program (*https://www.sap.com/partner/certify-my-solution.html*). This helps them to certify their business APIs and get an "SAP Certified" logo. API packages will also be listed as "SAP Certified" content on SAP API Business Hub. For integration packages, SAP Integration and Certification Center already supports certification scenarios, and these are already available on SAP API Business Hub.
5. **Publish to SAP API Business Hub**
   Publishing digital content on SAP API Business Hub can be visualized in four steps:
   - *Evaluate:* This phase involves validating the solution blueprint and use case and making sure it aligns with SAP's strategy.
   - *Legal/admin:* This phase involves signing the partnership and getting the contract status.
   - *Curate:* In this phase, the quality of the content is checked to ensure that it follows the guidelines.
   - *Value add:* If there is a change in content after a few months or years, SAP helps to manage the patches and, last but not least, gives tips and tricks to develop new use cases.

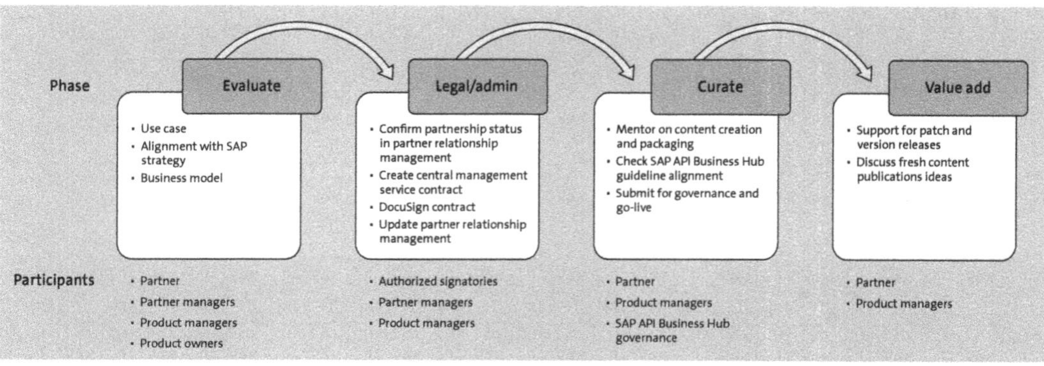

Figure 3.64 Phases of Publishing on SAP API Business Hub

6. **Engage with your consumers**
   Digital content published by partners can be consumed from SAP API Business Hub free of charge. The complete "launch to decommission" integration lifecycle can be managed via SAP API Business Hub.
7. **Sell via SAP App Center**
   Partners can choose to monetize their apps via the SAP App Center, while API/integration packages can be consumed from SAP API Business Hub.

3.5 Summary

In this chapter, you learned about the central catalog of digital content offered to SAP users by SAP and its selected partners. We focused on the significant value companies get from the out-of-the-box, predelivered content available on SAP API Business Hub. You can quickly build apps and extensions; integrate applications, business partners, and government agencies; and transform your business. You learned how to simplify and accelerate your integration journey by discovering, exploring, testing, and consuming different kinds of digital content bundled into logical packages on SAP API Business Hub. You can leverage high-touch and low-touch enablement through the vibrant community on SAP API Business Hub. As an application developer, you can build SAP Fiori applications, as detailed in this chapter, using predelivered APIs. Further, as an integration expert, you can configure predelivered integrations as described herein in your Cloud Integration tenant of SAP Integration Suite.

To help the partner ecosystem with digital transformation and move businesses at an ever-faster pace, partners can deepen their customer relationships and ensure they're delivering innovative solutions based on their customers' needs through deep partner engagement programs offered by SAP on SAP API Business Hub.

In the next chapter, we'll move on to discuss the API-led approach to hybrid integration.

Chapter 4
Open APIs

Application programming interfaces (APIs) are pivotal for real-time digital interactions. In this chapter, we'll take a first look at the API Management capability of SAP Integration Suite and learn how it helps to manage the complete lifecycle of APIs. We'll walk through the journey of enterprises, overcoming the complexities by leveraging SAP Integration Suite capabilities.

APIs are the foundation of digital transformation. They are the digital building blocks to create digital apps, integrations, and innovations. APIs help with openness and provide omnichannel experiences to businesses. Imagine you've walked into a meeting room for a seminar. There is a beautiful red carpet, but below it there are wires, sockets, and other types of equipment to help with the presentation. APIs can be compared to the red carpet, helping to hide the complexities of IT landscapes in order to provide a simple connected experience. When you book a trip online over the internet to travel from your home to your workplace, behind the scenes there are APIs providing a seamless experience. It may include tracing your car using location service APIs, using built-in logic to charge customers fees, paying using online wallets, watching your favorite movie while traveling, and the mobile app that gives you this beautiful connected experience.

APIs are the mainstream enablers for digital transformation for the following reasons, as shown in Figure 4.1:

- They empower the developers both within and external to your organization to build modern frontend applications like mobile apps, web apps, and chatbots.
- They foster collaboration between your partners and developers for omnichannel access by providing secure and governed access to your data and providing processes from your system of records as APIs.
- They give richer insights into API consumption and can be used for error reporting or monitoring.

In this chapter, we'll explore the API Management capability of SAP Integration Suite, create and deploy your first API, walk through the API management lifecycle, and see how to integrate with other capabilities.

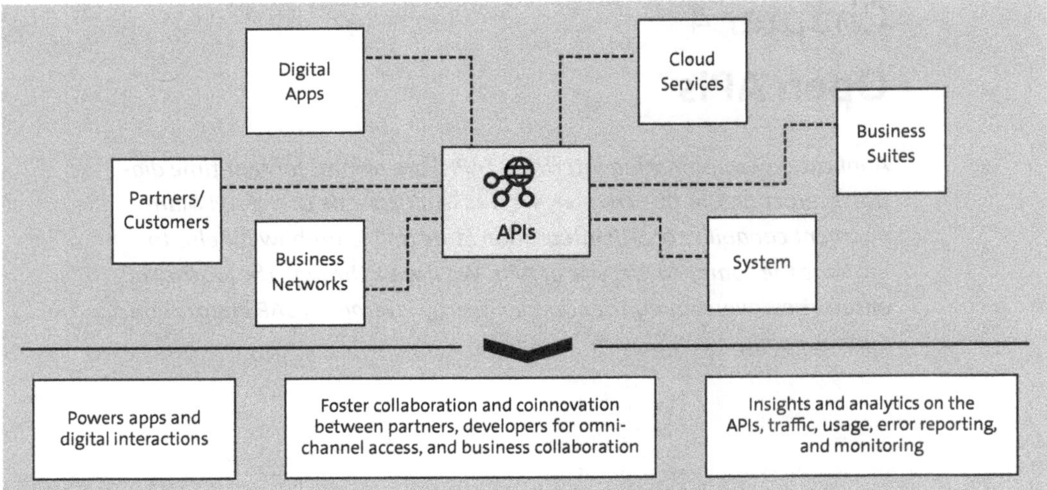

Figure 4.1 APIs Enabling Digital Transformation

4.1 Exploring API Management

When digital businesses have access to data and functions through APIs, it empowers them to drive innovations, deliver customer satisfaction, open channels for innovations, and identify new areas of collaboration with partners and customers. APIs are the largest enablers of digital transformation, and companies are using SAP Integration Suite's API Management capability for crafting and executing their API strategies by facilitating real-time open APIs.

In this section, we'll walk through the API Management capability, including customer use cases and the digital integration hub. We'll then teach you to create and deploy an API.

4.1.1 Use Cases

Companies are generating real business value by innovating with SAP Integration Suite capabilities across many diverse use cases, including application-to-application (A2A) integration, business-to-business (B2B) integration, and business-to-government (B2G) integration; API management and an API-first strategy; data integration; event-driven use cases; and IoT scenarios. They are primarily using the API Management capability for four different use cases:

- **API-driven business transformation**
  Underpinning any great digital experience is data. Almost every enterprise is involved in working with some form or another of data coming in from huge data lakes. Data is the nucleus powering digital growth, collaboration, and disruptive

complements. APIs allow enterprises to unlock digital capabilities in new and exciting ways.

Let's consider an example. One of the British multinational energy and services companies that supply gas and electricity to companies across the globe needed to transform its ability to meet their customers' digital expectations to transact in a 24/7 world with improved stability and response time. Its business was running on ten-year-old enterprise resource planning (ERP) and customer relationship management (CRM) backends, which did not offer the flexibility or agility to respond to company expectations and couldn't deliver the kind of service experience modern consumers expect. They didn't want to change their backends to switch their business, so they leveraged the API Management capability to quickly and easily build APIs to expose data from their backend systems and used SAP HANA as a service to cache the data necessary to deliver blazing fast performance. Now millions of their customers use the mobile app or the company website to handle their interactions, amounting to five to seven million interactions per day.

- **Omnichannel API access**
  With this unprecedented mindset shift, omnichannel is becoming more than just connecting systems and doing integrations. Enterprises are transforming beyond business transactions and into multichannel engagement to improve customer experience. An API-led approach helps the company to achieve greater organizational agility and greater control, develop a meaningful relationship, and build scalable and resilient infrastructure, especially during times of change.

  Let's consider an example. One global personal care cosmetics group faced challenges integrating SAP ERP with third-party tax solutions across cities, each with its own local and municipal taxes. It adopted a cloud-first and API-first strategy, and with the API Management capability, it quickly rolled out a mobile application to its local sales force, enabling it to comply easily with local market regulations.

- **Real-time integration via APIs**
  Real-time information is critical to today's digital experience, and APIs become the connecting tissue of great digital experiences. When you order groceries or apparel online, you can track your package by getting real-time data. While waiting for your ride, you can get the location of your car—another example of real-time updates.

  Let's consider an example. One of the largest Asian pharmaceuticals, with several medical facilities over different market segments, had a very ambitious vision of transforming a traditional, 100-year-old distribution company into a frontrunner in technological and digital health innovation. By leveraging the API Management capability, it was able to create an API ecosystem to improve company satisfaction throughout the end-to-end experience and reduce processing times by opening up APIs to various stakeholders.

- **Modernize IT and microservices**
  Gone are the days when monolithic architecture was commonly used to design applications. Today, organizations are taking a microservice approach to design their landscapes so that they have well-designed applications. Fine-grained functions (microservices) give them the flexibility to deploy in hybrid environments, thereby improving the infrastructure. Just as deployment has reached new speeds, with updates taking a few hours, fixes reach companies in a day or two and applications are available 24/7, so users' expectations have increased tenfold.

  Let's consider an example. One of the largest privately held family-owned spirits companies in the world was facing challenges with its increasingly heterogeneous landscape and increased complexity. This in turn led to an increased demand for integration. It built out an API-driven integration architecture to increase openness and reuse while driving down the cost and complexity of interface development. It already had quite a few integrations running in its on-premise SAP Process Orchestration platform, but realized it needed an effective and performant hybrid integration architecture that would let it use the best of what it already had while moving forward with API-driven integration to meet the demands of a modern digital enterprise. The company decided to create an easy-to-use on-demand CRM automation platform integrated into all markets, teams, and IT solutions as a necessary first step in the digital transformation journey. For the company, a hybrid integration approach with microservice architecture was the way to go, as it already owned SAP Process Orchestration on-premise.

4.1.2 API-Based Integration Architecture

The conventional API integration architecture, as shown in Figure 4.2, consists of the following:

- The system of records for application/data services exposes the data from your business suite, cloud services, or database.
- In the integration middleware, the orchestration, transformation, and mapping are done for exposing the process and data as APIs.
- In the API management layer, the API best practices for API governance like security or authentication policies and the traffic management policies like a quota or rate limit are applied. If required, policies for monetization like rate plan and the like can also be attached.

Finally, the APIs are made available to the system of engagements or channels. Here the APIs are shared with your partners/companies for omnichannel access or consumed within a digital application.

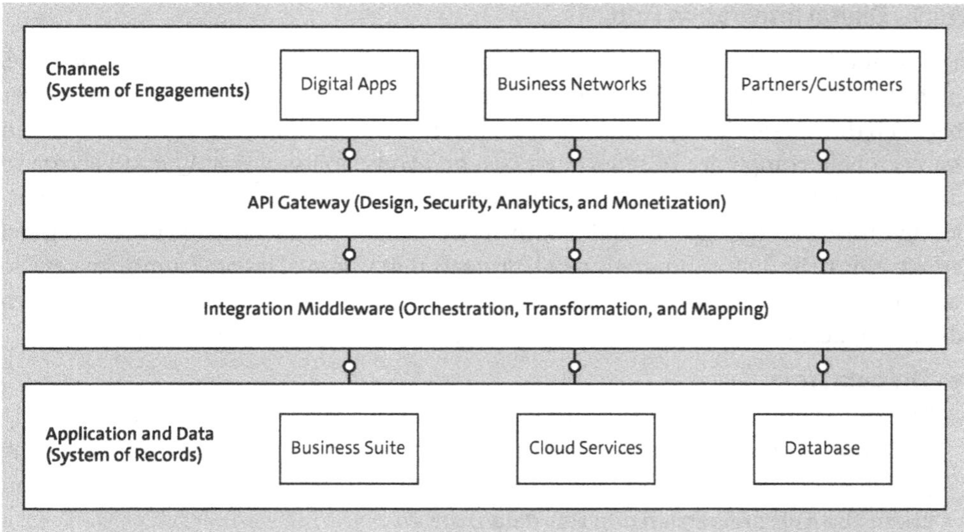

Figure 4.2 API-Based Integration Architecture

There are some challenges that go with an API-based integration architecture. Implementing APIs that provide fast and high-scale access to a backend system of records, applications, and data sources may imply the following:

- Exposing these legacy systems to potentially massive workloads puts stress on the backend systems to support additional workload generated by the frontend applications. Supporting these scenarios may be quite costly in terms of, say, hardware upgrades to the backend systems, required to scale the legacy systems to handle the massive API calls. The huge number of API calls potentially implies a massive workload that ultimately hits the system of record applications and the integration layer. This may entail significant costs to upgrade your infrastructure to support the additional workload. It may even be impossible to support this kind of workload altogether without radically redesigning the system.
- It would be difficult or a lot of effort to leverage cloud-native features of zero downtime and 24/7 high availability for the legacy system. If some backend systems are unavailable, such as during a maintenance window, then the integration layer cannot serve as the frontend for API services and can significantly impede business operations. This is increasingly unacceptable in a 24/7, always-on, global environment. A change in the backend systems (e.g., a modification of a field definition in a database table) implies some reengineering of the integration services or their complete redesign—for example, if a backend system is replaced with a different solution.
- The integration work needed to implement the frontend API services might prove substantially complex, especially if the data is fragmented across multiple backends and sometimes across various vendors (SAP or non-SAP).

4.1.3 Digital Integration Hub

The digital integration hub is an emerging architecture for implementing large-scale, high-throughput APIs by inserting a high-performance in-memory data store layer between the API service layer and the system of record. The emerging digital integration hub can help companies to reduce the cost and complexity of enabling API access to data held in the system of record applications for large-scale/high-performance scenarios. By storing an aggregated replica of the system of record data needed by the channel applications, the digital integration hub protects the system of record from an excessive workload while optimizing the data access latency and responsiveness for your modern digital application. The key components of the digital integration hub are as follows:

- **The data store**
  The high-performance data store stores the relevant subset of the backend data from one or more data sources in the cache—for example, information about the products, companies, prices, and so on. Sometimes even the metadata information about the APIs are persisted in this data store.
- **The frontend API services**
  These are the microservices that expose the data stored in the data store. Based on the principle of command query response segregation (CQRS), the read call would be handled by the in-memory data store, whereas the write scenarios can be performed directly on the system of record.
- **Integration layer**
  This layer is in charge of keeping the system of record data sources and the high-performance data store in sync. For the read-only scenario, the synchronization can be unidirectional; that is, changes from the backend system are propagated to the data store. In the case of conventional API architecture, the goal of the integration layer is to enable the frontend API services to synchronously call the backend to get data. In the digital integration hub pattern, the integration layer can be leveraged for the asynchronous event notifications pattern from the system of record to the high-performance data store layer.

The following are the benefits of a digital integration hub:

- Improved user experience and faster response time
- Decoupling the backend systems from frontend APIs to lower the cost of operations
- Protecting the backend system during peak loads from excessive frontend APIs
- Modernization of legacy system and microservice architecture to ensure rapid application management
- Normalization of APIs for certain domains, accessing data held in multiple data sources
- Real-time insights

The API Management capability is a core component of digital integration hub architecture from SAP, as shown in Figure 4.3. It facilitates the creation of a real-time digital access layer to insulate heterogenous transaction backends using in-memory data grids and leveraging microservice-based architectures. The API layer can be leveraged to modernize the legacy system and protect backends from heavy loads. It also allows for real-time insights.

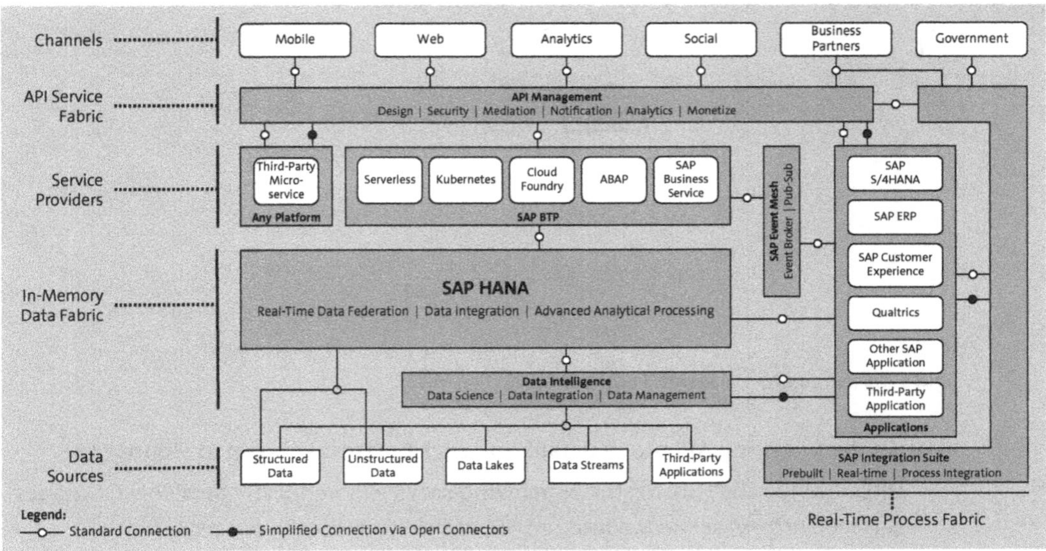

Figure 4.3 Digital Integration Hub and API Management

4.1.4 Creating Your First API

Let's get started with the API Management capability and let you create your first API proxy. After enabling the **Integration Suite** tile from your trial account and navigating to the SAP Integration Suite application (refer to Chapter 2, Section 2.3.3), follow these steps:

1. Click the **Design, Develop, and Manage APIs** tile, as shown in Figure 4.4.

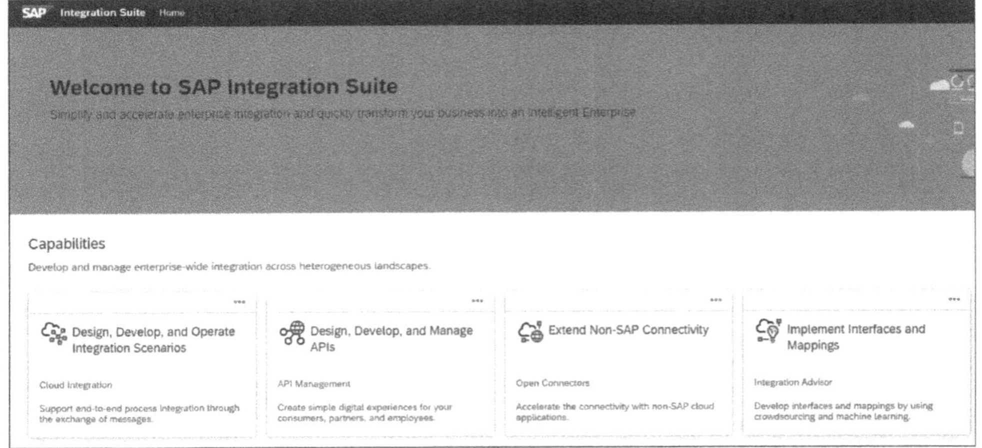

Figure 4.4 SAP Integration Suite Welcome Page

2. You will be taken to the API portal UI. Click the pencil icon, shown in Figure 4.5, to navigate to the **Develop** screen, and then click the **Create** button.

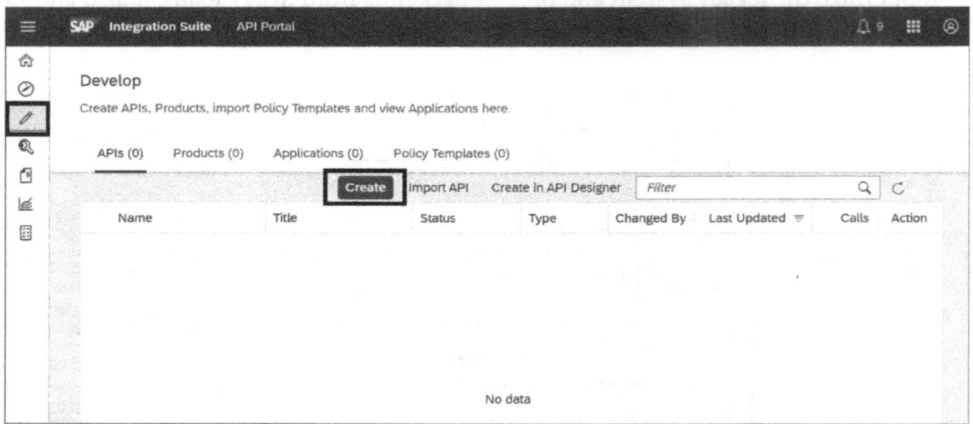

Figure 4.5 Navigate to Design Workspace

3. Fill in the following API proxy details and click **Create**, as shown in Figure 4.6:
   - **URL**: Provide the URL for the Northwind service (a publically available OData test service): *https://services.odata.org/V2/Northwind/Northwind.svc/*.
   - **Name**: This field is for the name of your API proxy.
   - **Title**: This field is for the title of your API proxy.

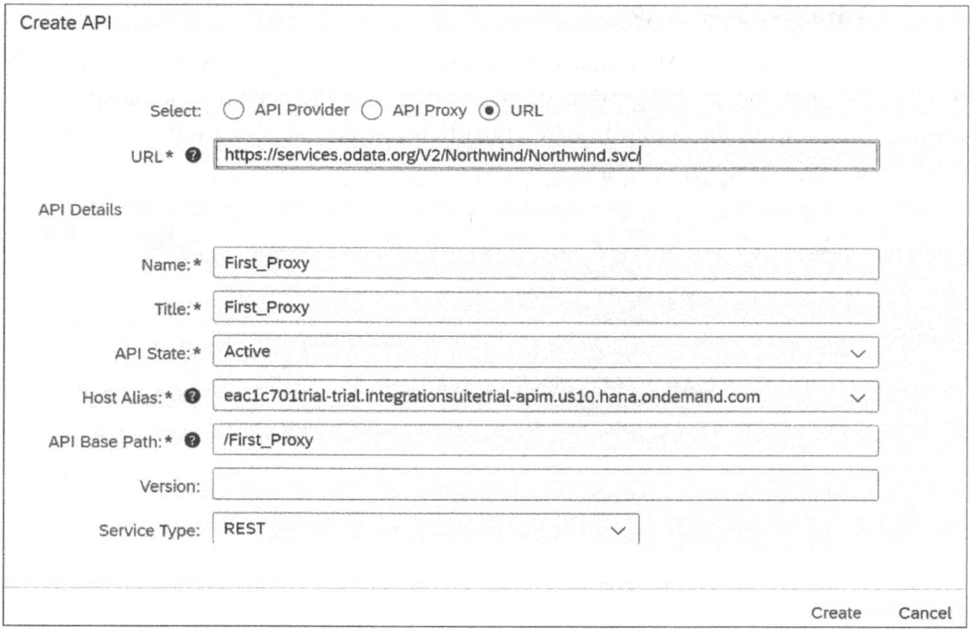

Create API

Select: API Provider API Proxy URL

URL * https://services.odata.org/V2/Northwind/Northwind.svc/

API Details

Name: * First_Proxy

Title: * First_Proxy

API State: * Active

Host Alias: * eac1c701trial-trial.integrationsuitetrial-apim.us10.hana.ondemand.com

API Base Path: * /First_Proxy

Version:

Service Type: REST

Create Cancel

Figure 4.6 Create API Proxy

- **API State**: This indicates the state of an API. In this case, it's **Active**.
- **Host Alias**: This is the virtual host that has to be provided while onboarding the API Management tenant so that the API proxies will be exposed with a proxy URL containing the same host.
- **API Base Path**: This is the base path; it should be unique for an API proxy.
- **Service Type**: This is the type of your API. In this case, it's the **REST** type.

4.1.5 Deploying and Testing APIs

Now, let's deploy and test your new API proxy. Follow these steps:

1. Click **Deploy** at the top of the screen, as shown in Figure 4.7.

Figure 4.7 Deploy Proxy

2. The API proxy will be registered successfully, as shown in Figure 4.8.

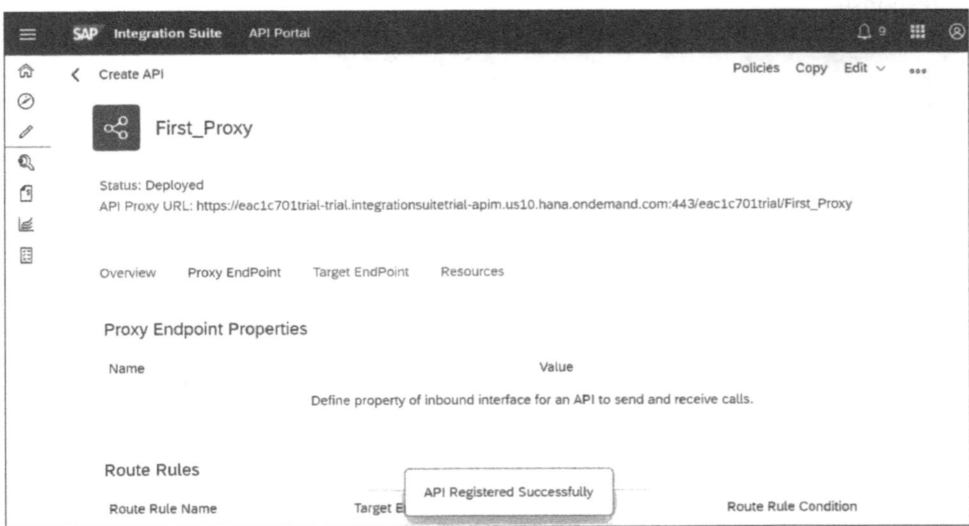

Figure 4.8 API Registered Successfully

3. Navigate to the API test console by clicking the last option on the left side, **Test Console**. You'll arrive at the screen shown in Figure 4.9, where the URL is autopopulated. Click the **Send** button.

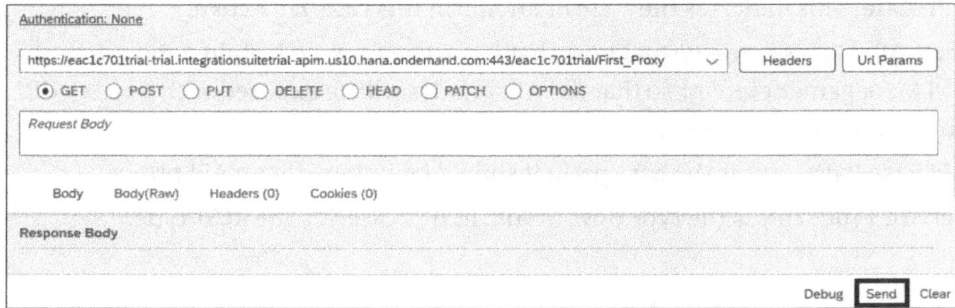

Figure 4.9 Test Console

4. You should get the response **200 OK**, as shown Figure 4.10.

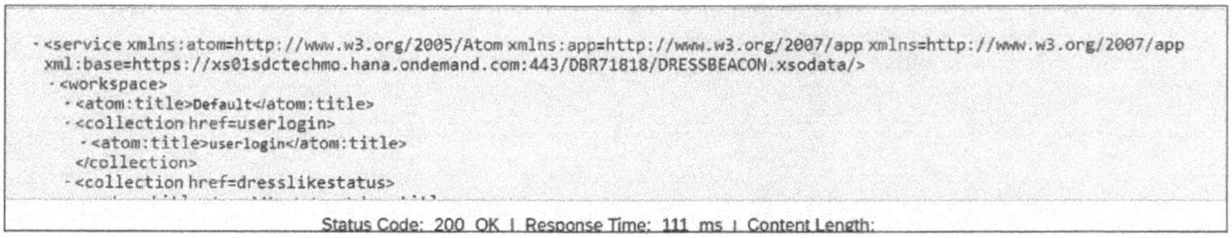

```
· <service xmlns:atom=http://www.w3.org/2005/Atom xmlns:app=http://www.w3.org/2007/app xmlns=http://www.w3.org/2007/app
  xml:base=https://xs01sdctechmo.hana.ondemand.com:443/DBR71818/DRESSBEACON.xsodata/>
  · <workspace>
    · <atom:title>Default</atom:title>
    · <collection href=userlogin>
      · <atom:title>userlogin</atom:title>
      </collection>
    · <collection href=dresslikestatus>
Status Code: 200 OK | Response Time: 111 ms | Content Length:
```


Figure 4.10 Test Response

5. Alternately, you can also test the API from your browser by copying the URL and pasting it into your browser, as shown in Figure 4.11. You should get the same HTTP response.

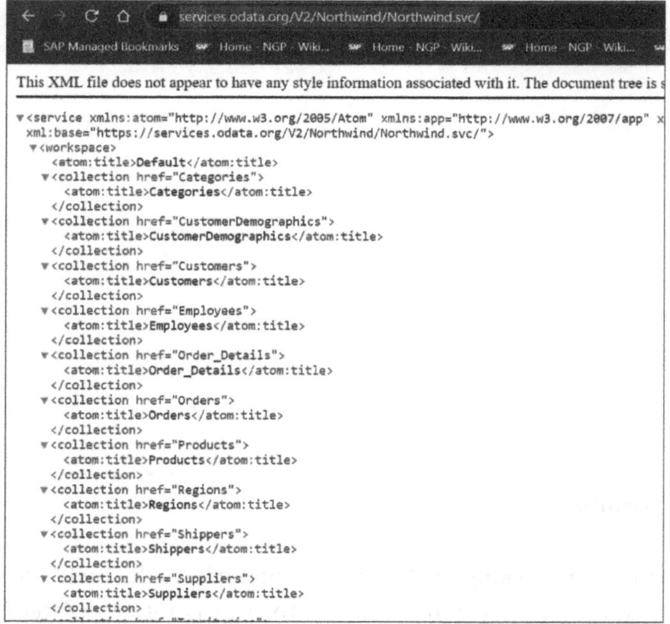

Figure 4.11 Browser Test

4.2 API Management Lifecycle

You've seen that APIs are becoming the de facto way to empower users to build cloud-native applications. Crafting and successfully executing your API strategy starts with defining the goal of what your organization wants to achieve with the API. Developing a well-thought-out API that aligns with your organization's needs is critical in achieving the desired outcomes. The API Management capability provides the ability to manage the full lifecycles of APIs. It guides you from an idea to realization, from designing and developing the right API to monetizing and engaging with the right audience. Let's look at the different stages in the API lifecycle in detail:

1. **Design**
   Once the API program and strategy are in place with rich semantics (defined by open standards such as OData and the OpenAPI Specification), users can easily design APIs using the API designer, as shown in Figure 4.12. Clicking **Create in API Designer** will launch the API designer.

   API Management provides an intuitive UI that makes API designing a breeze for architects, app developers, and integrators. The following are the key capabilities of API designer:

   - The API designer natively supports OpenAPI Specification but can easily import RAML and OData specifications and transform them into OpenAPI Specification.
   - The API designer allows the generation of client software development kits (SDKs) in Java, JavaScript, and Swift.
   - The API designer also allows the generation of server stubs in Java (JAX-RS and Spring Boot) and Node.js.

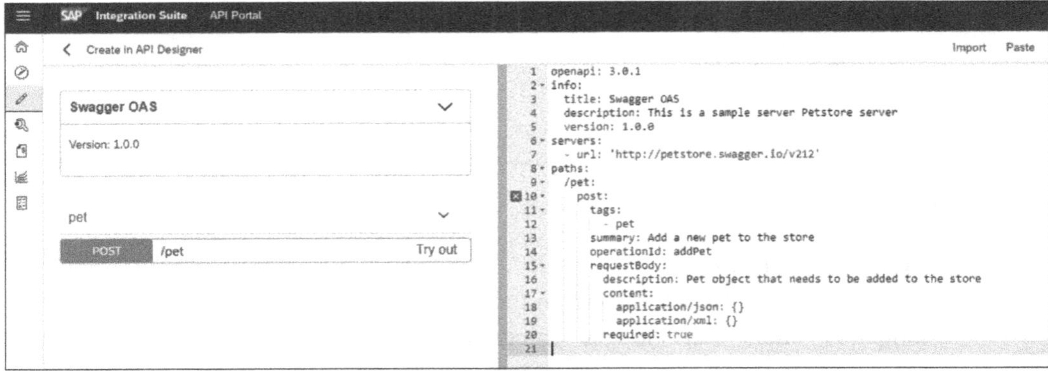

Figure 4.12 API Designer

2. **Develop**
   You can expose backend data from on-premise or cloud systems or even from any middleware in the form of APIs and leverage SAP Integration Suite's API Management capability to connect to your backend by creating an API provider. An API

provider abstracts the connection to your target system. We'll walk through creating a provider in Section 4.3.1.

3. **Manage**
   API Management provides a wizard-based approach to search, discover, and manage interfaces. It has primarily two components: design time and runtime. The runtime is an API gateway responsible for routing, policy enforcement (traffic management, security, and logging), and zero-downtime policy changes. The design time is integrated with the API gateway and empowers the user to achieve full lifecycle API management capability with a wizard-based UI.

   The API Management capability provides a rich set of out-of-the-box policies via the policy editor, as shown in Figure 4.13, for traffic management, security for all Open Web Application Security Project (OWASP10) threats, transformation, and mediation, as well as caching. For addressing specific requirements, you can also use JavaScript and Python scripts in a sandboxed environment to manipulate the request and response payloads.

   In addition to the policies, SAP also ships prepackaged digital content—API specifications, API connectors, and policy templates—that can be readily discovered and consumed, thereby accelerating the API development. Policy templates allow companies to standardize and reuse policies and best practices across APIs.

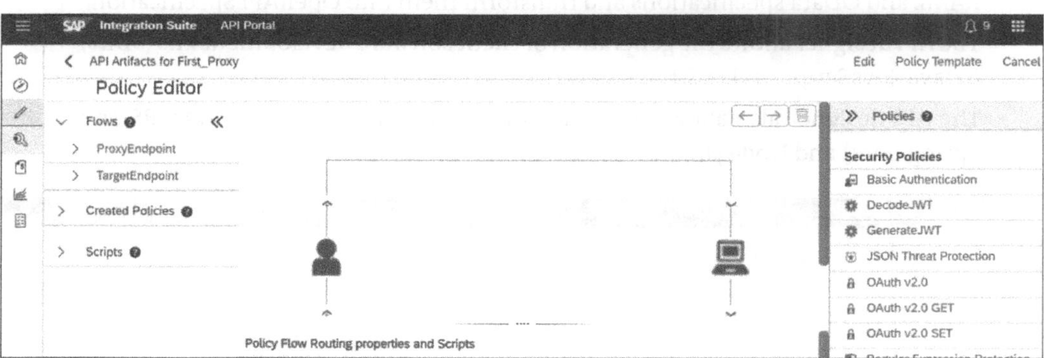

Figure 4.13 Policy Editor

4. **Meter and monitor**
   API analytics can help to meter and monitor the performance of your APIs. Both companies and API developers can benefit from API analytics, as shown in Figure 4.14. Companies can gain end-to-end visibility across API programs and drive decisions based on valuable insights. API developers can analyze how their APIs are performing by looking at metrics like traffic trends and spikes, latency, response times, and so on. These metrics can help you to drive proactive decision-making and protect your API from malicious attacks.

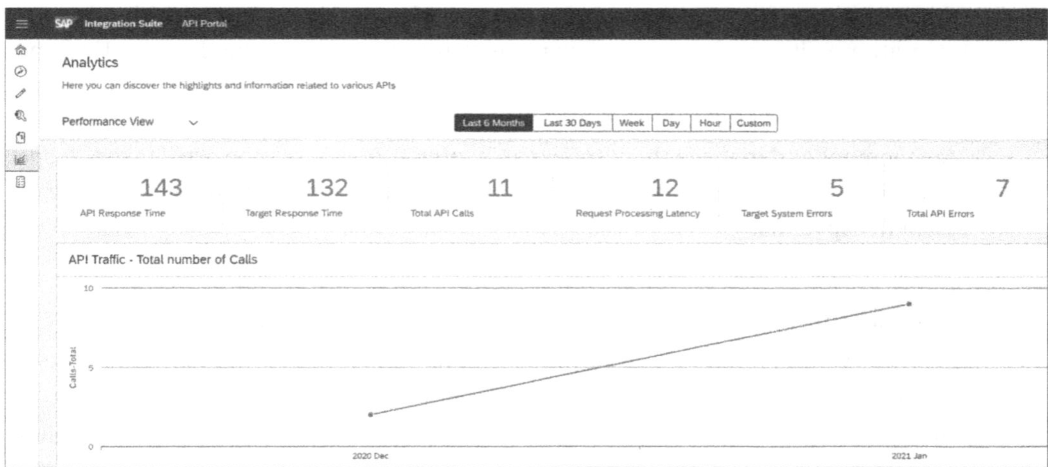

Figure 4.14 API Analytics

5. **Engage**
   Once your APIs are bundled into products, you can publish them to your API Business Hub Enterprise, as shown in Figure 4.15, so that developers can explore and consume APIs to develop applications. An API admin can define rate plans and add to API products to monetize them via a fixed or tier-based rate plan.

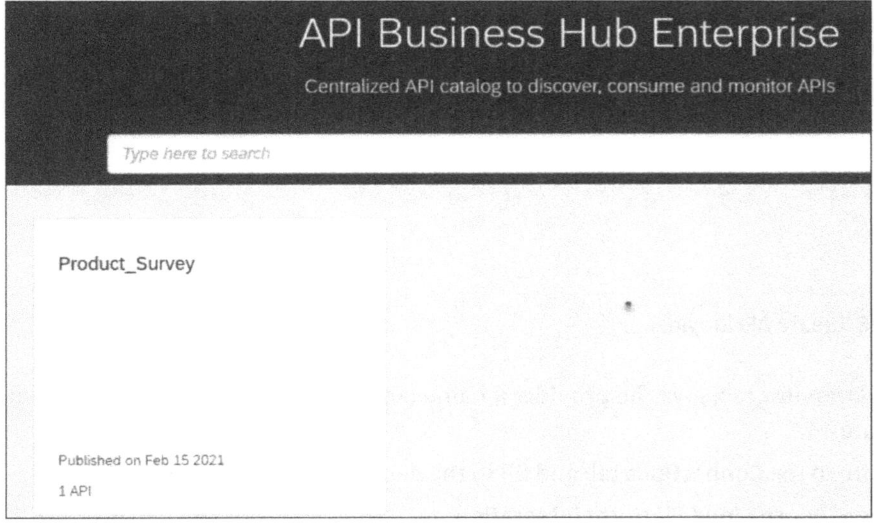

Figure 4.15 API Business Hub Enterprise

4.3 Integration with Other SAP Integration Suite Capabilities

Overall, the API Management capability is becoming a key driver for integration and extension scenarios, and therefore it's a core component of SAP Integration Suite. In addition, the API lifecycle management layer is also tightly integrated with other SAP Integration Suite capabilities; for example, you can manage a Cloud Integration endpoint in API Management by creating a provider system pointing to the Cloud Integration tenant. It also enables you to discover Open Connectors instances in API Management and manage them (refer to Chapter 7). In this section, we'll take a deeper look at discovering and managing the Cloud Integration endpoint using the API Management capability.

4.3.1 Creating an API Provider

In the API Management capability, backend systems can be represented by API providers. The configuration of the backend systems—port, host, and other details—is required to establish the connection from the API Management capability. Follow these steps to create an API provider:

1. Click the **Configure** icon in the left-hand menu and then click the **Create** button to create an API provider, as shown in Figure 4.16.

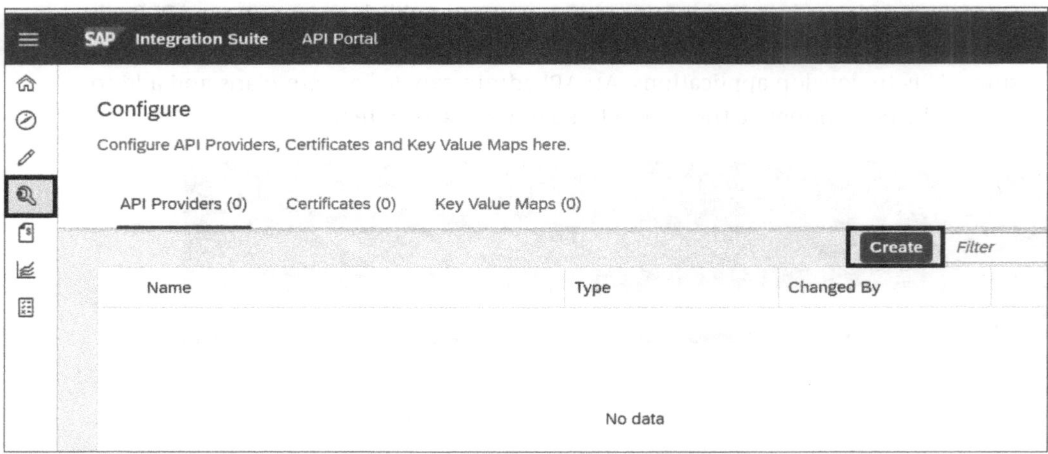

Figure 4.16 Create API Provider

2. In the **Overview** tab, give the provider a name (and optional description), as shown in Figure 4.17.
3. Navigate to the **Connections** tab and fill in the details as shown in Figure 4.18:
   - **Type**: Choose **Cloud Platform Integration.**
   - **Cloud Integration Management Host**: Enter your host for your Cloud Integration instance.
   - **Port**: Enter "443".
   - **Authentication**: Choose **Basic.**

- **Username**: Provide your Cloud Integration username.
- **Password**: Provide your Cloud Integration password.

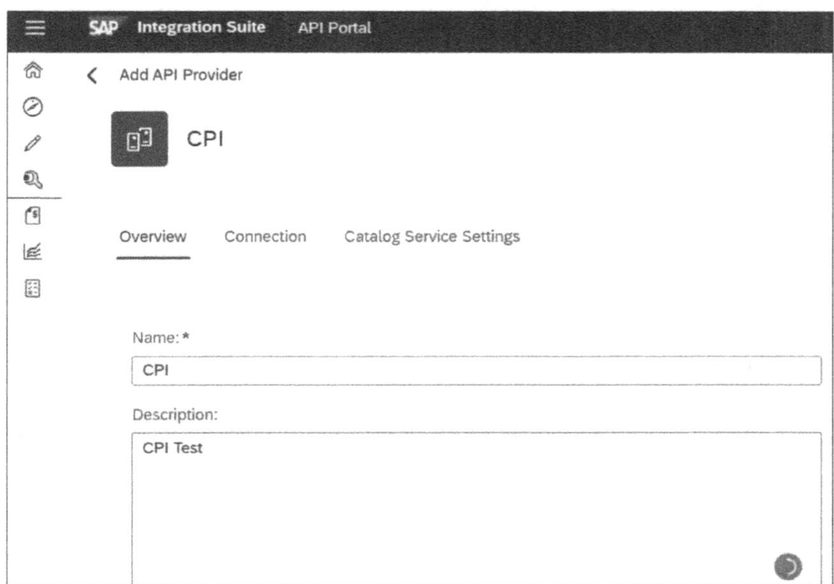

Figure 4.17 API Provider Overview

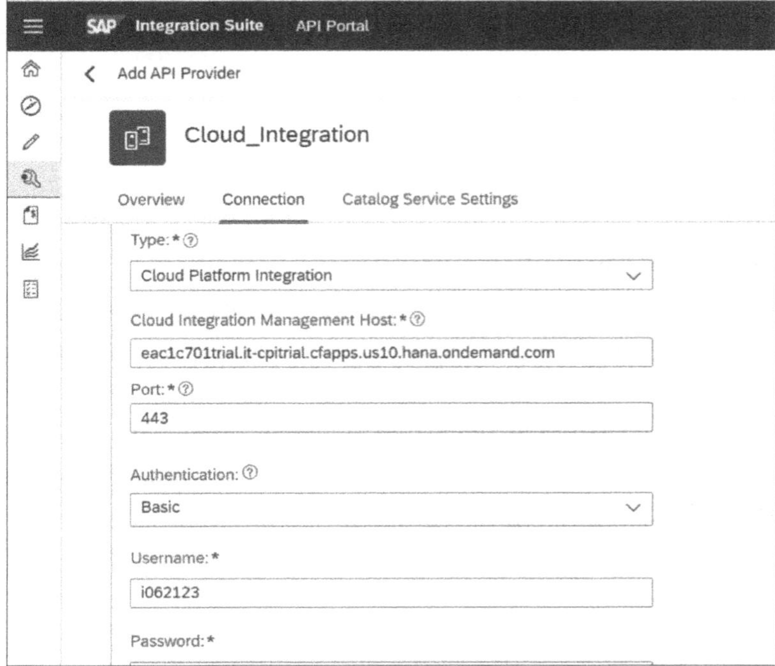

Figure 4.18 API Provider Connection Details

4. Save the provider by clicking **Save**, as shown in Figure 4.19.

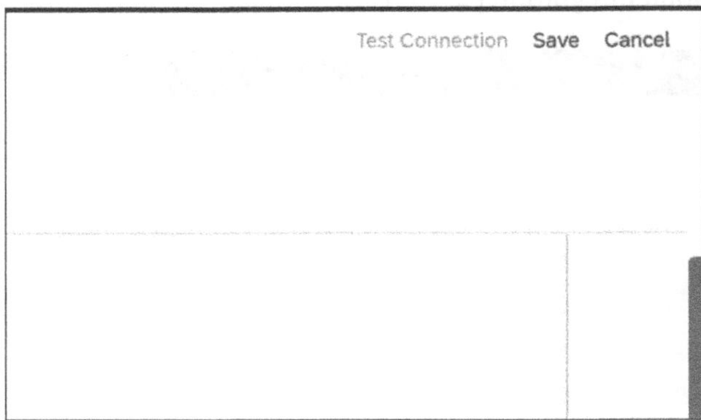

Figure 4.19 Save Provider

5. An API provider is created and appears on the **Configure** screen, as shown in Figure 4.20.

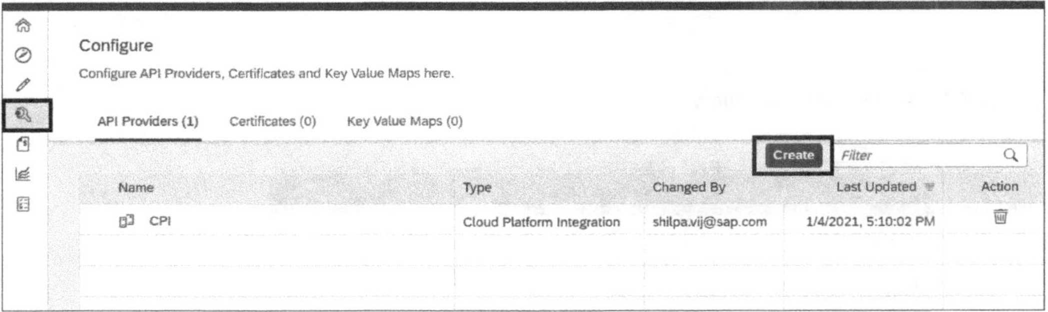

Figure 4.20 New API Provider

4.3.2 Discovering the Cloud Integration Endpoint

Once the connection is established to the backend system, the API Management capability facilitates browsing through the system and discovering APIs, be they in the cloud, on-premise, SAP, or non-SAP.

Follow these steps to discover OData endpoints from a Cloud Integration tenant:

1. Navigate to the **Develop** screen by clicking the pencil icon, and then click the **Create** button, as shown in Figure 4.21.

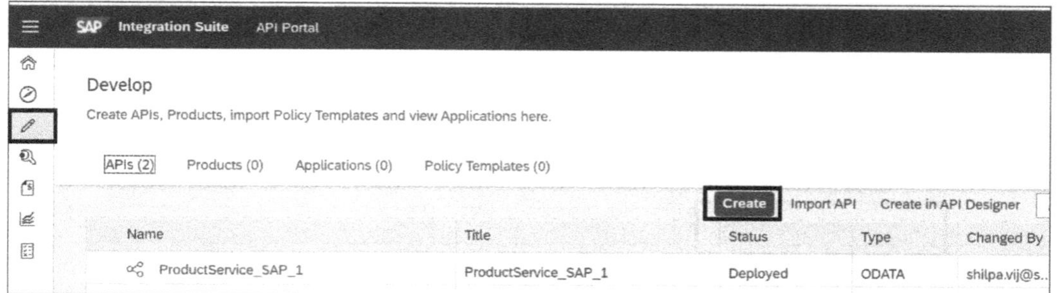

Figure 4.21 API Provider Discovery by Creating API

2. Select the **API Provider** that was created in the previous step. Click **Discover**, as shown in Figure 4.22.

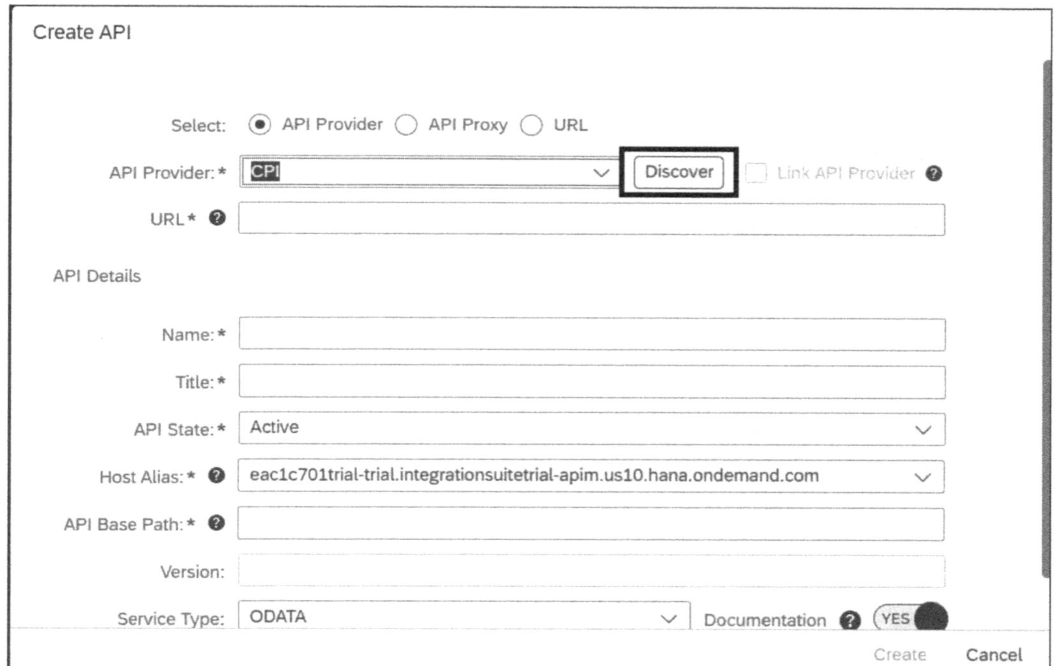

Figure 4.22 Discover APIs

3. Deployed OData endpoints will be listed as shown in Figure 4.23. Select the API and click **Next**.

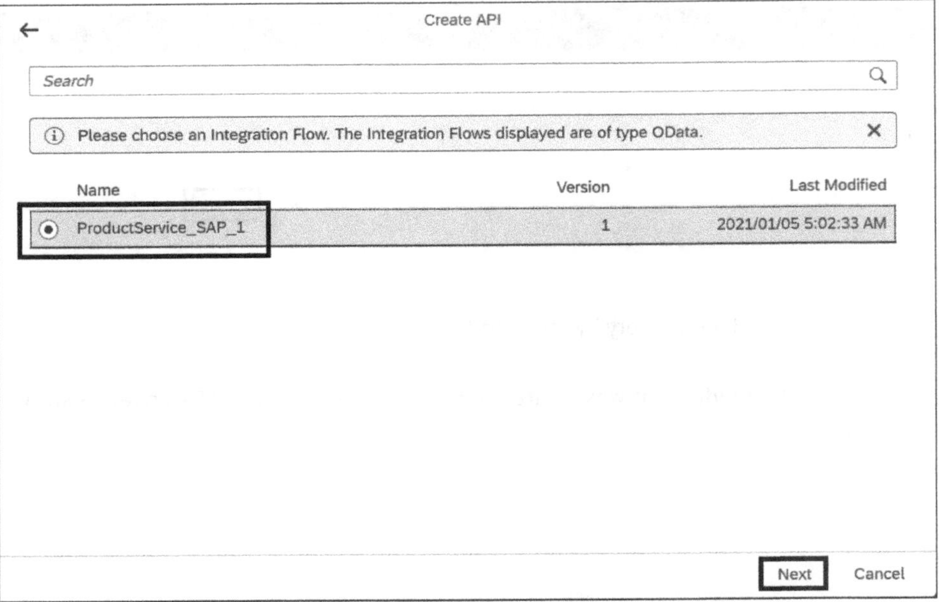

Figure 4.23 Select API

4. On the credentials popup, give the **Username** and **Password** for your Cloud Integration tenant, as shown in Figure 4.24, and select **Done**.

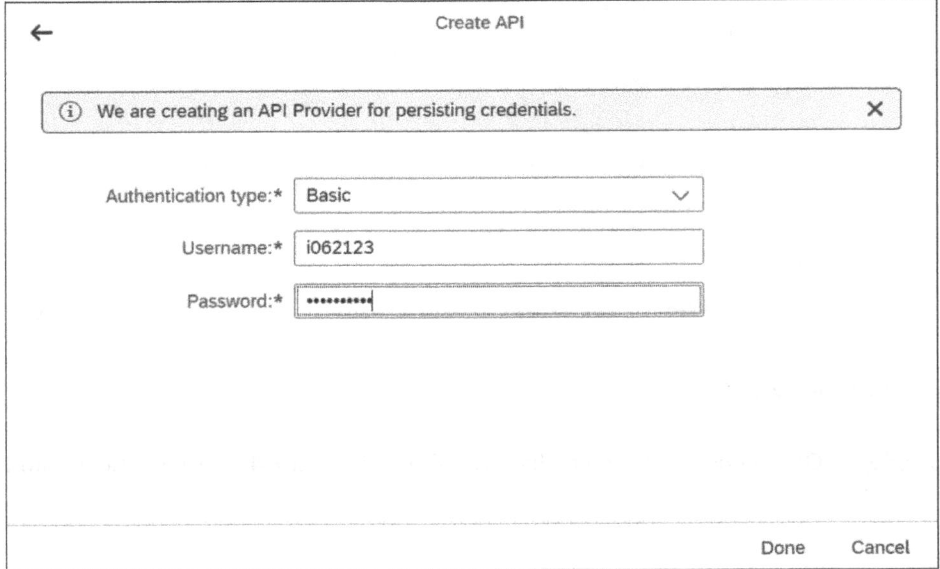

Figure 4.24 Tenant Credentials

5. The API proxy details will be automatically filled in, as shown in Figure 4.25. Click the **Create** button to create the API proxy.

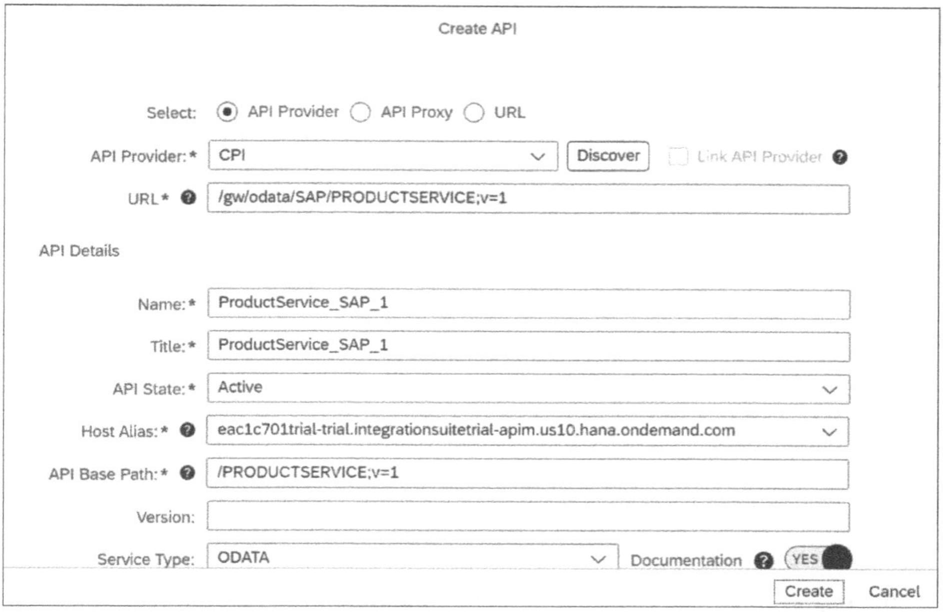

Create API

Select: API Provider API Proxy URL

API Provider:* CPI Discover Link API Provider

URL* /gw/odata/SAP/PRODUCTSERVICE;v=1

API Details

Name:* ProductService_SAP_1

Title:* ProductService_SAP_1

API State:* Active

Host Alias:* eac1c701trial-trial.integrationsuitetrial-apim.us10.hana.ondemand.com

API Base Path:* /PRODUCTSERVICE;v=1

Version:

Service Type: ODATA Documentation YES

Create Cancel

Figure 4.25 Create Proxy

6. Now you can deploy the API proxy and test it, as discussed in Section 4.1.5.

4.3.3 Managing the OData Endpoint

The API Management capability helps digital businesses manage the full lifecycle of APIs to easily enable open integrations, extensions, and app development. In this section, you'll learn how to manage the OData endpoint exposed from the Cloud Integration tenant against several requests using a quota policy. Follow these steps:

1. Open the API and navigate to the policy editor by clicking **Policies** in the upper-right corner, as shown in Figure 4.26.

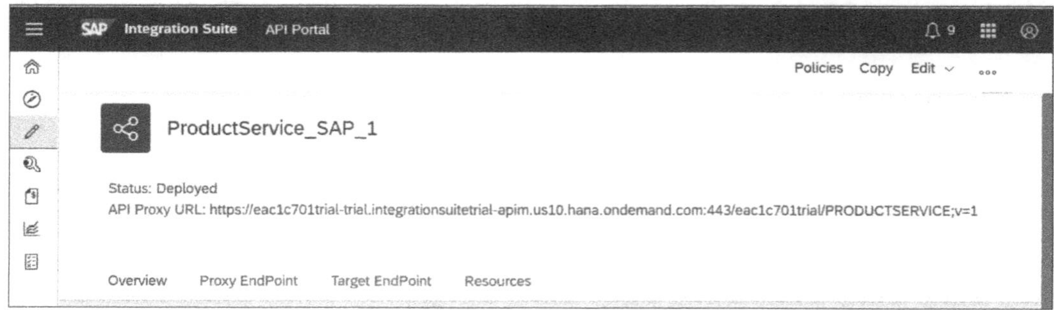

Figure 4.26 Policy Editor

2. Click the **Edit** button to edit the policy editor, as shown in Figure 4.27.

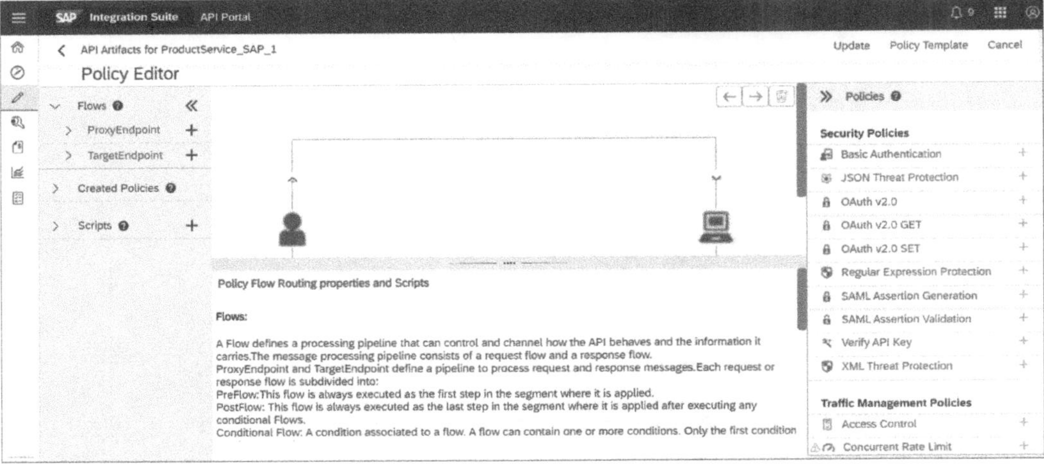

Figure 4.27 Policy Editor

3. Expand the **ProxyEndpoint** node on the left-hand side and select **Preflow**. From the **Traffic Management Policies** section on the right-hand side, add a **Quota** policy, as shown in Figure 4.28.

> **Note**
>
> The quota policy allows configuring the number of requests that an API proxy can allow over a time period. For this example, the count is 2, so this API proxy will allow two requests per minute.

4. Enter a **Policy Name** and click the **Add** button.

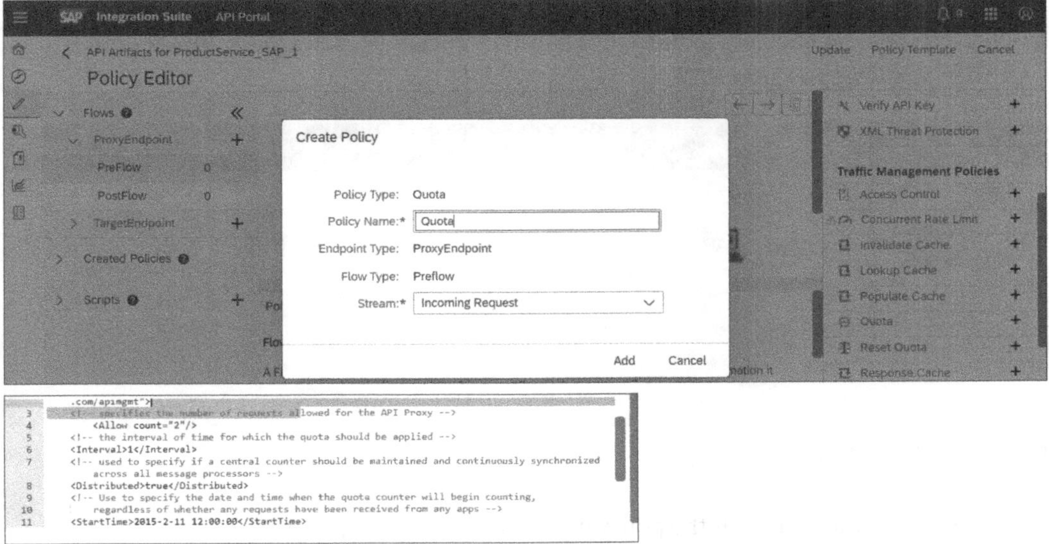


```
.com/apimgmt">
<!-- specifies the number of requests allowed for the API Proxy -->
    <Allow count="2"/>
<!-- the interval of time for which the quota should be applied -->
<Interval>1</Interval>
<!-- used to specify if a central counter should be maintained and continuously synchronized
    across all message processors -->
<Distributed>true</Distributed>
<!-- Use to specify the date and time when the quota counter will begin counting,
    regardless of whether any requests have been received from any apps -->
<StartTime>2015-2-11 12:00:00</StartTime>
```

Figure 4.28 Quota Policy

5. Click the **Update** button, as shown in Figure 4.29, and save the proxy.

Figure 4.29 Update Proxy

6. Navigate to the **Test Console** (refer back to Section 4.1.5) and try to send the request three times. Because the count mentioned in the quota policy is two, the third attempt will result in an error, as shown in Figure 4.30.

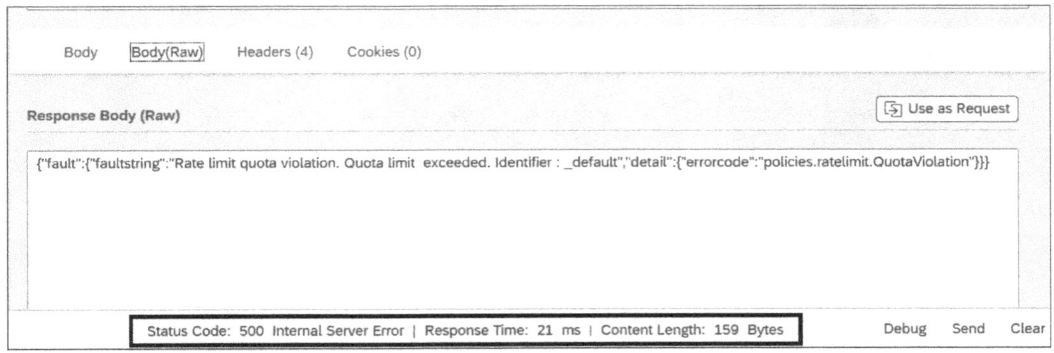

Figure 4.30 Quota Violation Check

4.3.4 Creating a Product and Exposing the API

Developers across the globe can easily and conveniently use APIs to build applications to innovate new edge scenarios for businesses to continuously adapt and advance.

API Business Hub Enterprise is a rebranding of the developer portal, which provides a central catalog to seamlessly search, discover, and consume the listed APIs. Companies can leverage multiple instances of the API portal to publish products to any API Business Hub Enterprise (the same or the other subaccount/tenant) as it is API gateway agnostic. These APIs can be from SAP or non-SAP; from different LoBs like SAP SuccessFactors, SAP S/4HANA, and SAP Ariba; and so on.

In this section, you'll see how a similar set of APIs can be bundled into a product and published to API Business Hub Enterprise. Follow these steps:

1. Navigate to **Products** under **Develop** and click the **Create** button, as shown in Figure 4.31.
2. Give a **Name** and **Title**, as shown in Figure 4.32, and navigate to the **APIs** tab.
3. Select the API you created previously in Figure 4.23, and click **OK**, as shown in Figure 4.33.

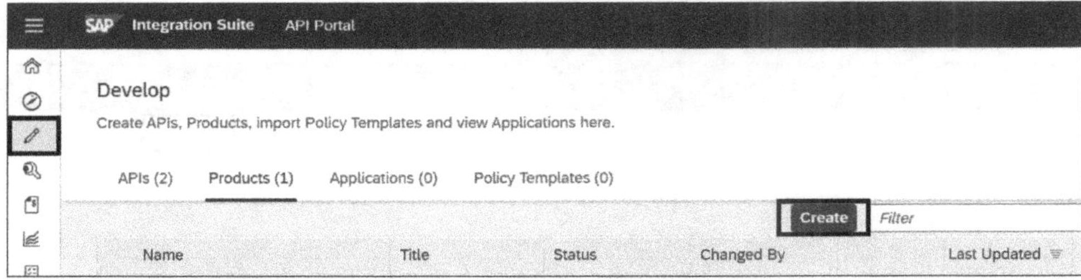

Figure 4.31 Create Product

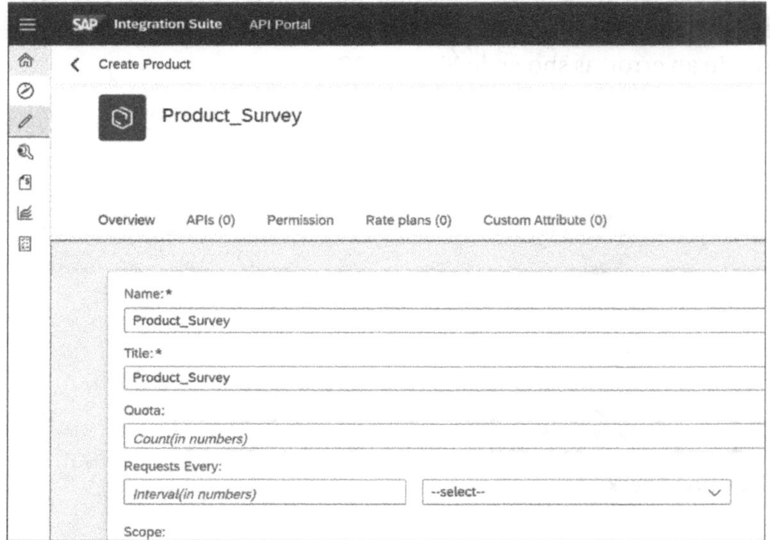

Figure 4.32 Product Details

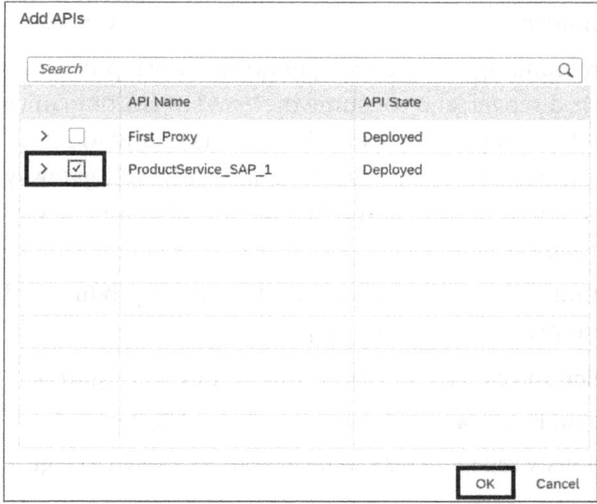

Figure 4.33 Add API

4. Click the **Publish** button, as shown in Figure 4.34.

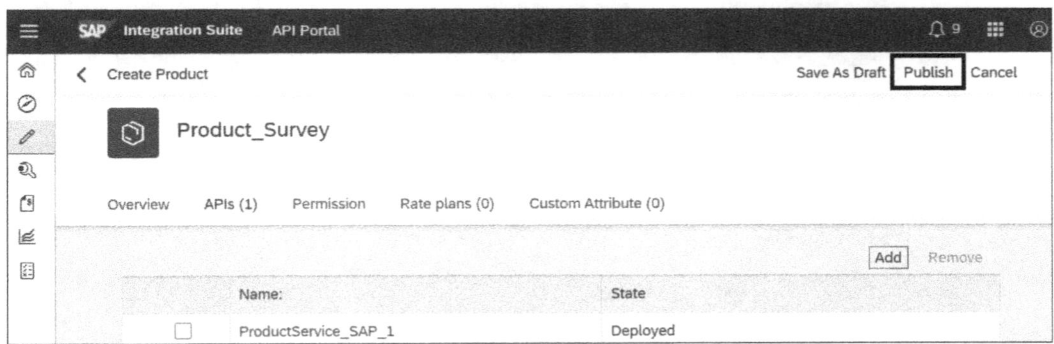

Figure 4.34 Publish Product

5. Navigate back to the **Integration Suite** tile from your trial account. Click **Manage Capabilities** to get to the provisioning screen. Navigate to API Business Hub Enterprise by clicking the **URL** under **API Business Hub Enterprise**, as shown in Figure 4.35.

Figure 4.35 Access API Management URLs

6. The published product will be available, as shown in Figure 4.36.

Now the APIs are available in your developer ecosystem to create applications and innovate with managed APIs. App developers can also leverage these APIs to collect, manage, analyze, monetize, and use information of all types to extend and connect to business systems.

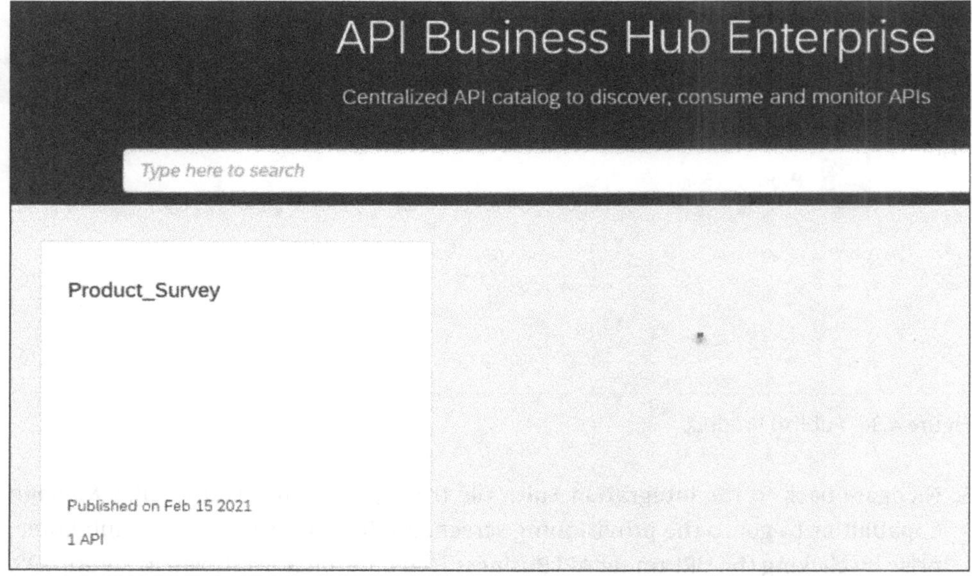

Figure 4.36 Published Product

4.4 Summary

In this chapter, you learned how organizations are using real-time API-based integrations to create omnichannel access and hide the complexity of underlying heterogeneous landscapes to improve company experiences. We focused on exposing open APIs managed via the API Management capability to integrate into marketplaces, build ecosystems through digital channels, and create new revenue streams by monetizing APIs. You can leverage agile data grids and microservices to support a digital integration hub that aggregates multiple backend systems and insulate them using in-memory data grids to create a low-latency and high-performance real-time digital access layer. You also learned how SAP API Management empowers the user to manage the complete lifecycle of APIs, which essentially binds different parts of the value chain. In the last section, you saw how API Management is a core component of SAP Integration Suite, which is the iPaaS offering from SAP that covers a broad range of integration styles, including API-led integrations.

In the next chapter, we'll move on to design integration scenarios using the Cloud Integration capability of SAP Integration Suite and will leverage prepackaged content to accelerate the integration journey.

5

Chapter 5
Designing Integration Scenarios

It's time for the next core capability of SAP Integration Suite: Cloud Integration. In this chapter, you'll learn how to consume prepackaged integration content, design integration scenarios, and run integration flows.

The Cloud Integration capability of SAP Integration Suite provides a platform to design, execute, and monitor integration scenarios. It provides many functionalities, such as transformation, routing, security, and multiple technical and application protocols for connecting to a variety of business systems in a hybrid environment via easy-to-use, drag-and-drop, browser-based tooling. It offers secure and governed access for different personas who develop, maintain, and monitor integration scenarios.

In this chapter, we'll go deeper into this capability and will describe the various components and building blocks of Cloud Integration. We'll start by exploring the main sections of the Cloud Integration capability before walking through the pipeline steps. We'll then examine connectivity options and end with the steps to create an end-to-end integration flow.

5.1 Exploring Cloud Integration

The Cloud Integration capability, which you can access via the **Design, Develop, and Operate Integration Scenarios** tile in your SAP Integration Suite tenant, provides a web-based workspace and three sections for end-to-end scenario designing and monitoring. All personas (developer, admin, and end user) can access web-based tooling and monitoring screens within these respective sections: **Discover**, **Design**, and **Monitor**.

Let's look into the details of these three sections.

5.1.1 Discover

As the name suggests, the **Discover** section (which you can navigate to via the left-hand menu) helps you discover the Cloud Integration prepackaged content shipped via SAP API Business Hub (see Chapter 3), as shown in Figure 5.1. It provides all the integration scenarios (bundled into logical packages) shipped by SAP and partner ecosystems for various business integrations. As of today, SAP provides 2,000-plus integration scenarios across lines of business (LoBs), such as SAP S/4HANA, customer experience, procurement, e-governance, human capital management (HCM), and third-party applications such as Salesforce, ServiceNow, Workday, and so on.

You can discover the content, get into the details, and view the integration scenarios here. You won't have any edit permissions in the **Discover** section. Once you have discovered the required integration scenario or package, you can **Copy** these packages into the **Design** section for further configuration, editing, and deployment. These packages are shipped in two modes:

- **Configure-only**
  Integration scenarios in these packages cannot be edited. They can only be configured with pre-exposed parameters.
- **Edit-only**
  Integration scenarios in these packages can be customized by companies, but after customization companies will not get new updates for these specific packages.

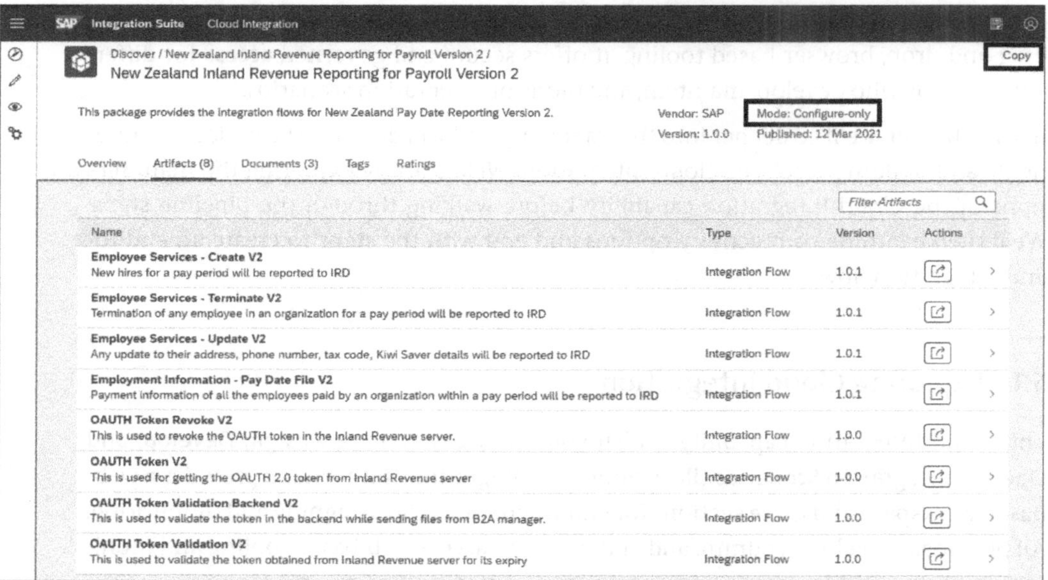

Figure 5.1 Discover Section View of Cloud Integration Capability

5.1.2 Design

The **Design** section provides capabilities to design and deploy integration scenarios with the available tooling. Once you enter the **Design** section via the left-hand toolbar, it displays all packages in a vertical view, as shown in Figure 5.2, with different attributes. It includes packages copied from the **Discover** section or created from scratch by companies.

Design /
Design

Packages (16)

Name	Mode	Version	Created By	Created Date	Description
New Zealand Inland Revenue Reporting for Payroll Version 2	Configure-only	1.0.0		Sat, 13 Mar 2021 16:18:54 GMT	This package provides the int Version 2.
SAP ERP or SAP S/4HANA Integration with SAP SuccessFactors Employee Central: Confirmation v2	Configure-only	3.0.0		Wed, 03 Mar 2021 10:48:07 GMT	Send confirmations to SAP Su SAP ERP or SAP S/4HANA is
SAP Financials Integration with SAP Concur	Configure-only	1.1.2		Tue, 02 Feb 2021 15:18:43 GMT	Integrate and automate the o your SAP Financial system
SAP Supply Base Optimization Integration with SAP ERP and SAP S/4HANA	Configure-only	1.0.2		Thu, 04 Mar 2021 13:31:28 GMT	This handles the transfer of m transactional data such as Or
demoflow	Editable	1.0		Mon, 14 Dec 2020 03:58:58 GMT	demo flow

Figure 5.2 Design Section View of Cloud Integration Capability

There are three options to create or use existing integration scenarios in the **Design** section, which we'll discuss next.

Configuring Prepackaged Integrations

You can configure prepackaged integrations copied from the **Discover** section, as shown in Figure 5.3, by clicking **Actions • Configure**. With this option, you can configure the available fields exposed by specific integration artifacts. These may be technical parameters such as the URL of the system, security credentials (as deployed in KeyStore), or functional parameters such as the company code or configuration of the date to execute a payroll run. These exposed parameters are different across integration artifacts and are dependent on business scenario configurations. Once you have entered externalized parameters, an integration scenario can be deployed directly. You cannot copy or edit the integration scenarios under a package that is defined as **Configure-only**.

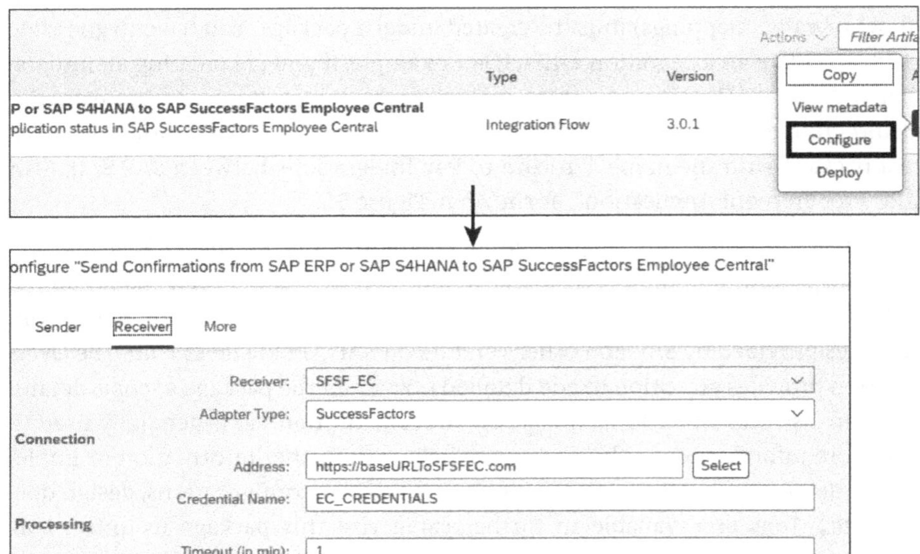

Figure 5.3 Configuring Existing Integration Flow

Editing Prepackaged Integration Content

Unlike the **Configure-only** option, there are integration packages that are shipped with editable options. Artifacts in these editable packages can be edited completely. They may or may not contain configurable parameters. If you edit the integration artifact using the **Edit** button, as shown in Figure 5.4, and customize the package, it is considered a customized object. Enhancements and bug fixes are shipped to integration packages regularly. If a user uses the **Configure** option to enter externalized parameters, updates will be available for such packages, but if a user uses the **Edit** button to customize, then it will be treated as a customization and further updates will not be provided.

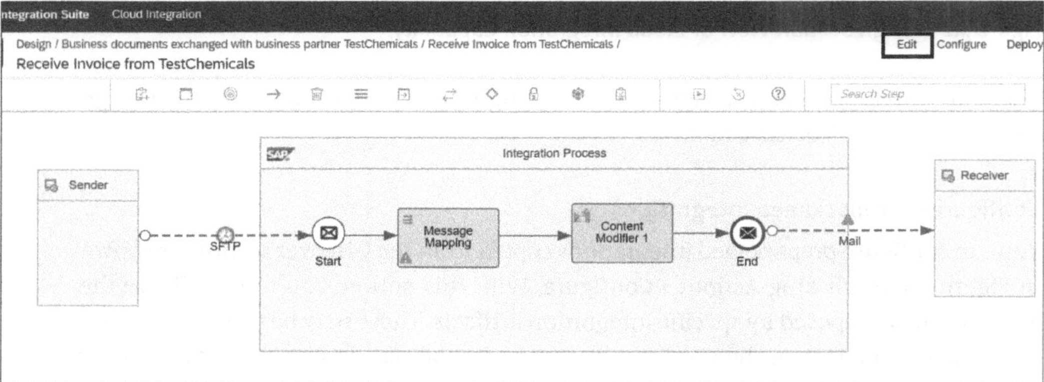

Figure 5.4 Editing Existing Integration Flow

Designing a New Package from Scratch

The final option is to create a custom integration package or scenario from scratch by clicking the **Create** button, as shown in Figure 5.5. Integration artifacts (such as integration flows or value mappings) must be created under a package. You have to go inside a package to create an integration artifact. For example, if you are creating an integration between SAP S/4HANA and a third-party procurement system, and there are different integrations scenarios for order, invoice, and shipment notification, you can create a package with the name "Procure to Pay Integrations between SAP S/4HANA and ABC Procurement Application", as shown in Figure 5.5.

While creating a package, you can provide a meaningful name for the package and add a short description, the version of the package, and a vendor name. The **Vendor** name is generally the name of the organization creating the package and differentiates it from the packages provided by SAP and other vendors on SAP API Business Hub. The **Overview** section provides an option to add detailed text about the package's scope/details, and you can also add an optional image. The **Documentation** tab is generally used to provide more information on the package to help with further information or implementation details (such as technical guides, best practices, configurations, design documents, etc.). **Tags** are available to further categorize this package (country, LoB, industry, and custom keywords) and also help during search in the **Design** section.

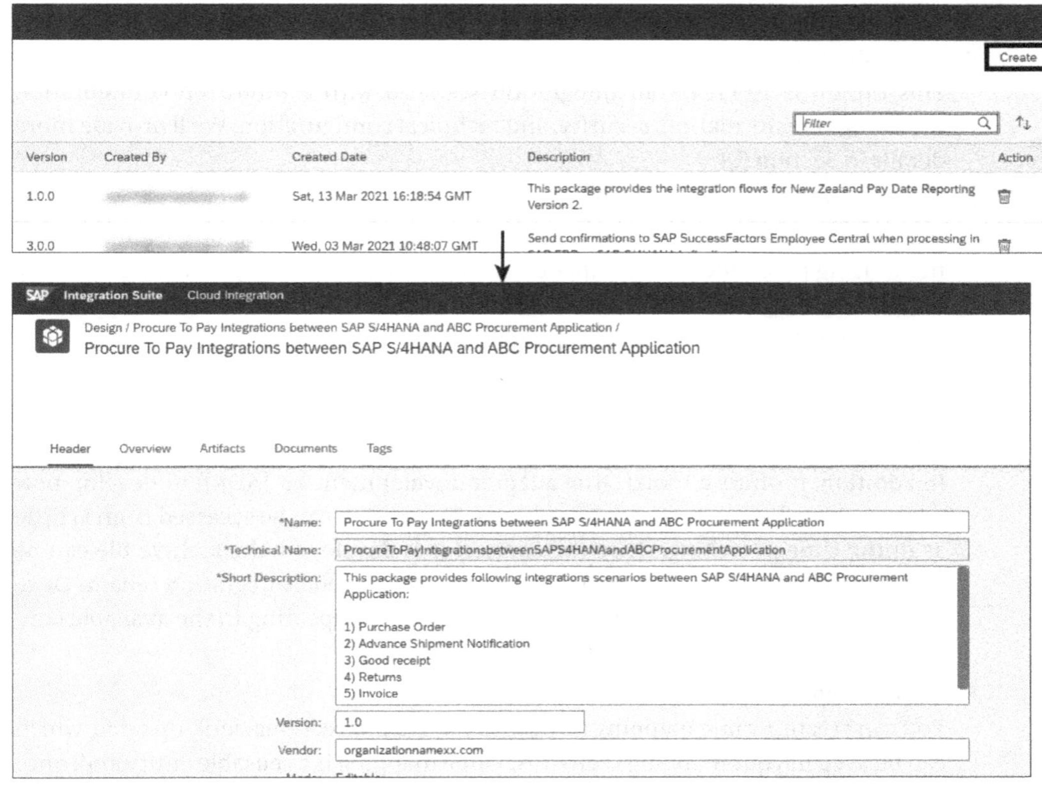

Figure 5.5 Creating Integration Package

Once you've designed a package, you can create different artifact types in the package by going to the **Artifacts** tab and clicking the **Add** button, as shown in Figure 5.6.

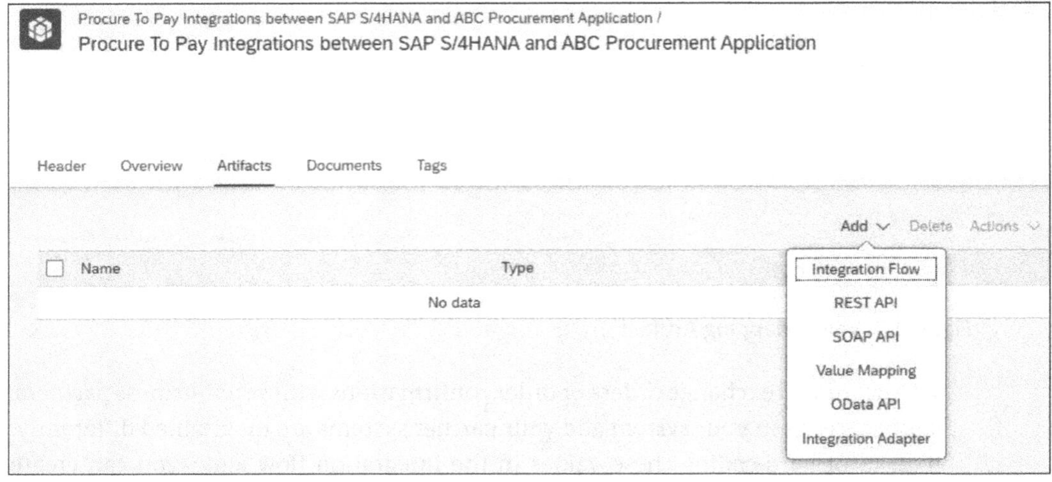

Figure 5.6 Creating Artifacts in Integration Package

The following artifacts are available:

- **Integration Flow**
  This option is to create an integration scenario with a multistep configuration including transformation, security, and technical configuration. We'll provide more details in Section 5.4.
- **REST API, SOAP API, OData API**
  This option helps with low-code application programming interface (API) development using predefined API templates. Once you select it, the system automatically creates the integration flow in the backend, which can be modified further based on requirements.
- **Integration Adapter**
  The Cloud Integration capability provides a lot of out-of-the-box technical and application connectors to connect to diverse systems in cloud and hybrid environments. In addition, it offers a toolkit (the adapter development kit [ADK]) to develop new adapters based on company requirements. This toolkit can be accessed from Eclipse as of the time of writing. Once the connector is developed, the archive file can be uploaded as an artifact and can be deployed on the Cloud Integration tenant. Once it's deployed, the connector will automatically start appearing in the available connectors list during integration flow development.
- **Value Mapping**
  You can create a value mapping artifact to act as a bidirectional look-up table, which can be used in your mapping scenarios. Value mapping is a reusable entity and, once deployed on a tenant, it can be used by different integration flows deployed on the tenant. It offers a capability to define key value maps with context that can accessed by message mapping in integration scenarios. Consider the following example, illustrated in Figure 5.7.

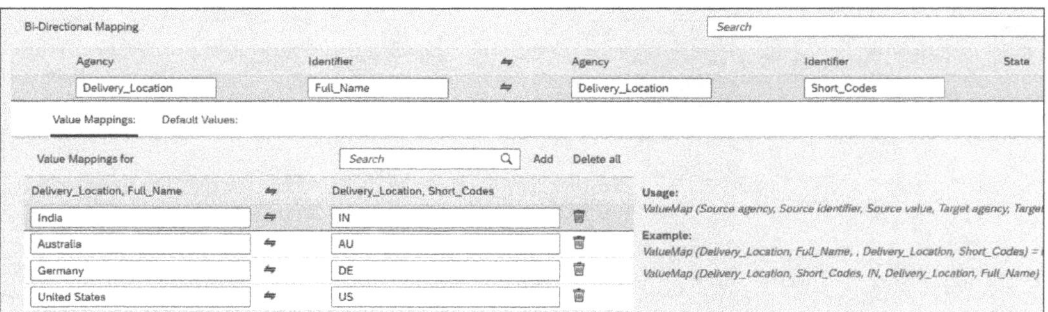

Figure 5.7 Value Mapping Artifact

Whenever you exchange orders or order confirmations with your business partners, country codes in your system and your partner systems are maintained differently. Instead of hard-coding these values in the integration flow logic, you can create these as independent reusable artifacts that can be used by multiple scenarios. If the country code is "Germany" for the shipment location in the order, you want to replace it with "DE" before you enter this order in your business system. Vice versa, once an order confirmation is generated from your business system, the shipment location "DE" needs to be replaced with "Germany" as per your partner's business requirements.

> **APIs**
>
> The Cloud Integration capability offers various REST/OData APIs to manage the operations of deployed integration content. It can be used for continuous integration/continuous delivery (CI/CD) use cases or for providing monitoring data to various other third-party dashboards. These APIs provide secure access for various activities, such as fetching or deploying/underlying integration artifacts; accessing message logs, attachments, and message stores; and managing security content and various operations via partner directory configuration.

5.1.3 Monitor

The **Monitor** section, shown in Figure 5.8, provides various views and activities related to monitoring deployed artifacts and the status of message queues, testing connectivity, deploying security artifacts such as certificates, and so on.

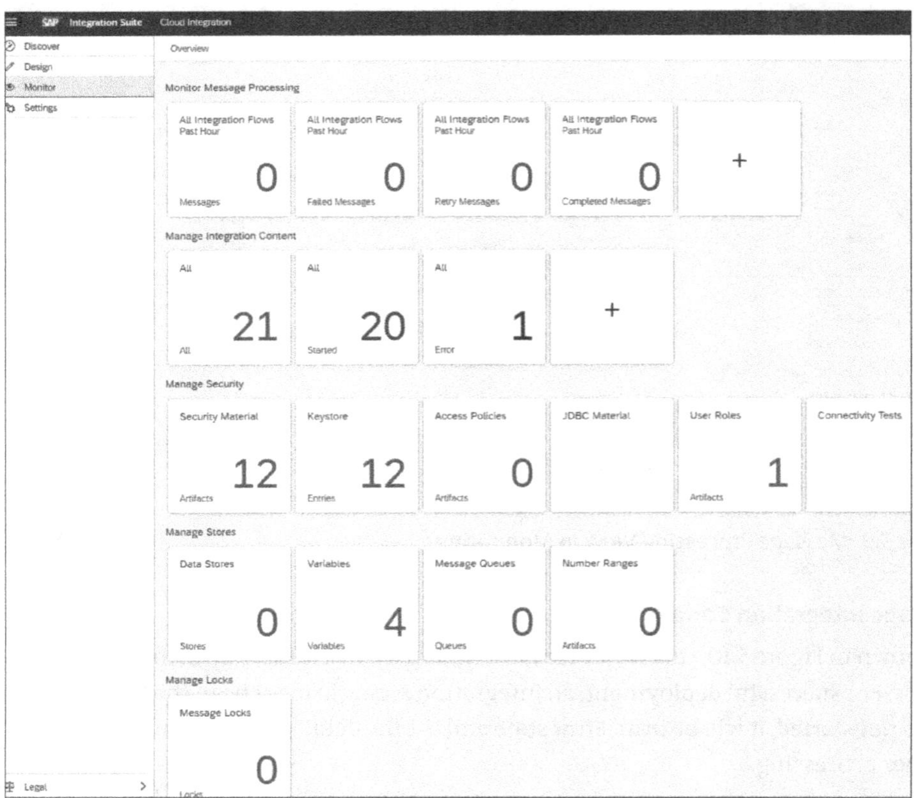

Figure 5.8 Monitoring View of Cloud Integration

Some of the tooling options in the **Monitoring** section are useful after deployment of integration scenarios, whereas others are mandatory as prerequisite steps depending on a scenario's requirements (e.g., certificates deployment).

Let's explore some of the key tiles available.

Monitor Message Processing

Once you trigger the deployment of an integration flow (in the **Design** section), it should be deployed successfully and will be ready to process messages. There may be multiple integration flows deployed. As an end user, you can monitor the message processing on all or specific integration scenarios, as shown in Figure 5.9. The monitoring will show you the status, such as **Completed**, **Failed**, and so on, along with additional details of errors encountered in case of failure status. If you have enabled **Trace** under **Log Configuration** (see next section), you can also look at detailed information about stepwise integration flow execution to trace the error and message content. There may be some business-critical processes that you need to monitor directly in the dashboard without selecting or filtering the integration scenarios. For this, you have the capability to create additional floating tiles per integration scenario, along with three tiles provided as default.

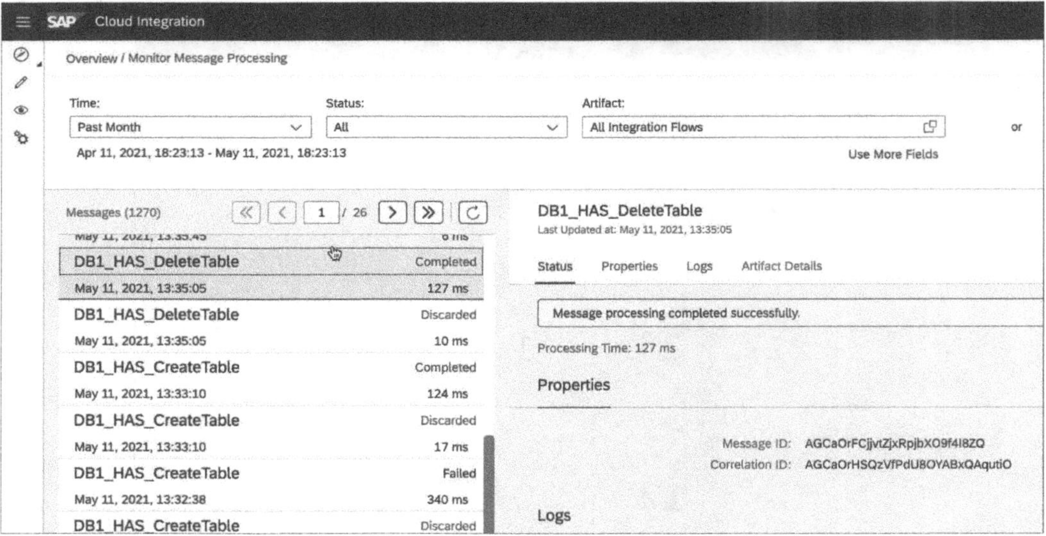

Figure 5.9 Message Processing View in Monitoring

Manage Integration Content

As shown in Figure 5.10, this view provides the deployment status of all integration scenarios. For successful deployment, an integration scenario must be in the **Started** state. If it is not started, it will be in an **Error** state and list the details of errors encountered for further processing.

Integration scenarios can use multiple different protocols, including SOAP, OData, and HTTP, and can also act as API listeners for specific protocols. APIs exposed by Cloud Integration scenarios are accessed by connecting a source system to make a call to the Cloud Integration scenario. **Endpoints** provide the full URL of the exposed API, which can be used by source system initiating the API calls.

You can see that the integration is deployed successfully (or that there are errors) in the **Status Details** tab. The **Artifact Details** tab provides a quick link to open the integration scenario in view-only mode without going to the **Design** section or an option to jump directly to the **Message Process Monitor** screen.

Log Configuration provides configuration for available **Log Levels**. The default is **Info**, which will display basic information related to successful or failed message processing. If you need further details, you can set the log level to **Debug**, which will display detailed error logs for message failures. The **Trace** option is used for interactive debugging; it shows the detailed stepwise execution of the integration scenario in a graphical view, with message details along with each step. You can monitor the stepwise execution of integration scenarios and look at values of an actual payload before and after each step's execution. Because it's resource intensive, this option is switched off by default and should be used only for debugging in case of errors or failures.

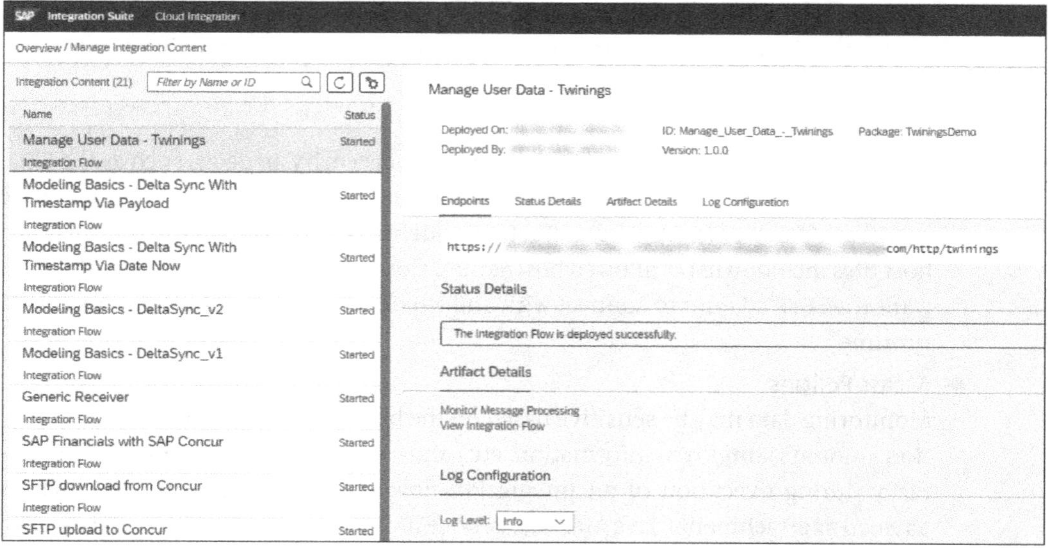

Figure 5.10 Integration Content Monitor

Manage Security

This view provides configuration options that are mainly used by administrators. Security is one of the key requirements for integration scenarios, as explained in Chapter 2, Section 2.3.2. Irrespective of the protocols being used, all calls must be secured. When a company gets an SAP Integration Suite tenant, there are preprovisioned certificate authority (CA)-signed certificates available for use by companies. Companies can use these certificates or upload and maintain new key pairs or certificates. The following tiles are available:

- **KeyStore**
  KeyStore is used to manage secure keys and certificates. It includes the following options:
  - *View SAP-provided keys*: Whenever a company activates a new SAP Integration Suite tenant, SAP provides a CA-signed key pair in the keystore that can used by companies in their integration scenarios.
  - *Generate a new key pair*: KeyStore also provides a tool to generate a new key pair (private/public key), but that pair is not signed by a CA.
  - *Create corresponding CA sign request*: You can create a sign request from the key pair and give it to a CA to sign. For the generated key pair, there is a process to export the request in certificate signing request (CSR) format, which then can be provided to a CA to sign.
  - *Get the response from CA*: Once you get a response file from a CA, you can upload it to KeyStore. Now you have your own CA-signed key pairs available in KeyStore, which include both public and private keys.
  - *Download public key/certificate chain*: You can download the key or certification chain and provide it to your integrating systems/applications so that secure communication can be designed.
- **Secure Store**
  Companies can deploy all non-KeyStore-relevant security artifacts, such as the SFTP (Secure Socket Shell [SSH] protocol) known_hosts file, PGP, OAuth credentials, basic username/passwords, SAP SuccessFactors credentials, and so on, in this view. Known host files include a list of allowed hostnames, along with public fingerprints that you want the SFTP adapter to connect with and which will be used for validation during runtime.
- **Access Policies**
  Monitoring data may be sensitive based on the business scenario (financials, quotation amounts, employee information, etc.) and can be designed to be stored temporarily during execution of an integration flow via several components, such as payload as attachments, Java Message Service (JMS) queues, tracing data, data stores, and the like. Organizations have a very strict policy for business data access and provide respective authorizations to a limited set of users. Cloud Integration provides secure access to business data using **Access Policies**. You can create a custom access policy and add different artifacts that require restricted access. A new **Role** needs to be created from the SAP BTP cockpit with the same name as the access policy. Once a role is created, you can add multiple users who need to access these artifacts. Once

the access policy is created, the access rule applies for both the **Monitor** user interface and for access via APIs.

- **JDBC Material**
  You can use a Java Database Connectivity (JDBC) adapter to connect to third-party databases such as MSSQL, Oracle, and so on. Some drivers are provided out of the box by Cloud Integration. For other databases, companies need to add valid JDBC drivers. Once drivers are uploaded, a JDBC data source is created with connection details such as the URL, user, and password. The name of this data source is used in the JDBC adapter during integration flow creation and is used during runtime to establish a secure connection. Uploading JDBC drivers is a task generally performed by administrators and does not require frequent changes.
- **Connectivity Tests**
  This is a very useful tool for developers/administrators during any project implementation. In any organization, there is a central security team that works on firewalls and ports that need to be opened or maintains access to various systems via a governed approval process. In addition, there are different personas who work on the Cloud Integration capability, like administrators who are responsible for managing keys/certificates and developers who focus on designing integration scenarios. It becomes quite challenging to manage different teams in an organization who deal with the overall connectivity and access requirements. This is also one of the reasons for delay in many implementation projects. SAP Integration Suite provides a very useful feature (as shown in Figure 5.11) to test the connectivity without deploying or creating any integration scenarios for various protocols. You can test it with or without security credentials as per your requirements. It helps you validate the connectivity with various systems with the provided security credentials and different protocols to diagnose the connectivity challenges.

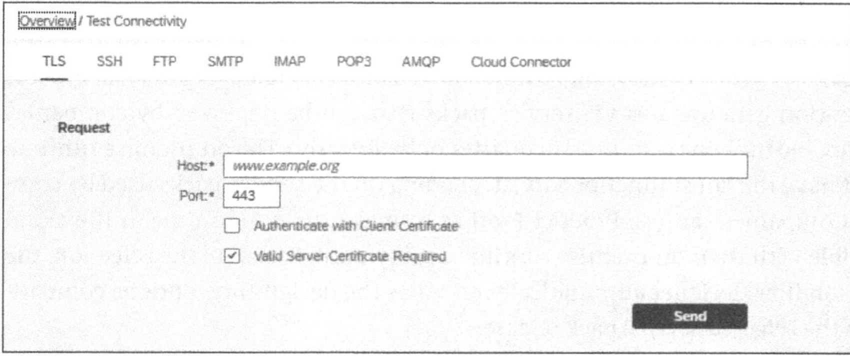

Figure 5.11 Connectivity Test UI from Monitoring Section

- **Manage Stores**
  Cloud Integration provides many steps that are used to store data temporarily during integration flow execution, such as data stores, JMS queues (to store data in

queues and process it reliably if the target system is not available with restart functionality), and global or local variables that can be defined to reuse required data values across integration flows or tenants.

Number ranges are used to generate unique sequential numbers within a given range. They automatically increase the count and provide the next number in the sequence during integration flow execution. One of their important use cases is to produce a unique number to identify an EDI interchange being sent to trading partner. You can create/edit a number range and also check the current value of a specific number range object.

Although there is no need to monitor the data regularly, there may be a need to check the current values in case of issues or for debugging purposes. Access to this data is strongly governed by access policies.

- **Managing Locks**
  Design-time artifact locks provide information on integration artifacts and packages in the tenant that are locked by the users of the tenant. Locks help avoid having multiple users editing same objects simultaneously. Message locks ensure runtime objects are locked to avoid duplicate processing via several different runtime nodes for applicable scenarios. These locks are automatically removed by the system once the required processing is over. In exceptional cases, objects may remain in a locked status. There is a self-service capability for the tenant administrators to access the active locks and unlock the artifacts if required.

5.1.4 Settings

This final section, **Settings**, provides various advanced-level settings required for Cloud Integration scenarios, as shown in Figure 5.12:

- **Product Profiles**
  SAP ships the Cloud Integration runtime on-premise with SAP Process Orchestration release 7.5. While Cloud Integration follows a monthly release cycle, SAP Process Orchestration gets updates via service packs that can be deployed by companies with a flexible timeline (e.g., once in quarter or half-yearly). The on-premise runtime may not have the latest functionality, depending on the service packs used by companies. Companies can use **Product Profiles** to make the design time in the cloud compatible with their on-premise runtime deployment. Based on the selection, the integration flow designer automatically provides the design-time options compatible with the selected service pack release.
- **Transport**
  These are the settings to configure integration package transport via various supported channels, such as the enhanced Change and Transport System (CTS+) and the SAP Cloud Transport Management service (as explained in Chapter 2, Section 2.3.1).

- **ES Repository**
  Cloud Integration provides a capability to reuse message mappings and operation mappings from the on-premise SAP Process Orchestration middleware. These mappings are stored in an enterprise service repository (ESR) component of SAP Process Orchestration. You can provide the connection details to connect to the ESR and can reuse these mappings during integration flow design.
- **Custom Tags**
  SAP provides tags that are associated with integration packages available in SAP API Business Hub. These tags can help with content categorization and search. You also can create custom tags that can be used by developers while creating integration packages in the **Design** section.
- **Malware Scanner**
  There are various types of project resources (documents, security material, key stores, drivers, integration artifacts, etc.) that can be uploaded in a tenant by developers. Such files may contain malware or malicious content that can compromise the security of a tenant. The **Malware Scanner** option enables file scanning during upload in a tenant and reports errors if any malicious content is found.

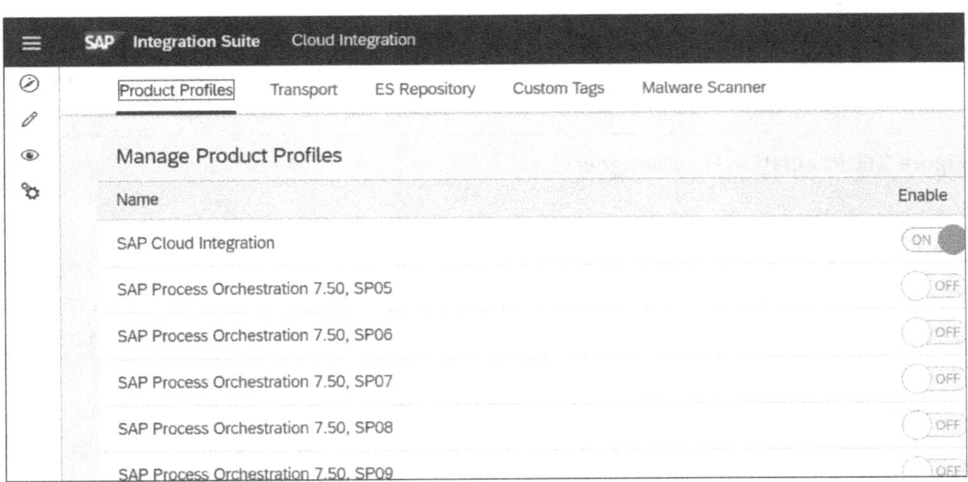

Figure 5.12 Advanced Settings View

5.2 Flexible Pipeline Steps

The Cloud Integration capability offers a process development environment based on Business Process Model and Notation (BPMN), as shown in Figure 5.13. You can see the pipeline steps in the menu at the top of the screen ❶ and can use it in the integration flow via a drag-and-drop approach. You can also see the source ❷ and target ❸ systems to be connected. You must select the required technical ❹/application ❺ connectors to be used to connect to the source and target systems. Finally, you can see the flexible pipeline ❻.

There are 40-plus different process steps available in the pallet that can be used to design a flexible multistep integration process. These process steps include security (encryption, signing, encoding, ZIP format, etc.), message routing (gateway, parallel and sequential routing, aggregator, filter, splitter, etc.), message transformations (content modifier, scripting, graphical transformations, predelivered functions, etc.), and various other components (validations, data stores, global and local variables, etc.). These out-of-the-box process components reduce the overall effort required to design and configure an integration process.

Now let's look at the different pipeline steps available in the integration flow designer. For a step-by-step integration flow example, see Section 5.4.

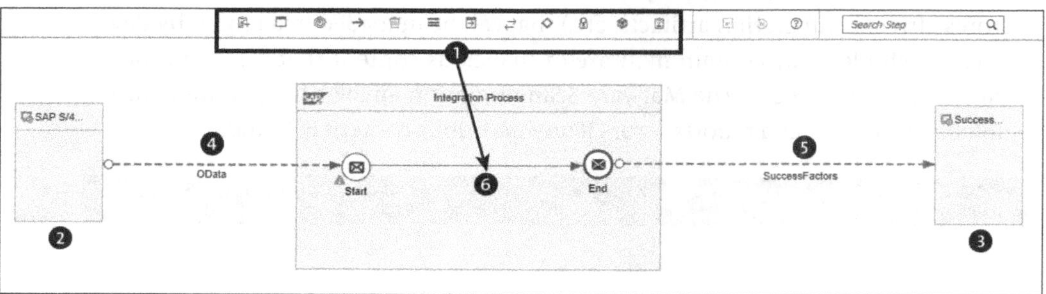

Figure 5.13 Integration Flow Designer

5.2.1 Participants

These are the entities that participate in integration scenarios. The sender represents the source system and the receiver represents the target system or intermediate system for the lookups. You can drag and drop to add any number of participants in your integration scenario. Every participant must be associated with an adapter. Adapters provide the required technical configuration for connecting to participants (e.g., SFTP, HTTP, SAP SuccessFactors).

5.2.2 Process

Cloud Integration provides various types of integration processes that can be designed in the integration flow designer. It is the entry point for modeling any integration flow. Let's look at the supported processes:

- **Integration process**
  An integration process is the default process necessary for any integration scenario. It is the main integration flow that controls the runtime execution of the business logic. It may include sender and receiver participants, multiple pipeline steps, and associated connectors.

- **Local integration process**
  A local integration process is an optional process that can be included in your main integration process. It helps break an integration scenario down into multiple small logical subprocesses. For example, say that one of the activities that the main process needs to do is a complex discount calculation based on incoming invoices and previous payment details. Instead of having this logic in the main process, making it bulky and complex, you can create a local integration process with discount calculation logic that can be called from the main process wherever it's required. Repetitive business logic can be reused or a big portion of integration logic can be modeled separately from a maintenance perspective.
- **Exception subprocess**
  An exception subprocess is used to handle exceptions in an integration scenario. In case of any issues within the main process, the main process may be terminated. If you use exception subprocess, then instead of the main process failing in case of any issues, control is passed to the exception subprocess and you can handle errors/exceptions correctly with the required business logic. For example, say that the main process is handling some critical sales order processing. Due to an invalid format that's received, the main process can't perform the transformations and raises an exception. Immediately, control is passed to an exception subprocess containing logic to inform the sales administrator of the order failure and send an error exception to the IT administrator so that he or she can look into it.

5.2.3 Mappings

Mappings are used to transform messages from one format to other. Let's look at the available mapping steps:

- **Message mapping**
  Message mapping provides a drag-and-drop graphical designer for message transformations. It offers multiple ready-to-use functions, such as arithmetic, Boolean, date, and so on. You can also create custom functions using scripts. In addition, it provides a simulation option that can be used to test the mapping transformations during design time. Once the mapping is created, it stores the artifacts in the integration flow project using Nmap format. This can be downloaded and reused across integration flows.

  Consider an example of replicating invoice details from system X to system Y, as shown in Figure 5.14. Each invoice that is received from the source system includes an amount associated with various line items, whereas the target system invoice format expects the total amount together instead of amounts associated with multiple line items. In addition, the field structure in target schema is also different for order numbers and invoice date order items. Most of the mapping use cases can be fulfilled via the graphical mapping editor.

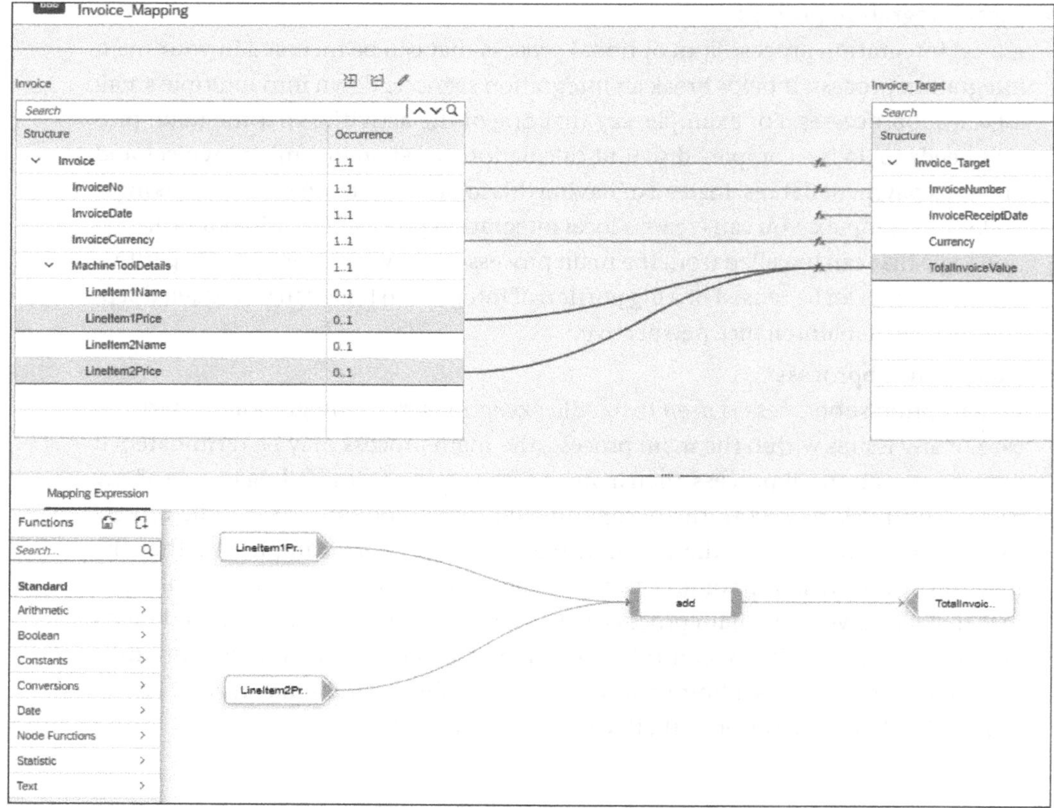

Figure 5.14 Message Mapping in Integration Flow Designer

- **Operations mapping**
Operation mapping provides a way to map two service interfaces, along with operations as defined in Web Services Description Language (WSDL) files. It can be asynchronous or synchronous and defines the operations signature. You can reuse the operations mapping defined in SAP Process Orchestration and can use it without any changes. It saves a lot of effort during migration.

- **Extensible Stylesheet Language Transformations mapping**
Extensible Stylesheet Language Transformations (XSLT) is used for transforming document formats such as XML. An XSLT transformation is called a stylesheet. Cloud Integration supports XSLT 3.0, the latest version of the standard. XSLT is a powerful tool for complex mapping scenarios but also requires skilled learning and careful usage: if it isn't designed properly, it may cause performance issues. You can import existing XSLT mapping from your files system, reuse it from your integration flow project resources, or create a new XSLT mapping. If you're creating a new mapping or importing one from a file system, it's automatically stored in your integration flow project resources.

5.2.4 Transformations

While message mapping provides an easy way to map two documents in XML format with the field mappings and conversion, the integration flow design editor provides many other transformation steps that reduce a lot of development effort by providing out-of-the-box conversion logic. Let's look at some of these transformation steps.

Content Modifier

As name suggests, the content modifier, shown in Figure 5.15, is used to modify the content of incoming messages. There are three sections available to read and update the message content. You can set or update the message header or property with content values or specific field values from incoming messages, constants, XML Path Language, and more. The difference between a header and a property is that header values can be lost after an external system call (as header values can be sent as is/set/reset by some receiver adapters, such as HTTP), whereas properties will be available for complete message execution and are not impacted by adapter calls. Both headers and properties can be used by other pipeline steps or receiver connectors by using syntax such as ${header.<Header Name>} or ${property.<Property Name>}. For example, you can set a sales order amount field in a property and can later use it for a condition routing step to check the amount and make conditional calls.

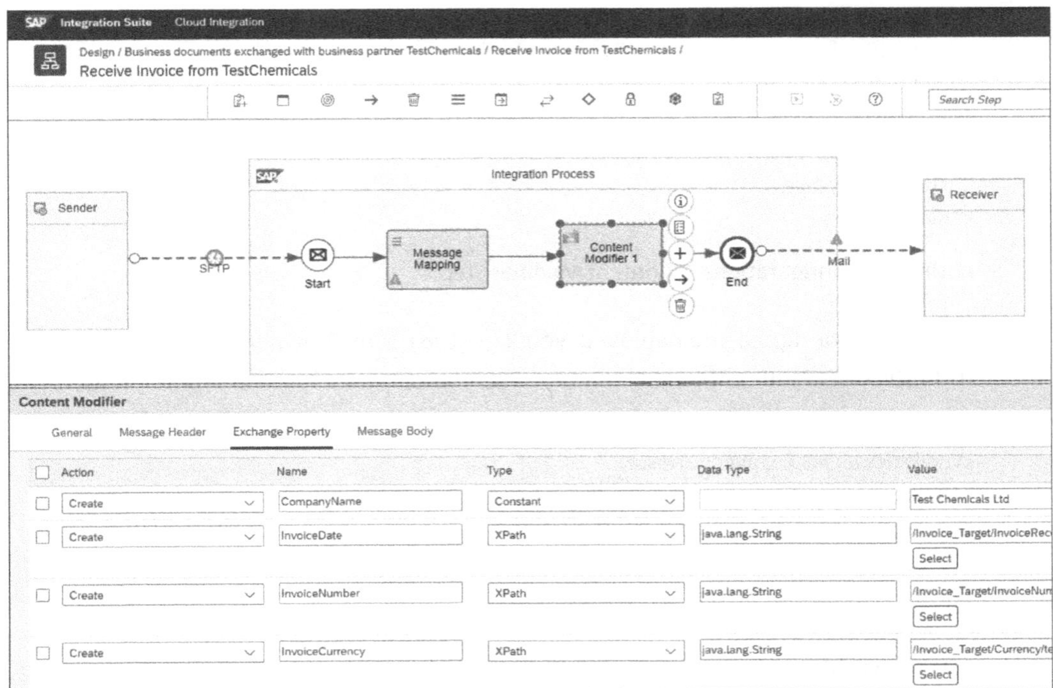

Figure 5.15 Sample View of Content Modifier

The **Message Body** section is used to change the message content partially or completely (to populate certain values from an incoming message or change the structure of a message by adding constants or dynamic values). After modifying the payload, it provides the modified payload for the next steps. Incoming payloads can be accessed by {in.body} elements. Let's consider an example. Say that you receive an invoice, but before sending it to the target system, you need to add additional vendor details and change the format so that the target system can understand it. Vendor details (values) to be added are already available in the pipeline header, exposed by previous steps of the integration flow. Listing 5.1 shows the input message for the content modifier step.

```
<Invoice>
<Invoice No>12345</Invoice No>
<InvoiceAmount>3456</InvoiceAmount>
<InvoiceDate>2021-21-03</InvoiceDate>
</Invoice>
```

Listing 5.1 Input Message to Content Modifier Step

Next, Listing 5.2 shows the configurations to be made in content modifier's **Message Body** tab. Assume that VendorID and VendorName are set as constants in the content modifier **Exchange Property** tab with the values **ABC122** and **ABCCorp**, respectively.

```
<InvoiceMaster>
<VendorID>${property.<vendorID>}</VendorID>
<VendorName>${property.<vendorName>}</VendorName>
${in.body}
</InvoiceMaster>
Outbound Message
<InvoiceMaster>
```

Listing 5.2 Configurations in Content Modifier Step

Once this is designed and deployed, you'll get the result shown in Listing 5.3 as output from the content modifier step.

```
<VendorID>ABC123<VendorID>
<VendorName>ABCCorpVendorName>
<Invoice>
<Invoice No>12345</Invoice No>
<InvoiceAmount>3456</InvoiceAmount>
<InvoiceDate>2021-21-03</InvoiceDate>
</Invoice>
</InvoiceMaster>
```

Listing 5.3 Content Modifier Configuration (Message Body Tab)

Non-XML Format Converters

Integration scenarios support different message formats, and often data exchange is in a non-XML format such as CSV, JSON, and EDI. This content needs to be converted to XML (and vice versa) for mapping or further processing. There are out-of-the-box connectors for the following formats:

- **CSV**
  Comma-separated values (CSV) is a plain-text file format that stores data in alphanumeric form, separated by delimiters and line breaks. This data can be read and easily represented in the form of tables. Each line consists of a data record with comma-separated fields. You can use a CSV converter to convert CSV files into XML (and vice versa). Listing 5.4 shows a CSV file with sample values. Once you use the CSV converter, the output from this step will be as shown in Listing 5.5.

  ```
  InvoiceNO, InvoiceAmount, OrderNo
  123,500,ABC
  456,600,DEF
  ```

  Listing 5.4 CSV File

  ```
  <InvoiceList>
  <Invoice>
  <InvoiceNo>123</InvoiceNo>
  <InvoiceAmount>500</InvoiceAmount>
  <OrderNo>ABC</OrderNo>
  </Invoice>
  </Invoice>
  <InvoiceNo>456</InvoiceNo>
  <InvoiceAmount>600</InvoiceAmount>
  <OrderNo>DEF</OrderNo>
  </Invoice>
  </InvoiceList>
  ```

  Listing 5.5 XML Conversion

- **JSON**
  JavaScript Object Notation (JSON) is an open standard data interchange format. It offers a lightweight way to transport data in an easy-to-access and organized manner. Cloud Integration provides a JSON converter to convert JSON format into XML and vice versa.

- **EDI**
  Electronic Data Interchange (EDI) format is used to exchange business data between trading partners. These messages are created according to different EDI specifications (X12, EDIFACT, EANCOM, etc.) for message types used in different regions and industries. Cloud Integration provides an EDI converter to convert EDI format into XML and vice versa.

- **Base64 encoder and decoder**
  One of the main use cases for Base64 is transferring binary data (images, PDFs, etc.). Sending raw binary data via different protocols may result in data corruption as different protocols may interpret binary data for different purposes. Converting binary data to Base64 encoding represents it in ASCII strings and makes it easy to transfer via different protocols. Base64 is also widely adopted for sending email attachments. Cloud Integration provides capabilities to both encode and decode Base64 messages.
- **ZIP and GZIP encoder and decoder**
  ZIP and GZIP are used to compress message content. As of the time of writing, Cloud Integration supports ZIP archives with single entry only. The difference between GZIP and ZIP is that ZIP is used for both file archiving and compression and GZIP only for compression; GZIP relies on TAR for archiving. Cloud Integration provides both capabilities to compress and decompress data.
- **Multipurpose Internet Mail Extensions multipart encoder and decoder**
  This transforms the message content into a Multipurpose Internet Mail Extensions (MIME) multipart message. A multipart/mixed MIME message consists of different types of data together. A MIME attachment may contain text data and other nontext attachments (images, video or audio files, etc.). The content type in the header is multipart/mixed if it contains both text and nontext data.
- **EDI extractor**
  EDIs include business transaction details (orders, invoices, shipments, etc.) and header parameters such as sender identification ID, recipient ID, type of business transaction, EDI format version, and more. The EDI extractor picks all the relevant header parameters to add to the message exchange so that they can be utilized by other steps for further processing (e.g., monitoring or acknowledgement processing).
- **Filter**
  Filters help to take a section of an incoming XML document and provide the filtered data to next steps. You can provide XPath information to extract a specific value of a node element or a whole node list that can be used for further processing. For example, if you are receiving an employee record with complete information (personal details, address, payroll information, etc.) and there is a need to extract the address data and store it in a separate target system, you can use the filter step to extract address details from the employee record and send it to the next steps for further processing.
- **Message digest**
  Message digest is used to maintain the integrity of the message. It converts the XML document into canonical format and calculates a hash value on the document. This hash value (digest) is added to the message header. The message recipient can read this digest value and check the integrity of the received message. Canonicalization

skips nonsignificant elements from an XML document before calculating a digest that makes it possible to compare it with other XML documents. Message digest provides options to select algorithms for calculating hashes and to add them to the message header.

- **Script**
  There are lot of steps provided in the pipeline for designing integration scenarios with a no-code approach. But users still have the option to write custom logic using Groovy or JavaScript to create complex customized business logic. Groovy is a Java syntax–based programming/scripting language. Using its scripting elements, you can also extract the header, property, and payload details and manipulate the messages.

5.2.5 Call Steps

Call steps are used to make external calls to third-party systems or local calls to other integration processes from the main integration flow, as shown in Figure 5.16. We'll walk through both in the following sections.

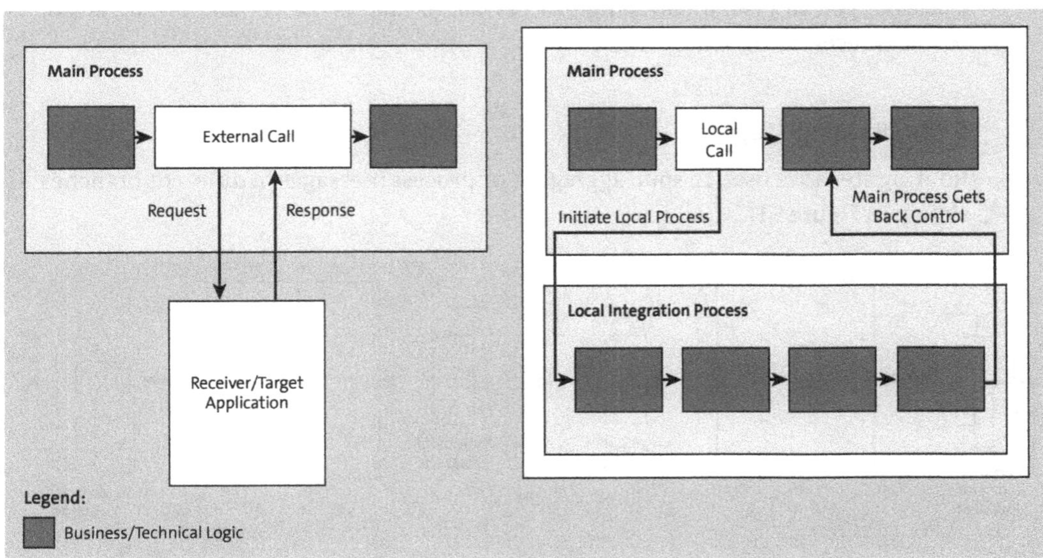

Figure 5.16 Call Step Variants Supported in Integration Flow Design

External Calls

External calls are used as intermediary calls to external systems before sending the message to the actual target system. They may be required for data enrichment or for looking up some business data to add required fields to the message payload before sending it to the actual receiver. For example, say you've received an order confirmation that needs to be sent to the manufacturing planning system. But before you send it, you need to add shipping location information, which is available in another third-party logistics system. The content enricher step is used to make external calls using receiver adapters and it appends the received responses from the third-party system to the original payload. Request reply is used to process only the response message. The send step helps send the required information to the target system without expecting a response. You can use only receiver adapters with the send step. As an example, once the shipment information is received, a quality status notification is sent to the logistics company to update the status before shipment details are shared with the company information system for an end user to consume.

Local Calls

You can also use the ProcessDirect adapter to call local integration processes, as explained in Section 5.2.2. A local integration process helps to break the business logic into small chunks that are easy to maintain and operate and can be called multiple times from the main process. For example, say you've received a batch of invoices. The local integration process calculates the overall tax amount of multiple line items in a single invoice and sends the updated invoice payload back to the processing step. In this case, you can use the looping process call to update the tax amount details of all invoices received in a batch file.

5.2.6 Routing Steps

Routing steps are used to split, aggregate, or process messages in different branches, as shown in Figure 5.17.

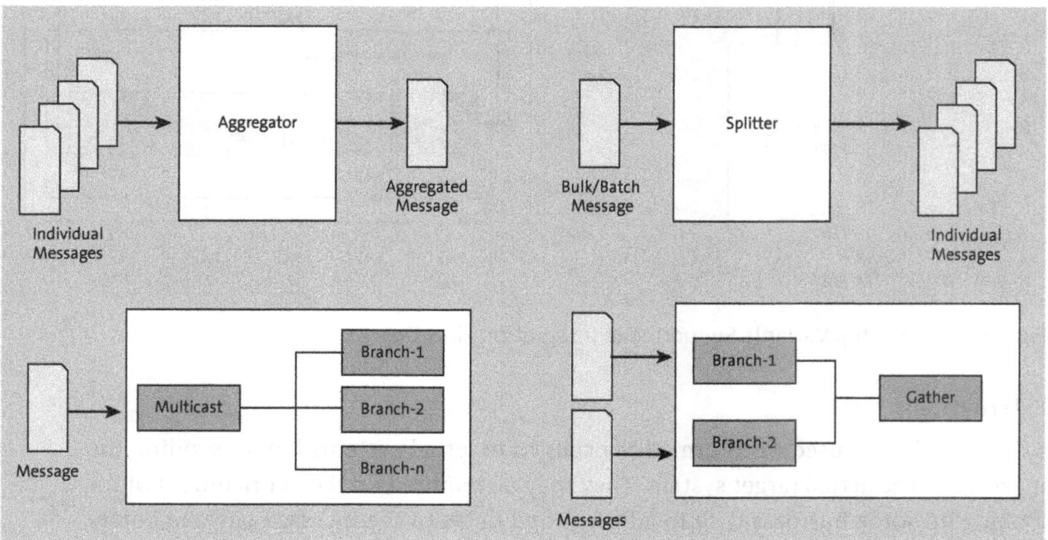

Figure 5.17 Routing Step Variants Supported in Integration Flow Design

Let's walk through the available routing steps:

- **Aggregator**
  The aggregator step helps to aggregate multiple messages. Assume you are receiving multiple sales orders from the source systems, but you need to create a batch before sending it to the target systems either to optimize the communication bandwidth or due to a business requirement that all orders need to be processed together. You can have a condition-based aggregation (e.g., until order 100 arrives) or a time-box aggregation (e.g., all orders that arrive within 60 minutes), which can be configured in the aggregator step.
- **Gather**
  The gather step combines messages from multiple branches and can support aggregation of plain text and XML (same or different) messages. You can provide an exact location in an XML document in which incoming messages need to be concatenated. It also provides the option to convert concatenated messages into ZIP or TAR files.
- **Multicast**
  This step can be used to send the same messages to different branches in an integration scenario. Parallel multicast allows parallel processing of different branches in an integration flow simultaneously, whereas sequential multicast provides an option to execute different branches in sequence. For example, say you've received invoices with multiple line items. You need to calculate the total amount for all line items, and execution logic is mentioned in branch 1. Once the total amount is calculated, control will automatically be sent to branch 2 to calculate the tax amount, which is different for various line items, and the final tax amount will be sent to the target system.
- **Router**
  The router step helps with conditional routing. Based on the different conditions that can be configured, it routes the control to respective branches in an integration flow. For example, if the sales order amount is less than 100K, it needs to be posted directly to the target system (logic from branch 1). But if it is greater than 100K, it needs to be sent for two levels of approvals (logic written in branch 2) before it's posted to the target system.
- **Splitter**
  The splitter step helps split various types of aggregated or composite messages. It provides flexible conditions to split different types of messages, such as EDI, XML, IDocs, ZIP and TAR files, and so on. It also provides a PKCS #7 splitter to split actual content and signatures from a digitally signed incoming message to process it separately. Say that a trading partner sends a batch of messages that consists of different types of business transactions bundled together (orders, shipment notifications), but the target system can only accept one business transaction at a time. A splitter is needed to split messages, map them to target system formats via message mapping, and post the converted messages one by one in the target business system.

5.2.7 Security

Secure communication (HTTPS, SMTPS, SSH, etc.) is available on transport-level transmissions, but message-level security adds another layer of security on top and further secures messages being transmitted on the communication layer. These messages can be encrypted and signed, as shown in Figure 5.18, using PGP, X.509, XML digital signatures, and other supported techniques. PGP encryption is a common requirement for transferring business-sensitive data (e.g., financial messages for bank integration). Digital signatures are required to confirm your identity to know that the message is coming from the indicated sender. Encryption secures the content so that no one can read it except the intended recipient, using required keys to decrypt it. There are two different steps available in the integration flow designer for encryption and digital signatures. While using the encryption step, optionally, you can digitally sign the message in the same step. The keys for encryption and signatures should be present in the Key-Store (see Section 5.1.3).

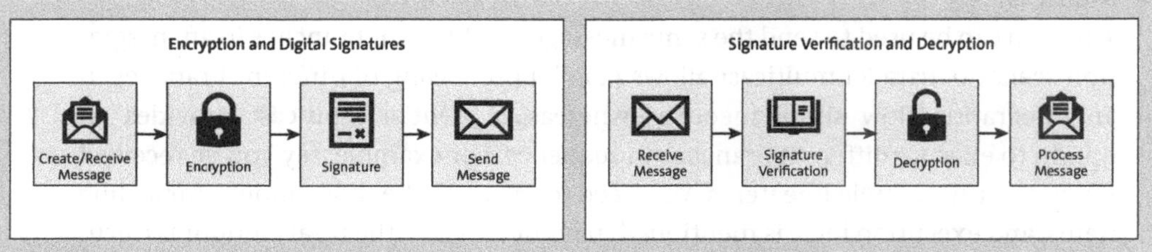

Figure 5.18 Security Step Variants in Integration Flow Designer

5.2.8 Validator

The validator step helps to validate the incoming messages against standard schemas. These schemas are based on open specifications with a controlled syntax and semantics, like XML and EDI. Before processing the document, there is a need to check the semantics of the message. EDI is validated against the EDI schema, and XML documents are validated against XSL Schema Definition (XSD).

5.2.9 Persistence

There are many integration scenarios in which temporary persistence is needed, as shown in Figure 5.19. For example, say you're getting purchase orders and shipping location details separately from different applications and via separate transactions. While processing shipping details, you need to include incoming purchase order details that are already received by Cloud Integration a day before. You can store purchase orders in data stores and retrieve them later whenever they're needed—in this case, when preparing the shipping details. You can also delete the data from the data store once business integration requirements for that process are completed and you no longer need the temporary storage.

Another persistence step is *write variables*, which provides a way to populate the message headers with the incoming message values that you would like to reuse in integration scenarios or across multiple integration scenarios in same tenant. The *persist message* step helps you to persist required messages that can be accessed via APIs (e.g., if you want to store incoming or outgoing messages from or to business partners).

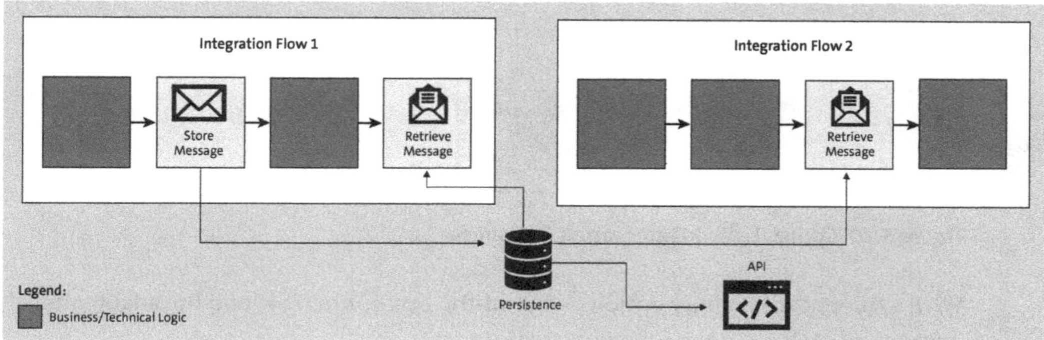

Figure 5.19 Persistence Step Variants Available in Integration Flow Designer

5.3 Connectivity

The Cloud Integration capability of SAP Integration Suite offers a wide variety of protocols to connect to cloud and on-premise applications. It supports technical and application-level connectivity to many different technical and business systems that are required to be connected in an integration scenario. Further, on-premise systems and applications can be securely connected via the cloud connector without any need to open ports in a firewall (as explained in Chapter 2, Section 2.3.2). These connectors are used along with integration flow participants as source, target, or intermediary calls, as shown in Figure 5.20. These business applications either support open specifications for technical connectivity such as HTTP, OData, SSH and the like or follow certain proprietary mechanisms that are deviations from the standards to provide enhanced communication such as enriched security and message exchange procedures. Security is a very important aspect of message exchange and is tightly associated with these protocols, such as transport-level security (SSH for SFTP, SMTPS, POPS for mail communication, HTTPS for OData, SOAP, REST, etc.) and message-level security (encryption, digital signatures, PGP, X.509, XML digital signatures, etc.). Technical adapters provide implementation of open protocol specifications, and application adapters include technical protocols, application-specific deviations, and business API access for enriched and easy connectivity.

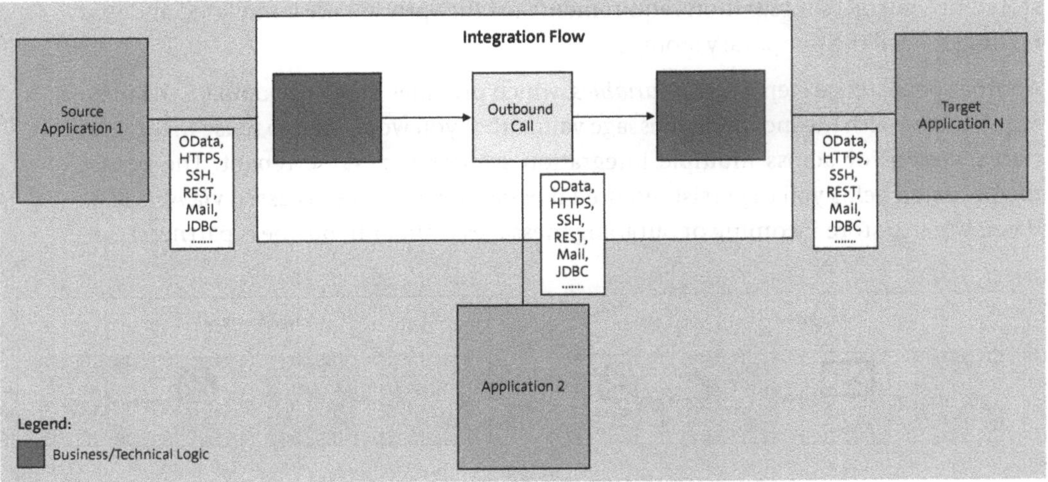

Figure 5.20 Connectivity in Integration Scenarios

We'll explore connectivity options—out-of-the-box connectors and the adapter development kit (ADK)—in the following sections.

5.3.1 Out-of-the-Box Connectors

Let's look at the different technical and application adapters supported by the Cloud Integration capability of SAP Integration Suite, as described in Table 5.1.

Adapters	Short Description
HTTP(S)	Allows you to connect with HTTP-based scenarios and to make plain HTTP or REST calls.
OData	OData provisioning and consumption is used to connect to any SAP or non-SAP-based OData service. Low-code API development is available for provisioning OData APIs easily with required business logic at the backend.
SOAP	Helps to provision or consume SOAP-based web services. You can also connect to an SAP ERP backend by exposing remote function calls (RFCs)/business application programming interfaces (BAPIs) as web services. Low-code API development is available for provisioning SOAP APIs easily with the required business logic in the backend.
FTP	Used to connect FTP servers for transferring files using an FTP/FTPS protocol.

Table 5.1 Out-of-the-Box Technical and Application Adapters Provided by Cloud Integration Capability

Adapters	Short Description
SFTP	Used to connect to SFTP servers for both reading and writing files via a secure client-based certificate authentication mechanism. This is used to connect to SFTP servers in order to transfer files using the SSH protocol.
Mail	Used to send and receive emails using SMTP(S) and IMAP(S) and POP3(S). It supports attachments and can also be used to send email alerts based on different business/technical statuses of the integration process.
JDBC	Helps connect to databases such as SAP HANA, Oracle, MSSQL, and so on, both in the cloud and on premise. It helps store/read data to/from various SAP and third-party databases.
Kafka	Can connect to external Kafka broker and can send or fetch records from multiple Kafka hosts.
AMQP	The Advanced Message Queuing Protocol (AMQP) adapter is used to transfer messages or events via message brokers. It can act as both provider and consumer of messages. SAP Event Mesh can also be connected using the AMQP adapter.
AS2	Applicability Statement 2 (AS2) is an HTTP-based protocol used to securely transport business messages to trading partners using encryption and digital signatures in a reliable manner. It's mainly used to communicate business-to-business (B2B) integration messages, such as EDI.
AS4	Applicability Statement 4 (AS4) is an HTTP- and SOAP-based protocol and uses XML Encryption and XML Digital Signatures. It's based on the ebMS 3.0 specification and is mainly used for B2B integration scenarios.
RFC	Used to connect to on-premise ABAP systems (e.g., SAP ERP) using the RFC protocol. It uses the cloud connector to securely connect to on-premise systems.
XI	Provides compatibility with on-premise systems that support XI 3.0, such as SAP Process Orchestration.
IDoc	Can be used to connect to SAP Business Suite for transferring IDocs over HTML in XML format. This is the easiest option for companies to open HTTP ports in their firewalls considering security guidelines.
JMS	Used to store and retrieve messages from internal asynchronous queues and helps process messages a in a reliable manner.

Table 5.1 Out-of-the-Box Technical and Application Adapters Provided by Cloud Integration Capability (Cont.)

Adapters	Short Description
LDAP	Lightweight Directory Access Protocol (LDAP) is an open specification for retrieving and maintaining distributed directory services over the internet. The LDAP adapter provides insert, modify, delete, and search operations. One of its common use cases is to connect to Microsoft Active Directory for authentication services using usernames and passwords.
ProcessDirect	Used to call the local integration process from the main integration scenario. It is an internal adapter used within Cloud Integration to call business logic written in subprocesses.
Open Connectors	SAP Integration Suite provides 160-plus third-party connectors that can be accessed via normalized APIs. Cloud Integration provides direct connectivity via the Open Connectors adapter to communicate with third-party systems.
Salesforce	Used to connect to the Salesforce system. It supports multiple APIs with over 50 operations and provides XSD generation capabilities. It uses OAuth2 as its authentication mechanism and can exchange data in both JSON and XML formats.
Workday	Used to connect to the Workday HR application.
ServiceNow	Used to connect to the ServiceNow application.
Facebook	Used to extract information such as keywords and user data from Facebook (e.g., you can use it for social marketing activities).
Twitter	Can be used to extract information from Twitter such as keywords and tweets (e.g., to check market sentiments for a new product launch).
SAP SuccessFactors	Used to connect to the SAP SuccessFactors application using the SFSF API (based on SOAP) or OData for Employee Central and talent suite components. The REST protocol in the SAP SuccessFactors adapter can be used to connect to SAP SuccessFactors's learning management solution. The adapter also supports a delta sync option to optimize the number of records fetched from SAP SuccessFactors and ad hoc query support to work with multiple entities.
SAP Ariba	Used to connect to Ariba Network. This adapter supports messages being exchanged in cXML format and transfer them using an HTTP MIME-based protocol.
ELSTER	Elektronische Steuererklärung (ELSTER) is used by German fiscal management to process tax declarations over the internet.

Table 5.1 Out-of-the-Box Technical and Application Adapters Provided by Cloud Integration Capability (Cont.)

Adapters	Short Description
Microsoft Dynamics CRM	Used to connect to the Microsoft Dynamics CRM system. It provides simplified discovery of Microsoft Dynamics CRM entities, respective properties, and XSD generation. It uses OAuth2 as its authentication mechanism and can exchange data in both JSON and XML formats.
Amazon AWS	Used for connecting to Amazon Web Services (AWS), and supports the following connectivity options: ▪ Simple Cloud Storage (S3) ▪ Simple Queue Service (SQS) ▪ Simple Notification Service (SNS) ▪ Simple Workflow Service (SWF)
Sugar CRM	Used to connect to the Sugar CRM system. It provides simplified discovery of Microsoft Dynamics CRM entities, respective properties, and XSD generation. It uses OAuth2 as its authentication mechanism and can exchange data in both JSON and XML formats.

Table 5.1 Out-of-the-Box Technical and Application Adapters Provided by Cloud Integration Capability (Cont.)

Note

Connectivity adapters follow multiple respective security specifications. These security configurations require various certificates (public, private, X.509, PGP) for security operations. These certificates need to be stored in the central KeyStore, as explained in Section 5.1.3, and can be referred to in adapter configurations with an alias.

In addition to the adapters in Table 5.1, SAP Integration Suite provides 160-plus third-party connectors. Companies can perform message transformation in cloud connectors, but the Open Connectors adapter can be used to exchange data with these 160-plus third-party connectors available within SAP Integration Suite. See Chapter 7 for more details.

5.3.2 Adapter Development Kit

In addition to the out-of-the-box technical and application adapters in Table 5.1, the Cloud Integration capability offers a public ADK that can be used to develop new connectors based on the ADK framework. Once an adapter is developed, it can be deployed on the Cloud Integration tenant and will automatically appear in the list of connectors available for configuration. Currently, new adapters can be developed in Eclipse-based development environments, with ADK available as a plug-in. (There is a plan to move this development experience soon to a cloud-based development environment.) ADK includes components for connectivity functionality, logic to interpret messages exchanged using adapters, logging, security, on-premise connectivity APIs, and so on. You can develop any technical and application connectors using the ADK framework. From a skillset perspective, you need knowledge of Java and Apache Camel components. There are many partners offering adapters developed using ADKs, such as HubSpot, Microsoft SharePoint, Microsoft Azure, and more.

Partner Directory

There may be various scenarios in which the business logic defined in an integration flow is the same but you want to process messages in a different manner based on certain input parameters. In this case, you won't want to deploy multiple integration scenarios with these different values configured in various pipeline steps because maintenance and operations will become difficult. Instead, you can use partner directory APIs provided by Cloud Integration to define partner-specific types of data that can be utilized by integration flows during runtime for dynamic scenario execution. Various pipeline steps in the integration flow can query the partner directory based on values that can be extracted from the input payload, such as partner ID and scheme. This is a very common use case for B2B integration scenarios in which you need to connect to multiple trading partners over a single connection from a value-added network (VAN) provider but need to execute the same integration flow with different parameters and conditions based on different trading partner agreements. SAP plans to provide a user interface for partner directory APIs via trading partner management (TPM) functionality in the future.

5.4 Creating an End-to-End Integration Flow

Now that we've gone through the various elements and functionalities of the integration designer, you're ready to create an integration scenario from scratch. We'll walk through the steps in this section.

5.4.1 Business Requirements

Let's start with the business requirements for our example scenario. A company, Test Solvent Corporation, wants to launch a new product in the market. For manufacturing this product, a new vendor, Test Chemicals Ltd., is onboarded with due diligence so it can provide raw materials for the new solvent product. Test Solvent can't create a new vendor in its business system due to some technical issues and decides to proceed by providing documents over email to Test Solvent's team. It plans to do manual processing for the next one to two months because the new product launch is critical. The business team provides the requirement to the IT team to get the invoices from Test Chemicals and send the business transaction documents with the required information in emails to Test Solvent's finance team for manual processing. The IT team is utilizing SAP Integration Suite for developing this integration. Test Chemicals provides access to business transactions over an SFTP server. For Test Solvent's business system, Gmail will be the email server used to send email notifications to the finance team.

5.4.2 Prerequisites

There are security prerequisites to meet for this scenario, which will be configured by tenant administrators, for connecting to the SFTP and mail servers. We won't show these security settings in the next steps and will assume that these configurations are maintained by tenant administrators. Let's look at the requirements:

- **SFTP**
  For connecting to an SFTP sever, we need to add the required SFTP server hostname and public fingerprints to the known_host file and deploy it via a secure store. This hostname and public certificate will be checked during execution of the integration scenario. If it doesn't match, the scenario execution will fail. This is done to ensure we are connecting to the required SFTP servers allowed for the scenario. In addition, we'll be connecting to the SFTP server using basic authentication. The username/password needs to be deployed in the secure store with a unique name that can be used in the SFTP adapter settings.
- **Mail**
  For connecting to Gmail in a secure way, you need to import the right Gmail certificates in the KeyStore. Then you need to provide your username (email ID) and password and deploy it in the secure store with a unique name. This name will be used in the mail adapter while configuring the security parameters. In addition, you need to configure your Google account to allow non-browser-based access to your email account. These are the security requirements Google has for the security of your email account (as these security requirements may be changed by Gmail, be sure to follow the latest documentation for Gmail access via APIs). You can use any other email server as well.

5.4.3 Creating an Integration Package and Artifact

Now, let's walk through the steps to create an integration package:

1. You need to activate the SAP Integration Suite offering as described in Chapter 2, Section 2.3.3. Once you have it activated, use the SAP Integration Suite landing page to start developing scenarios, as shown in Figure 5.21.
2. Click the **Design, Develop, and Operate Integration Scenarios** tile. That will open the workspace to design the Cloud Integration scenarios.
3. By default, it opens to the **Discover** section. Because you're planning to create a custom integration scenario without utilizing prepackaged integrations, you can skip the **Discover** section and move to the **Design** section, as shown in Figure 5.22.

Figure 5.21 SAP Integration Suite Landing Page

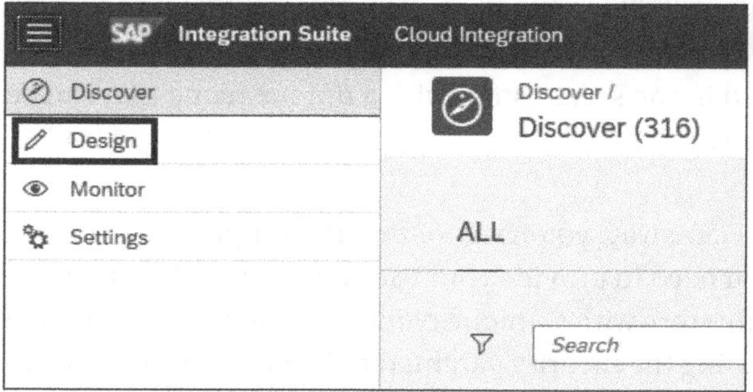

Figure 5.22 Design Section of Cloud Integration

4. Click the **Create** button to create a new integration package, as shown in Figure 5.23.

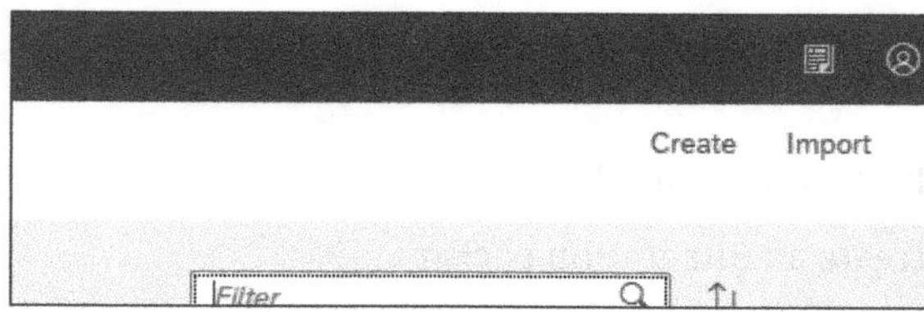

Figure 5.23 Creating New Integration Package

5. Create a package with "Business documents exchanged with business partner Test-Chemicals" as the **Name** and enter the details as shown in Figure 5.24 (referring back to Section 5.1.2). Optionally, you can also provide additional details in the **Overview**, **Artifacts**, **Documents**, and **Tags** tabs, but it isn't necessary for this example. Click the **Save** button.
6. Once created, the package will appear in the **Design** section, as shown in Figure 5.25.

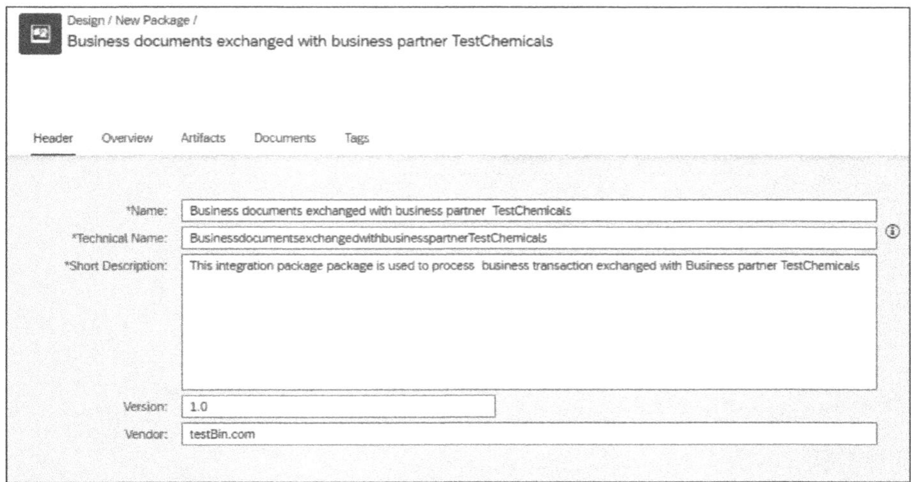

Figure 5.24 Providing Details for New Integration Package

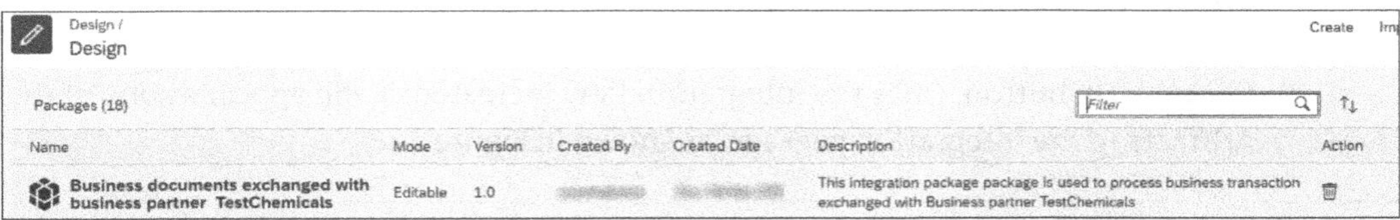

Figure 5.25 Package View in Design Section

7. The next step is to create an integration flow artifact in the newly created package. Edit the package again, then go to the **Artifacts** tab and choose **Add • Integration Flow**, as shown in Figure 5.26.

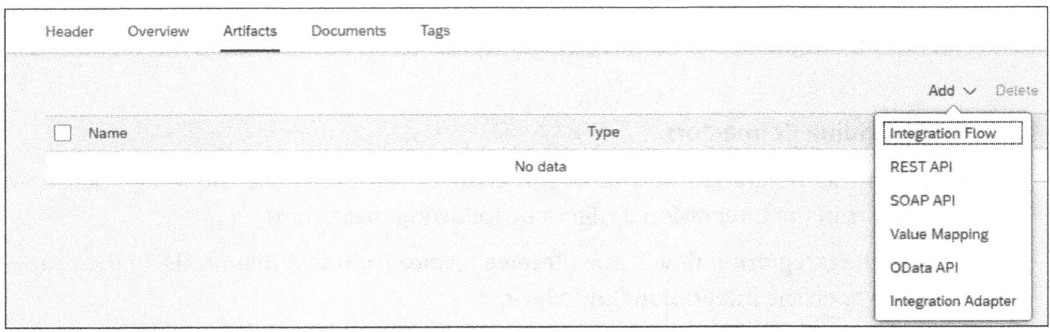

Figure 5.26 Creating New Integration Flow Artifact in Existing Package

8. Enter the required details as shown in Figure 5.27, including a **Name, ID,** and **Product Profile.** You can select the **Sender** and **Receiver** systems from the provided list or add a custom name. We have listed the adapters that we will be using in the sample integration scenario: **SFTP** and **MAIL**.

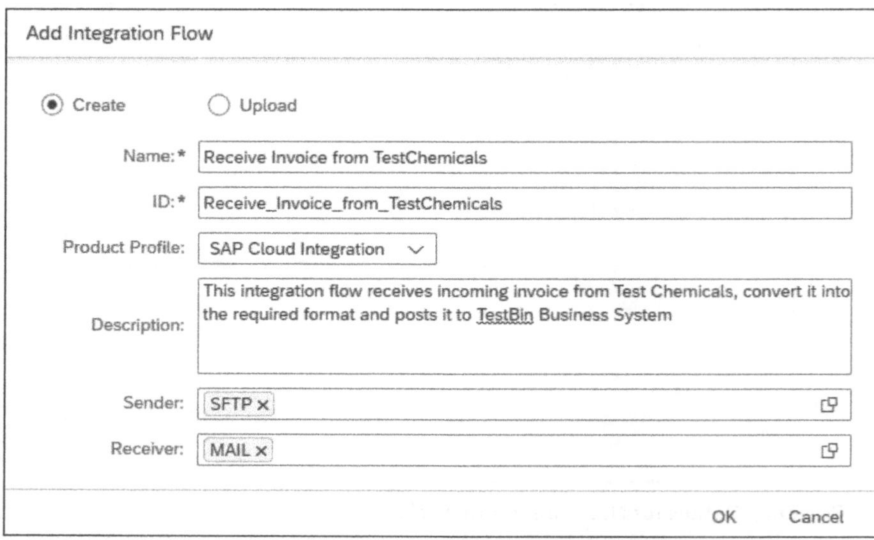

Figure 5.27 Creating Integration Artifact

9. Click the **OK** button. Once the integration flow is created, it will appear as one of the **Artifacts** in the integration package, as shown in Figure 5.28.

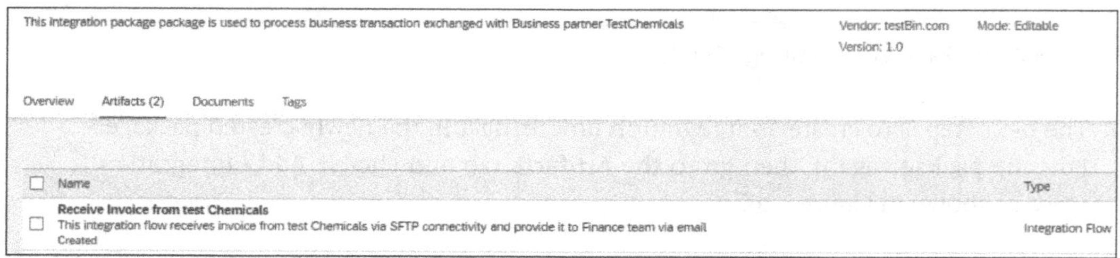

Figure 5.28 Integration Flow in Integration Package

5.4.4 Adding Connectors

Now that the integration flow artifact is created, you can create the integration flow connectors in the integration designer by following these steps:

1. Click the integration flow name (**Receive Invoice from test Chemicals**, in the example) to open the integration flow editor.
2. You need to edit the integration flow and add the required connectors and business logic for the next steps. Drag the arrow from the **Sender** participant to the **Start** message icon and select **SFTP** from the adapter list, as shown in Figure 5.29.

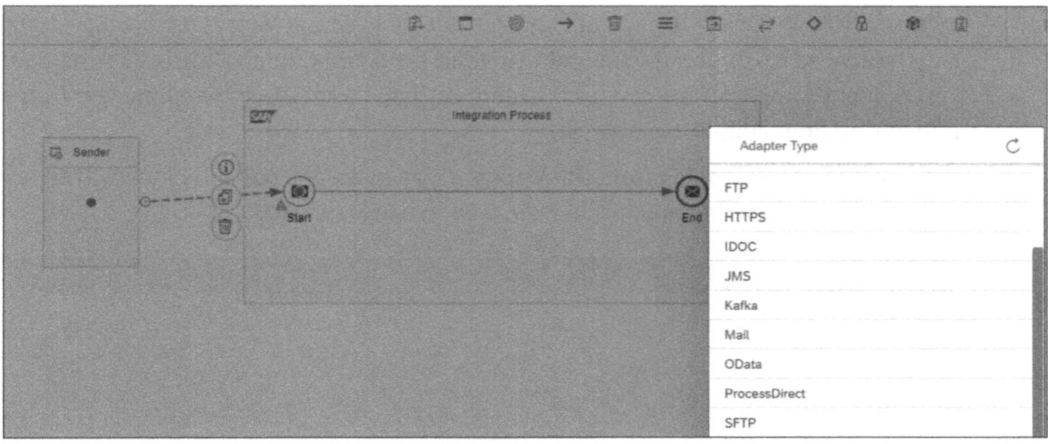

Figure 5.29 Selecting SFTP Adapter for Source Connectivity

3. Similarly, drag the arrow from the **End** message icon to the **Receiver** participant and select the **Mail** adapter, as shown in Figure 5.30.

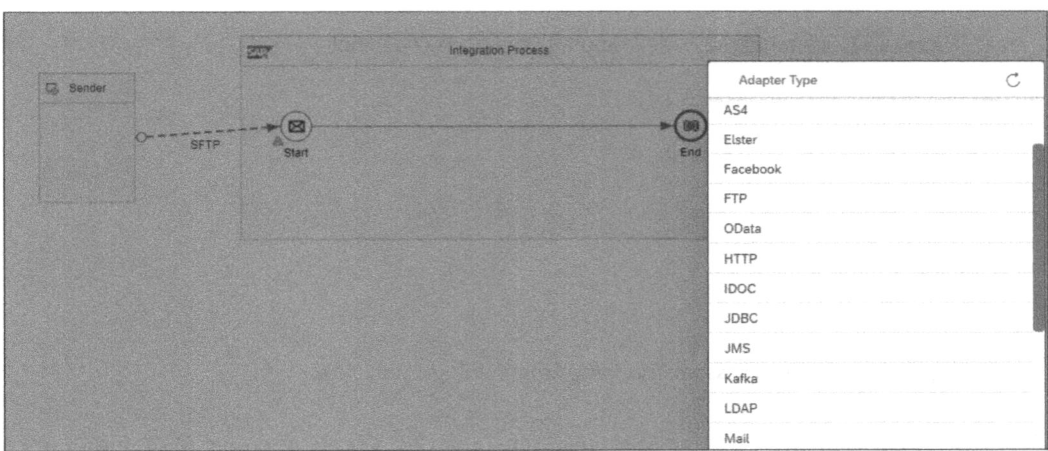

Figure 5.30 Selecting Mail Adapter for Target System Connectivity

4. After selecting both connectors, your integration flow will look as shown in Figure 5.31.

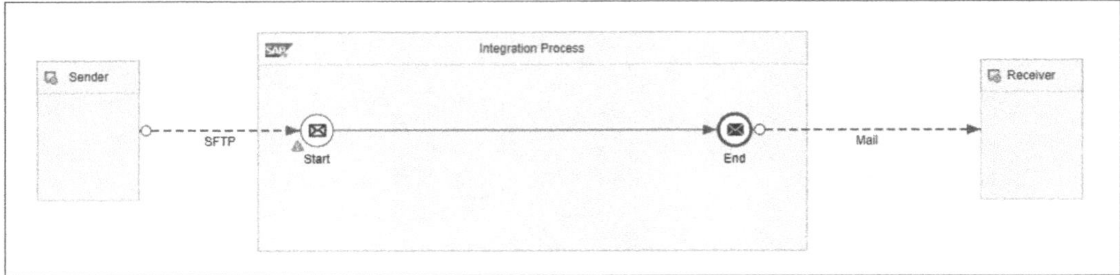

Figure 5.31 Integration Flow View after Adding Adapters

5.4.5 Message Mapping

Next, you need to convert the incoming invoice from Test Chemicals into a format that Test Solvent's applications can understand. Let's add a message mapping step in the integration flow. Follow these steps:

1. Select **Message Mapping** from the **Mapping** icon in the top panel, as shown in Figure 5.32, and add it in the integration flow as shown in Figure 5.33.

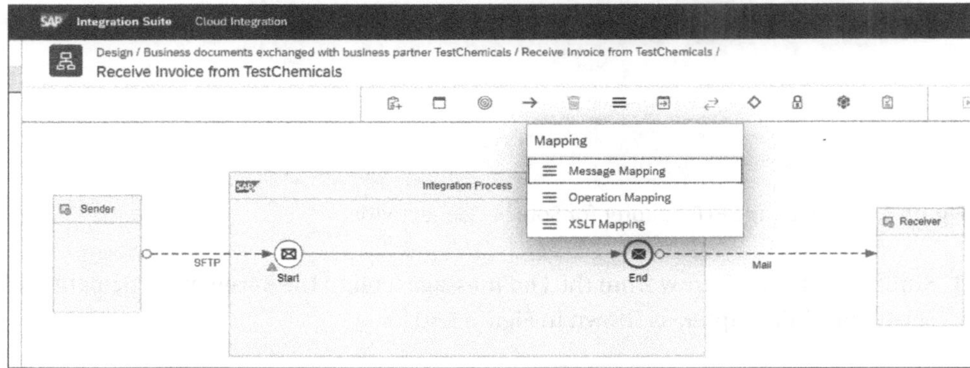

Figure 5.32 Selecting Message Mapping Step from Top Panel

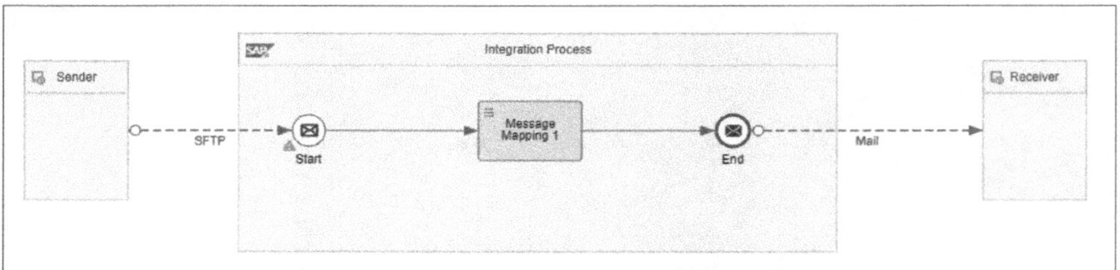

Figure 5.33 Adding Message Mapping Step in Integration Flow

2. Right-click the **Message Mapping** step and select **Create Message Mapping** to open the popup shown in Figure 5.34.

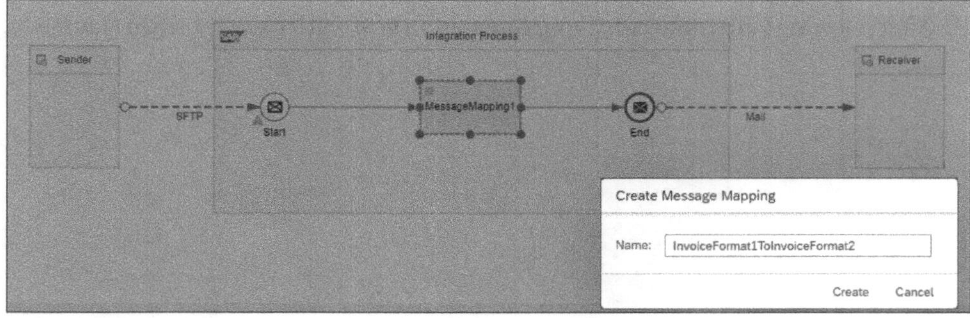

Figure 5.34 Creating New Message Mapping

3. Click the **Create** button to open a new window, as shown in Figure 5.35, with options to select the source and target message structures for message mapping.

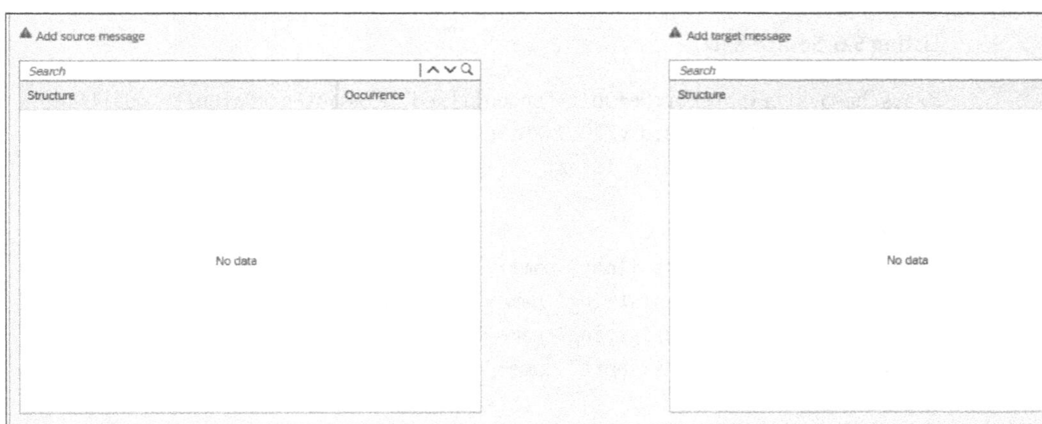

Figure 5.35 New Message Mapping View

4. Upload source and target XSD structures from your file system by clicking the **Add source message** and **Add target message** links, respectively, as shown in Figure 5.36. These XSD structures are schemas for XML documents (provided by Test Chemicals and Test Solvent's internal teams) and are used to exchange messages with business systems. Listing 5.6 shows the source XSD structure for the example, and Listing 5.7 shows the target XSD structure.

```
<xs:schema attributeFormDefault="unqualified" elementFormDefault="qualified"
xmlns:xs="http://www.w3.org/2001/XMLSchema">
 <xs:element name="Invoice">
  <xs:complexType>
   <xs:sequence>
    <xs:element type="xs:float" name="InvoiceNo"/>
    <xs:element type="xs:string" name="InvoiceDate"/>
    <xs:element type="xs:string" name="InvoiceCurrency"/>
    <xs:element name="MachineToolDetails">
     <xs:complexType>
      <xs:choice maxOccurs="unbounded" minOccurs="0">
       <xs:element type="xs:string" name="LineItem1Name"/>
       <xs:element type="xs:short" name="LineItem1Price"/>
       <xs:element type="xs:string" name="LineItem2Name"/>
       <xs:element type="xs:string" name="LineItem2Price"/>
      </xs:choice>
     </xs:complexType>
    </xs:element>
   </xs:sequence>
```

```
    </xs:complexType>
  </xs:element>
</xs:schema>
```

Listing 5.6 Source XSD

```
<xs:schema attributeFormDefault="unqualified" elementFormDefault="qualified"
xmlns:xs="http://www.w3.org/2001/XMLSchema">
  <xs:element name="Invoice_Target">
    <xs:complexType>
      <xs:sequence>
        <xs:element type="xs:float" name="InvoiceNumber"/>
        <xs:element type="xs:string" name="InvoiceReceiptDate"/>
        <xs:element type="xs:string" name="Currency"/>
        <xs:element type="xs:short" name="TotalInvoiceValue"/>
      </xs:sequence>
    </xs:complexType>
  </xs:element>
</xs:schema>
```

Listing 5.7 Target XSD

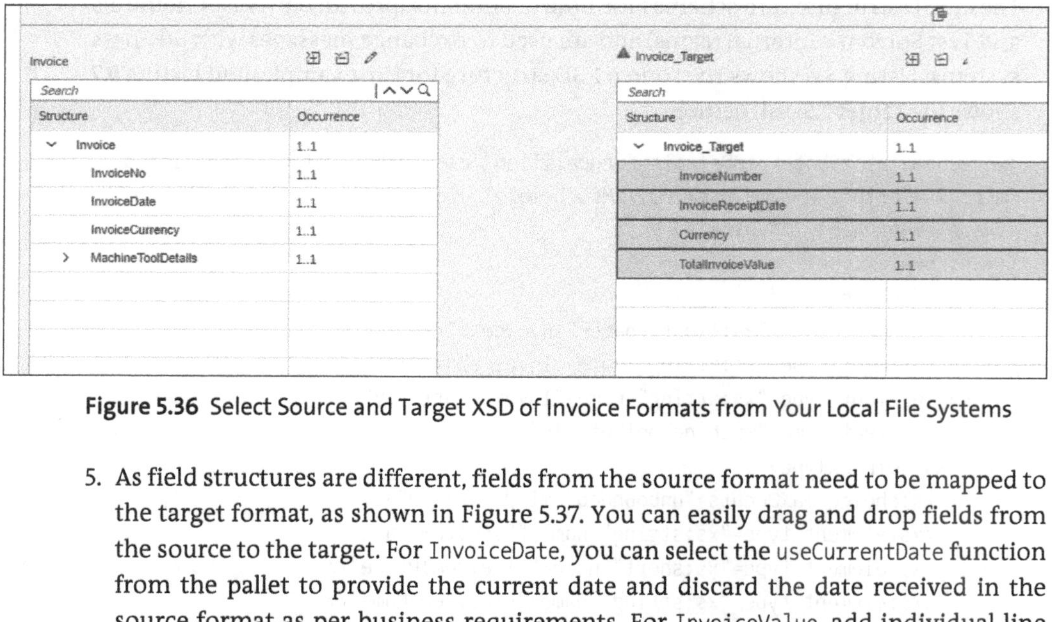

Figure 5.36 Select Source and Target XSD of Invoice Formats from Your Local File Systems

5. As field structures are different, fields from the source format need to be mapped to the target format, as shown in Figure 5.37. You can easily drag and drop fields from the source to the target. For `InvoiceDate`, you can select the `useCurrentDate` function from the pallet to provide the current date and discard the date received in the source format as per business requirements. For `InvoiceValue`, add individual line items' price values from the source format and use the arithmetic `add function` to provide a sum to the `InvoiceValue` field of the target format.

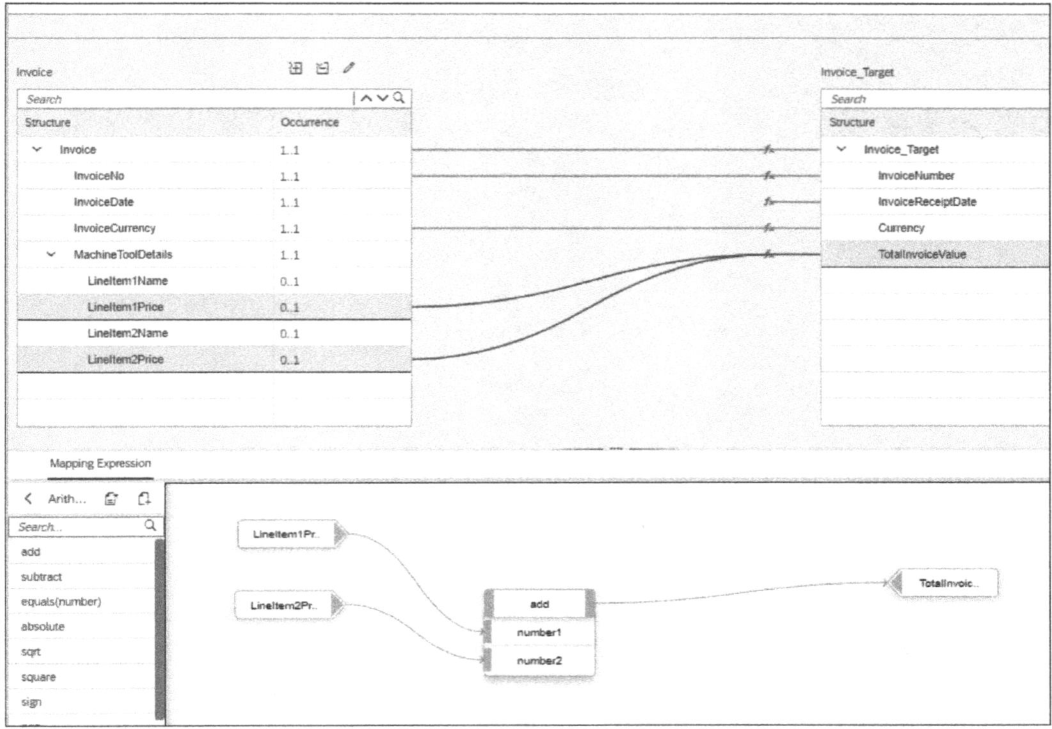

Figure 5.37 Message Mapping between Source and Target Format

6. Once mapping is done, you can quickly simulate the mapping using the **Simulate** option, as shown in Figure 5.38. You can **Browse** for and upload the test payload that matches the incoming XSD format of the source structure and then click **Test**. This will convert the payload into the target format as per the defined mapping logic, as shown in Figure 5.39, and you can validate the values.

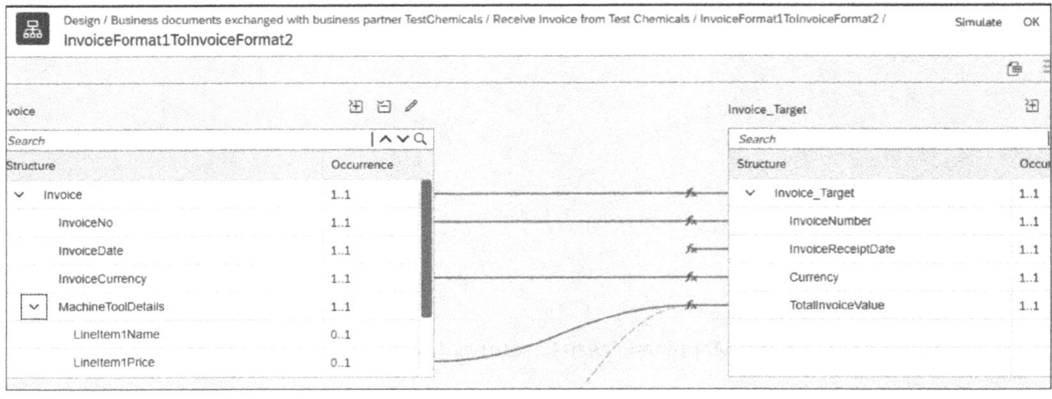

Figure 5.38 Simulate Message Mappings

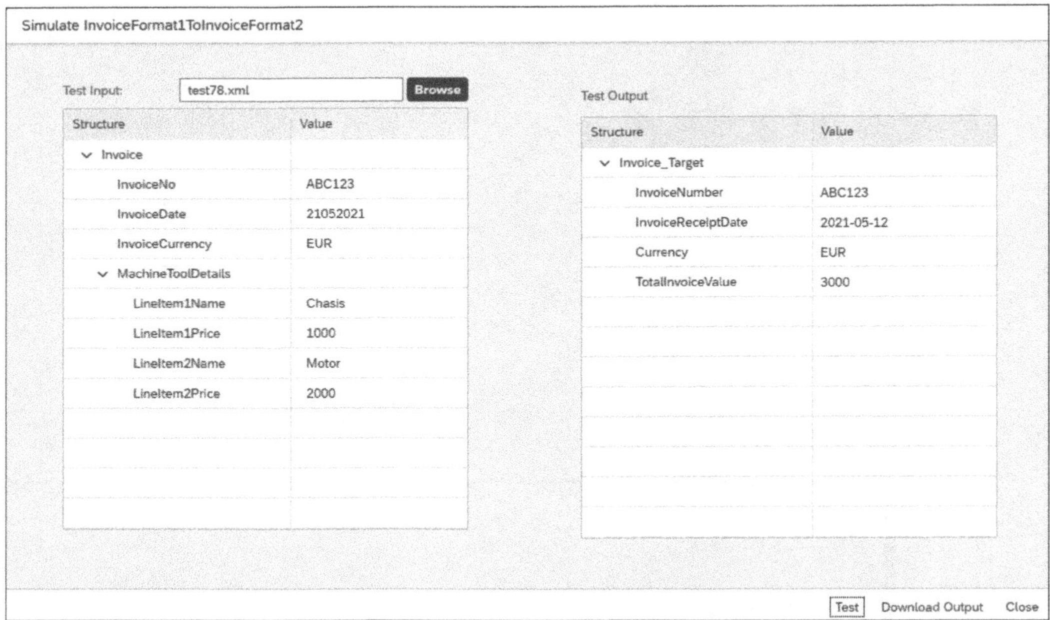

Figure 5.39 Simulation Result with Target Structure

7. Once the message mapping configuration is over, return to the main integration flow screen and click the **Save** button. You can also navigate to the **Resources** tab and see that that XSDs and mapping files used in the message mapping are automatically added to the **Resources** tab of the main integration flow, as shown in Figure 5.40.

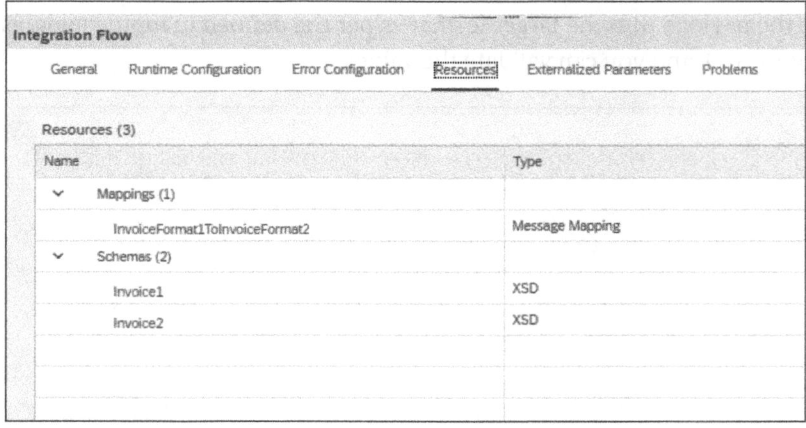

Figure 5.40 XSDs and Mapping Files in Resources Tab

8. Return to the main integration flow, where you can optionally change the name for the **Message Mapping** step in the **General** tab. For this example, change the name to "Invoice Structure Mapping", as shown in Figure 5.41. Save the integration flow.

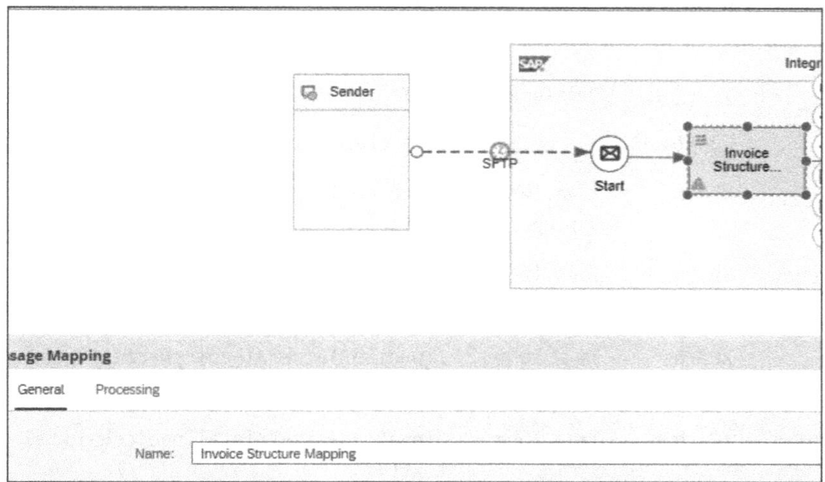

Figure 5.41 Updating Name of Message Mapping Step

5.4.6 Creating a Custom Text Format

Message mapping is complete, but the Test Solvent business system isn't ready yet. You need to create a simple text format from the invoice XML that can be sent in an email. To do so, follow these steps:

1. Select the **Content Modifier** step from the top panel, as shown in Figure 5.42.

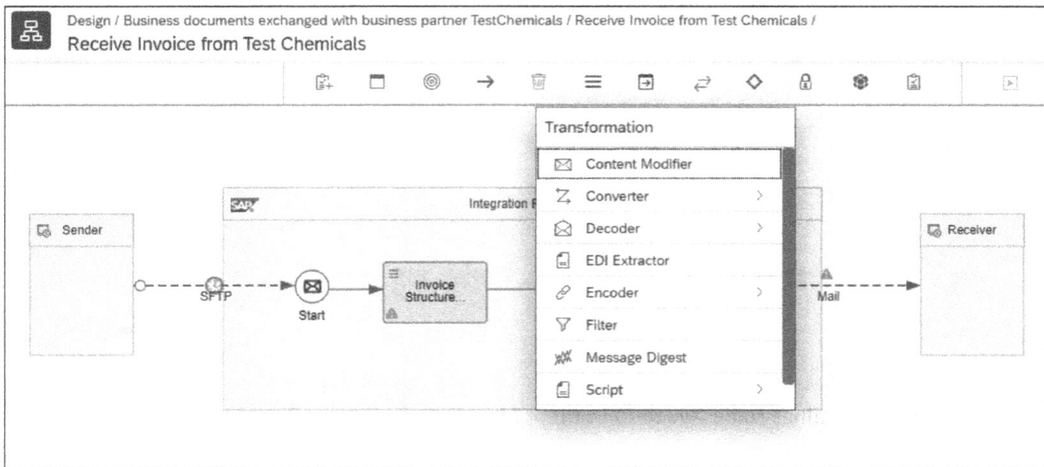

Figure 5.42 Select Content Modifier Step from Transformation Category

2. Add the **Content Modifier** step in the main integration flow and change the step name to "Email Message Creation" within the **General** tab.
3. You need to extract values from the target XML format that can be used in email templates. Add the values under the **Exchange Property** tab, as shown in Table 5.2 and Figure 5.43.

Name	Type	DataType	Value
CompanyName	Constant		Test Chemicals Ltd
InvoiceNo	XPath	Java.lang.String	/Invoice_Target/InvoiceReceiptDate/text()
InvoiceCurrency	XPath	Java.lang.String	/Invoice_Target/InvoiceNumber/text()
InvoiceDate	XPath	Java.lang.String	/Invoice_Target/InvoiceReceiptDate/text()
InvoiceAmount	XPath	Java.lang.String	/Invoice_Target/TotalInvoiceValue/text()

Table 5.2 Values to Be Added in Content Modifier's Exchange Property Tab

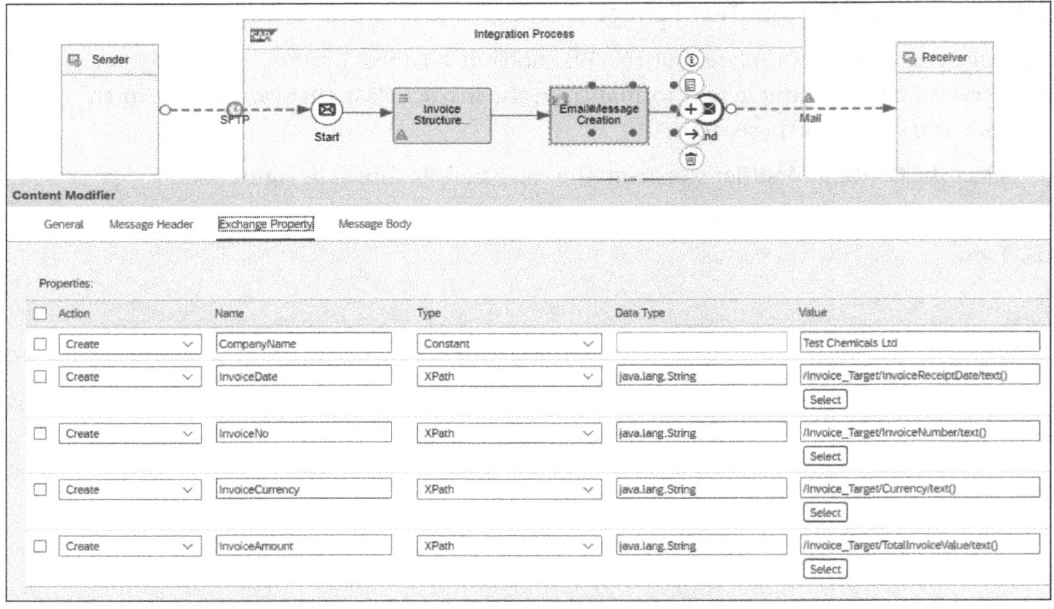

Figure 5.43 Values to Be Added in Content Modifier's Exchange Property Tab

4. Once the required values are extracted from the XML, you need to create a message structure that can be sent in an email via the Mail adapter. You can go to the **Message**

Body tab in **Content Modifier** and add the text shown in Listing 5.8 to create a custom message from dynamic values, as shown in Figure 5.44. These dynamic parameters will be replaced with actual values during integration flow execution.

```
Dear Finance team,
A new invoice has been received from ${property.CompanyName}.
Please check the details in the email below. Payment terms are 30 days.
Invoice No.    ${property.InvoiceNo}
Invoice Date    ${property.InvoiceDate}
Invoice Currency ${property.InvoiceCurrency}
Invoice Value    ${property.InvoiceAmount}
Regards,
Administrator
```

Listing 5.8 Custom Message Configuration in Content Modifier

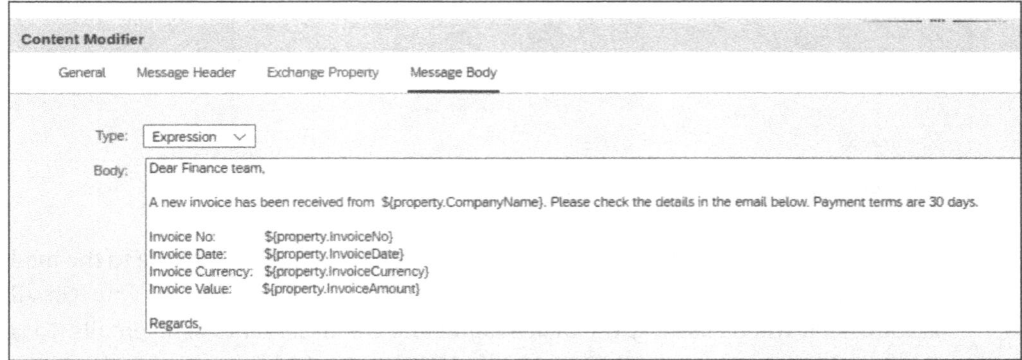

Figure 5.44 Configure Message Body Tab with Custom Message

5.4.7 Configuring Adapters

Configuration of the content modifier steps are completed. The next step is to configure the sender and receiver adapters. Follow these steps:

1. Click the arrow to configure the Mail adapter, as shown in Figure 5.45.

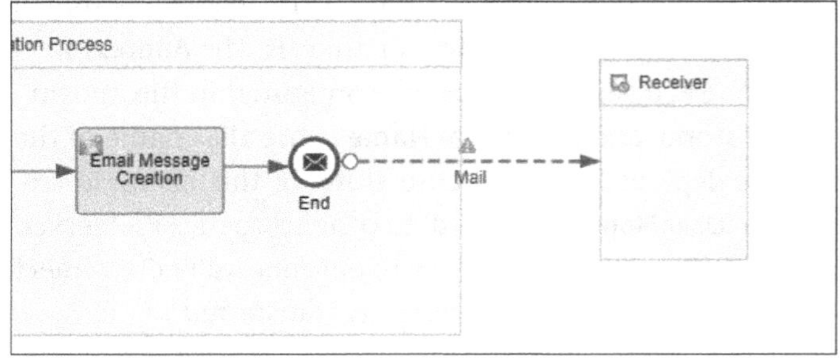

Figure 5.45 Select Mail Adapter for Configuration

2. Provide configuration details for email server connectivity in the **Connection** tab, as shown in Figure 5.46. The **Credential Name** is the exact alias for the user credentials deployed in the secure store. It must match exactly. `smtp.gmail..com` is the hostname for the Gmail server connectivity, as defined in the **Address** field. Make sure that you have configured your Gmail account to receive emails from nonweb applications.

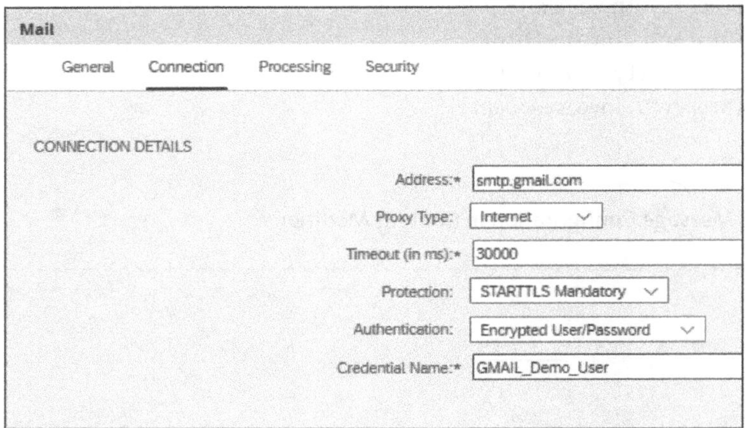

Figure 5.46 Configure Mail Receiver Adapter

3. You need to provide further **Processing** settings for sending the message to the mail server, as shown in Figure 5.47. **From** describes the email address of your Gmail account that will be used by the Mail adapter and Gmail server to send emails. **To** is the email address where you want to send emails. In this example, it's the email address of Test Solvent's finance team. **From** is the administrative email account provided by Test Solvent's IT team to send emails. **Subject** needs to be customized; you can configure to read it from dynamic values. Here, use the `${property.CompanyName}` parameter to get the actual value of the company name that was set up via the content modifier step. Because you've created the email template already via the content modifier, there's no need for further customization and you can use the final payload from content modifier as is via the `${in.body}` attribute.
4. Similarly, you need to provide connectivity settings for the SFTP adapter, as shown in Figure 5.48, to read the invoice file provided by Test Chemicals. The **Address** field represents the hostname of the SFTP server (it must be configured in the known_host file present in the secure store), and **Credential Name** is the alias name of the SFTP server secure credentials deployed in the secure store. In this example scenario, use basic authentication (**User Name/Password**) to connect to the SFTP server. You can select the **Automatically Disconnect** checkbox to automatically disconnect the session with the SFTP server once the file is successfully transferred.

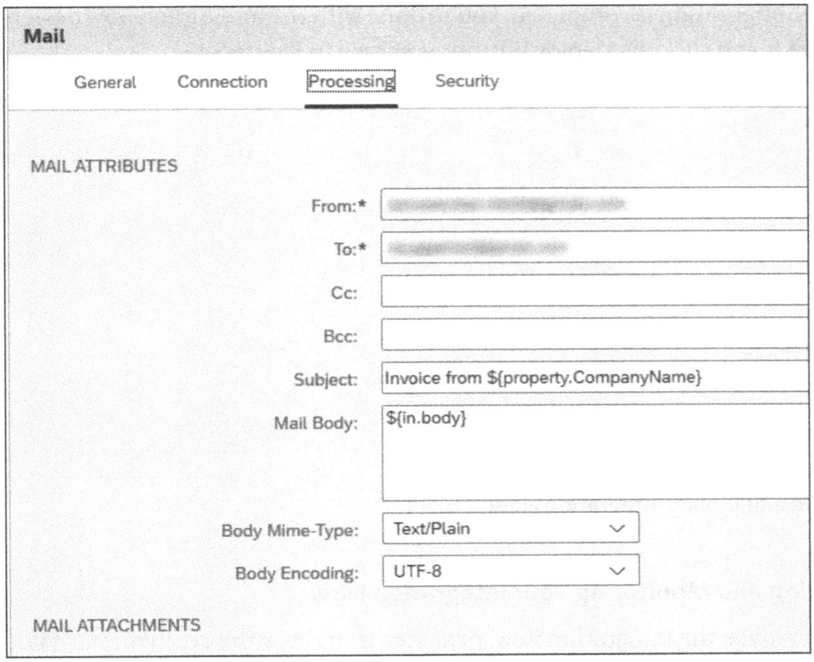

Mail

General Connection Processing Security

MAIL ATTRIBUTES

From:*

To:*

Cc:

Bcc:

Subject: Invoice from ${property.CompanyName}

Mail Body: ${in.body}

Body Mime-Type: Text/Plain

Body Encoding: UTF-8

MAIL ATTACHMENTS

Figure 5.47 Configure Processing Settings of Mail Adapter

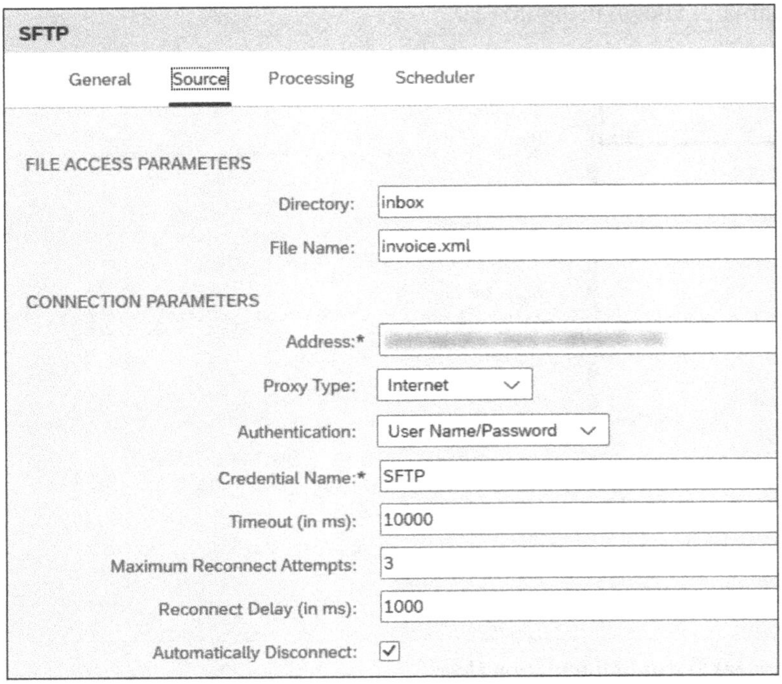

SFTP

General Source Processing Scheduler

FILE ACCESS PARAMETERS

Directory: inbox

File Name: invoice.xml

CONNECTION PARAMETERS

Address:*

Proxy Type: Internet

Authentication: User Name/Password

Credential Name:* SFTP

Timeout (in ms): 10000

Maximum Reconnect Attempts: 3

Reconnect Delay (in ms): 1000

Automatically Disconnect: ☑

Figure 5.48 Configure SFTP Adapter

5. Once SFTP configuration is completed, you're done with the integration flow design. You can **Save** it and click the **Deploy** button, as shown in Figure 5.49.

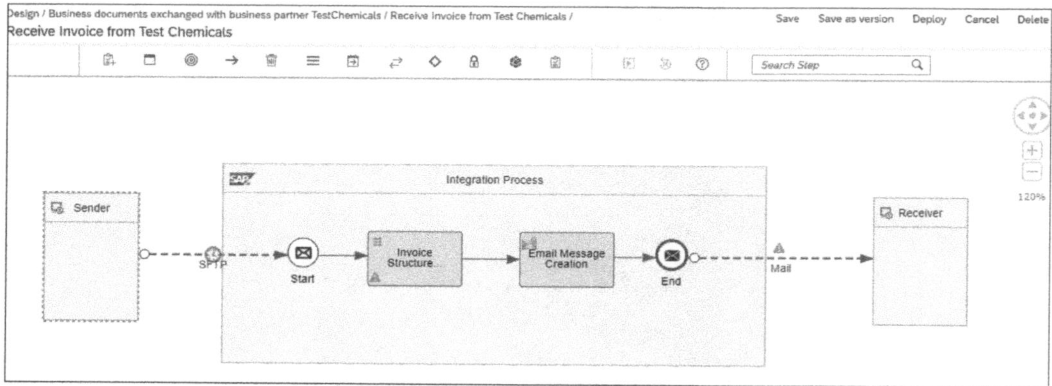

Figure 5.49 Save and Deploy Integration Flow

5.4.8 Executing and Monitoring Your Integration Flow

Once you've deployed the integration flow, next step is to check the deployment status in the **Monitor** section (introduced in Section 5.1.3). Follow these steps:

1. Go to the **Monitor** section and select **Monitor Integration Content** to check the integration flow status, as shown in Figure 5.50.

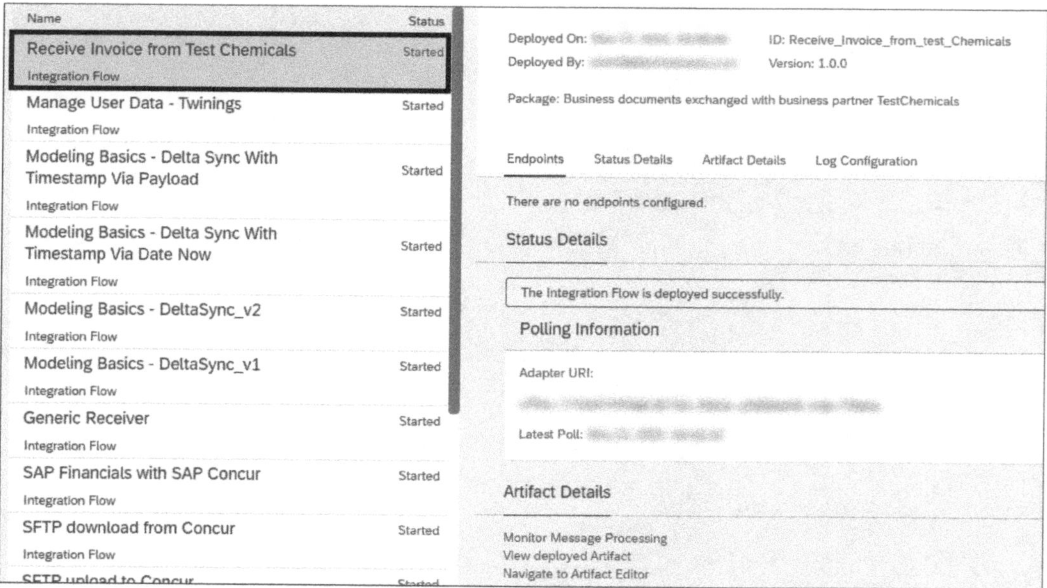

Figure 5.50 Deployment Status of Integration Flow

2. Now you need to test the integration scenario by providing an input invoice XML payload in the SFTP server's inbox folder. You can use any SFTP client. In this example, FileZilla is used, as shown in Figure 5.51. You can connect FileZilla to the SFTP server and upload the invoice.xml file shown in Listing 5.9 in the inbox directory.

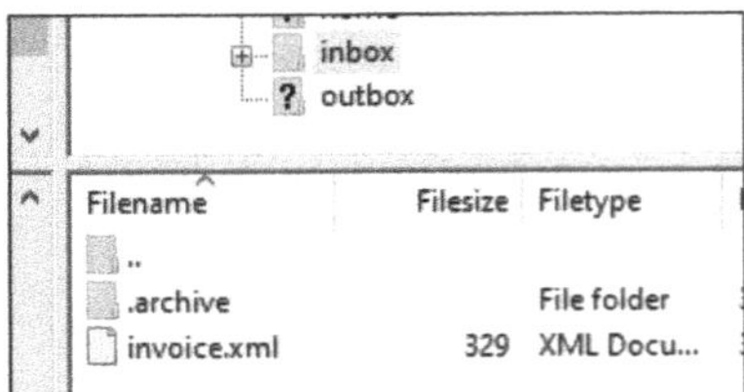

Figure 5.51 invoice.xml File in SFTP Inbox Directory

```
<Invoice>
<InvoiceNo>INV123</InvoiceNo>
<InvoiceDate>21052021</InvoiceDate>
<InvoiceCurrency>EUR</InvoiceCurrency>
<MachineToolDetails>
<LineItem1Name>Chasis</LineItem1Name>
<LineItem1Price>1000</LineItem1Price>
<LineItem2Name>Motor</LineItem2Name>
<LineItem2Price>7000</LineItem2Price>
</MachineToolDetails>
</Invoice>
```

Listing 5.9 Content of invoice.xml File

3. The SFTP adapter is polling this directory per your defined logic in the integration flow. The file will be picked up and a successful email message sent to the configured email account. **Monitor Message Processing** must show that there was successful processing, as shown in Figure 5.52.

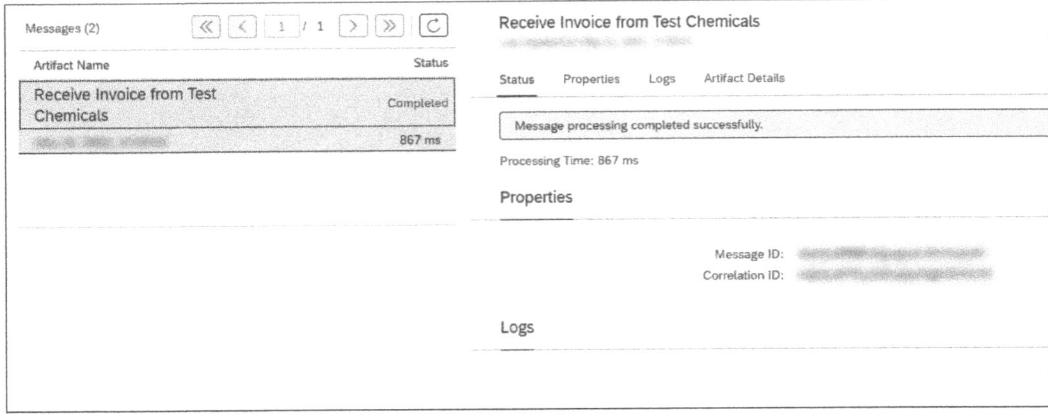

Figure 5.52 Message Process Status in Monitor Message Processing View

4. The last step is to check the successful email message in your Gmail account, as shown in Figure 5.53.

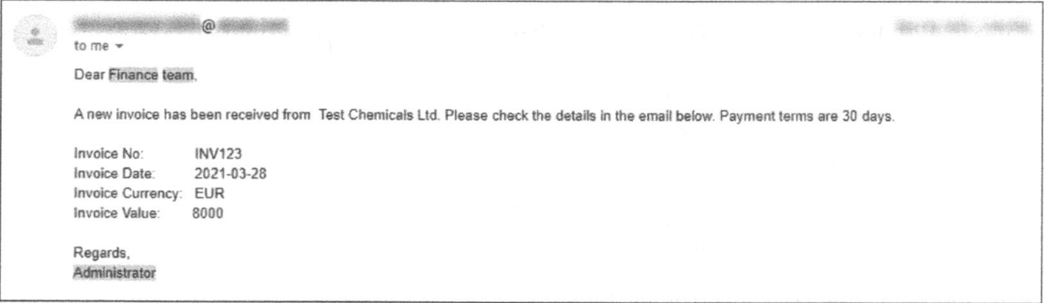
to me
Dear Finance team,

A new invoice has been received from Test Chemicals Ltd. Please check the details in the email below. Payment terms are 30 days.

Invoice No: INV123
Invoice Date: 2021-03-28
Invoice Currency: EUR
Invoice Value: 8000

Regards,
Administrator

Figure 5.53 Message Received Successfully in Email Account

This completes the end-to-end scenario.

5.5 Summary

In this chapter, we went through the details of the Cloud Integration capability of SAP Integration Suite. We looked at sections of the web-based workspace, including **Discover**, **Design**, and **Monitor**. We took a detailed look at various integration processes, pipeline steps available for designing integration scenarios, and connectivity options supported by Cloud Integration. The **Monitor** section provides lot of details for integration flows and message monitoring, including various debugging options and access to temporary data stores. We also looked at administrative sections to store certificates, the secure store, JDBC drivers, and connectivity test options. Finally, we created a custom integration flow covering various components and several operations aspects.

In subsequent chapters, we'll take a detailed look at other SAP Integration Suite capabilities. We'll continue with AI-driven and open B2B integration in the next chapter.

Chapter 6

AI-Driven and B2B Integration

Now that we have an understanding of how to design standard integration scenarios, let's go a step further and discuss artificial intelligence-driven business-to-business integration scenarios. Using the Integration Advisor, a machine learning-based tool, you can reduce implementation efforts and quickly design your integration scenarios.

SAP Integration Suite provides an integration platform for connecting various business scenarios, such as application-to-application (A2A), business-to-business (B2B), and business-to-government (B2G). These business transactions need to follow specific structures and align with business partner agreements. While designing these scenarios, developers have to invest a lot of time during integration flow creation, and business consultants have to understand the business requirements, align them with definitions from connecting systems, and prepare a functional specification. These functional specifications are converted to technical designs for message interfaces and business mappings, which are required to exchange business data between different systems. Artificial intelligence (AI)-powered design tools and proposal services help reduce the implementation effort and time and guides both functional and technical experts by providing proposals for selecting the right business interface, structure, and mappings.

In this chapter, you'll discover the need for and importance of B2B integrations, the requirements of core B2B elements, and how the Integration Advisor as a machine learning-enabled intelligent tool helps with B2B integration.

6.1 Intercompany Integration

Integration spans multiple applications and stakeholders involved in a business process across industries and organizational boundaries. There is a need for open frameworks, automation to avoid manual errors, and compliance to ensure the integrity of business messages. In this section, we'll discuss various business networks that run on SAP Integration Suite and involve diverse collaboration from intercompany applications, followed by a detailed discussion of the Electronic Data Interchange (EDI) framework and important components required for B2B integration.

6.1.1 Open Integration

Companies that are starting their cloud journeys have multiyear roadmaps and assorted application landscapes, which include SAP S/4HANA, SAP ERP, and many third-party applications deployed on premise, in the cloud, in private data centers, or hosted on different infrastructure providers. In addition, business transactions need to be exchanged with business partners and government agencies.

SAP provides an open integration strategy, as shown in Figure 6.1, with SAP Integration Suite to connect to various systems via application programming interfaces (APIs) and open interfaces. These APIs and associated prepackaged integration content are published on SAP API Business Hub (refer to Chapter 3) and in third-party applications with 160-plus connectors (explained in Chapter 7). The open integration strategy with APIs from different line of business (LoB) solutions published in SAP API Business Hub creates harmonized access to various business functions.

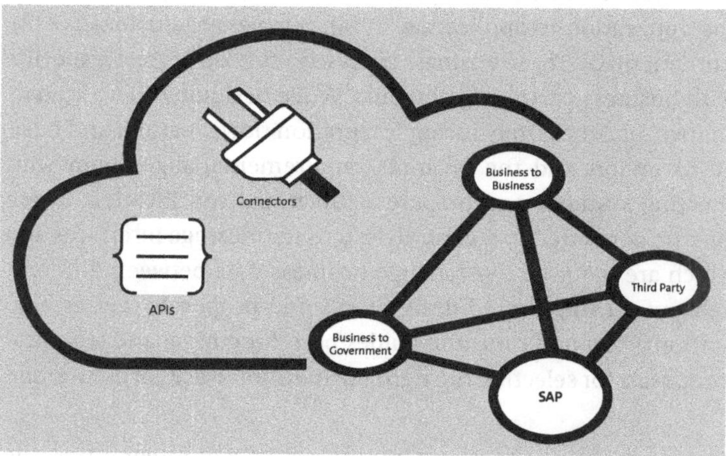

Figure 6.1 Open Integration with SAP Integration Suite

One of the main aims of an open integration platform is to provide an easy and secure way to connect to business partners. SAP Integration Suite provides a stable platform for companies to integrate their trading partners, and that same platform is used underneath many business network solutions (as shown in Figure 6.1) provided by SAP to provide end-to-end business scenarios. The solution strategy for business networks provides interlinked enterprise-centric business scenarios enabled by SAP applications: on-demand and on-premise across LoBs and industries such as supplier integration, carrier integration, transportation tendering, and invoicing. It provides added value for companies via a connected business ecosystem, enabling companies to focus on expanding business and getting end-to-end visibility into their supply chains. Business networks form a community of stakeholders with quick onboarding mechanisms, platforms to collaborate, and the ability to exchange real-time information with the community, ensuring reliable and secure connectivity.

Let's look at some of the network solutions enabled by SAP Integration Suite, as shown in Figure 6.2:

- **SAP Ariba Cloud Integration Gateway**
  SAP Ariba is the procurement and supply chain collaboration solution that helps organizations accelerate their business and visibility into their spend. SAP Ariba Cloud Integration Gateway provides seamless onboarding and integration experiences for buyers and sellers on the Ariba Network. It helps simplify the integration process by providing out-of-the-box business mappings to Ariba Network with backend systems such as SAP ERP and SAP S/4HANA. It provides citizen integration with a self-service, wizard-based framework that lets companies select their trading partners and respective business transactions to create automatic technical integrations via SAP Integration Suite.

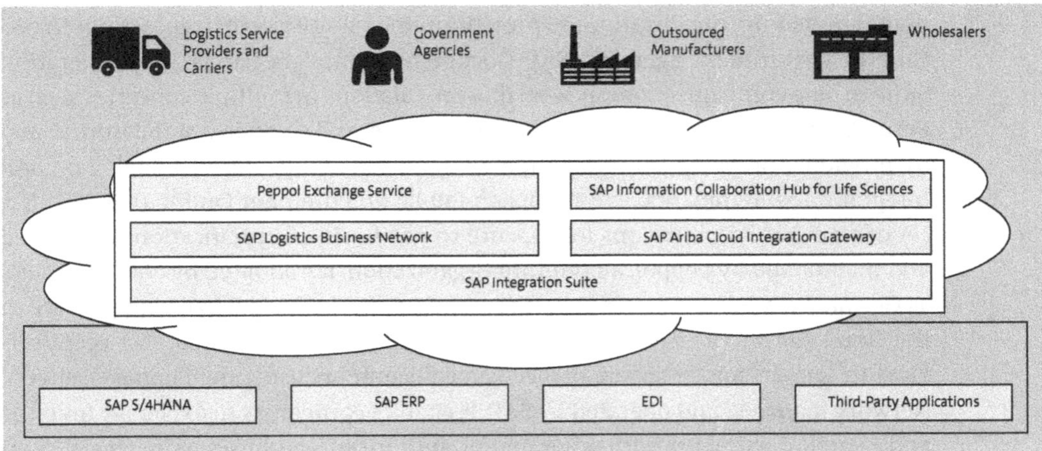

Figure 6.2 Network Solutions Enabled by SAP Integration Suite

- **SAP Information Collaboration Hub for Life Sciences**
  Drug counterfeiting is becoming a major issue for pharmaceutical companies and health divisions of many governments. The pharmaceutical supply chain stakeholders, such as contract manufacturers, wholesalers, logistics providers, and government agencies, must ensure tracking of individual product units using unique identifiers to avoid counterfeiting. This can enable patients and healthcare professionals to easily perform product verification for a secure supply chain network to manage and operate end-to-end tractability. SAP Information Collaboration Hub for Life Sciences is a cloud-based solution that connects pharmaceutical organizations and supply chain partners through a secure cloud service owned and managed by SAP. It supports pharmaceutical manufacturers to manage the unique marking and identification of packages as per defined guidelines. It also enables tracking and tracing activities for the data entering the warehouse and supply chain operations with reporting capabilities for the government.

- **SAP Logistics Business Network**
  SAP Logistics Business Network is an open network that connects business partners for intercompany logistics collaboration and insights. It allows companies to manage logistics transactions via freight collaboration and tracking capabilities and gain insights across the complete value chain. Logistics collaboration is a must in today's supply chain environments for reducing total costs and improving service to end customers. It securely connects multiple business partners and provides integration via out-of-the-box standardized scenarios such as contracting, dock appointment scheduling, milestone reporting, settlement, and insights into freight orders/booking and container tracking. It can also connect into the scattered logistics network space for transaction visibility and collaboration.
- **Peppol Exchange service**
  Statutory reporting requirements are evolving from legacy manual processes and mandates require organizations to provide online submission of transactions in real time to government agencies. SAP Document Compliance uses SAP Integration Suite to help companies comply with these regulations in multiple countries such as Peru, Chile, India, Australia, Hungary, Italy, Greece, and so on in a standardized way by providing prepackaged integration content that companies can deploy on SAP Integration Suite tenants. Pan-European Public Procurement Online (PEPPOL) is a set of standard specifications for procure-to-pay business specifications developed and maintained by Peppol, a nonprofit organization. It's adopted by countries such as Germany, Sweden, Austria, Singapore, and so on. The Peppol Exchange service in the cloud edition of SAP Document Compliance provides Peppol access point services to senders and receivers of electronic documents using the Peppol eDelivery Network managed and operated by SAP. It enables companies to exchange invoices and related documents with government authorities and business partners in an electronic format and utilizes SAP Integration Suite as its underlying technology layer.

6.1.2 B2B and EDI Integrations

Organizations have been communicating across their company boundaries for many years in processes such as procure-to-pay and order-to-cash, working with trading partners, financial (tax, wages, etc.) services, information exchanges with governments, customs and shipment information with logistics providers, and other widely used business transactions. Business trends are changing rapidly, and there is an increased need to collaborate with suppliers, partners, and outsourced vendors to innovate, sell, market, and compete. Accurate and fast execution of transactional data like orders, inventory status, demand forecast, and other metrics allows for better inventory management and supports production-critical processes like just-in-time (JIT) delivery. Considering such business requirements, organizations want support for integrating with multiple business partners and quick onboarding mechanisms, as shown in Figure 6.3.

Bulk business transactions across organizational boundaries are defined as B2B. B2B integration is costly and time-consuming. Only large companies have the expertise and the transactional volume to justify the investment in improving operational costs. This becomes more challenging when the connecting business partners are enormous. It's very tedious to execute, maintain, and audit these bulk transactions using manual mechanisms such as paper, which result in a lot of time being spent on communication. Such mechanisms also delay the execution of business processes affecting procurement, sales cycles, shipping, payments, invoices, and revenue. Adherence to generic standards is necessary to reduce the complexity caused by different message exchange formats and communication mechanisms used by different trading partners. At the same time, enforced standards increase the entry barriers for business partners and add to the level of complexity. There is a huge volume and various types of transactions that are executed between trading partners, and it isn't possible to keep track of them manually.

Organizations have already invested a lot in their hardware infrastructure and business software solutions, like enterprise resource planning (ERP) and customer relationship management (CRM) to enhance efficiency and productivity throughout their shared network. While planning B2B integration, companies look forward to extending and leveraging their existing technology landscapes for utilizing their existing skillsets, minimizing the need to train their IT and business users, and avoiding adding new systems to their landscapes. Companies want to have a single platform in their landscape for defining, managing, and monitoring business processes (A2A, B2G, and B2B). This helps them to enhance business efficiency and reduce the overall total cost of ownership (TCO) of the complex integration requirements. Organizations need to have transparency in business communications, which is also required from legal and compliance perspectives.

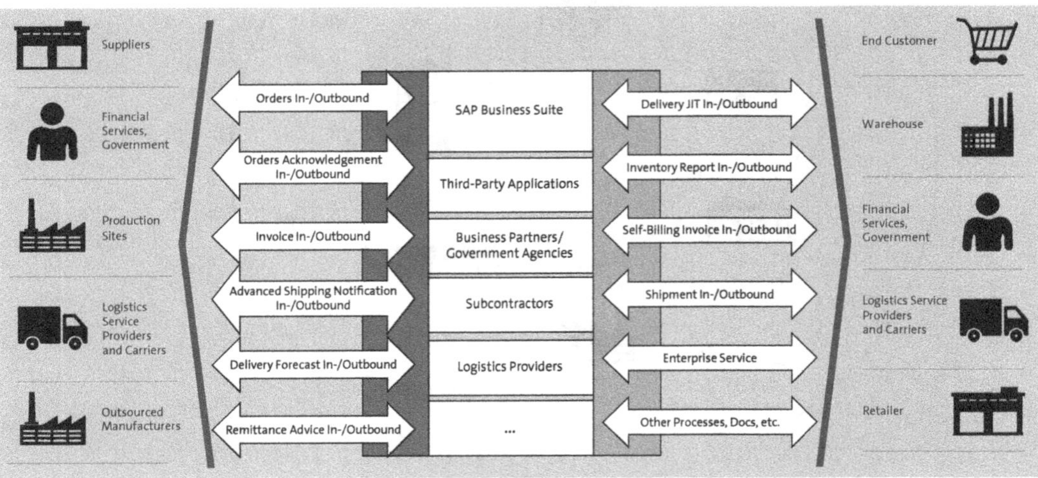

Figure 6.3 B2B and EDI Integration Requirements for Enterprises

Companies want to reduce cost and business errors by replacing manual processes with automated tools. Another challenging task is to make partners adhere to an organization's internal processes, which may be different from the partners' processes. Open standards help to ensure this interoperability yet limit the flexibility of the business processes. The majority of organizations want to have real-time process visibility for sales, inventory, logistics, procurement, and production. This can help organizations make quick business decisions, ensure valid forecasts, set up an effective supply chain, and solve inefficiencies in their business processes. They want to reduce the complexity of bulk transactions by integrating with their trading partners and sharing business information in a fast and accurate way.

EDI is particularly used in B2B communication. There are many standard defining bodies and nonprofit organizations (GS1, OASIS, ISO, DISA, UN/EDIFACT, WCO, Rosettanet, etc.) that focus on improving electronic business capabilities. The specifications published by these organizations include technical and functional protocols. These protocols are developed with inbuilt flexibility to ensure fast and accurate integration between trading partners, to drive an easy way forward for business collaborations, and to describe new e-business capabilities that can be used globally or for a specific industry vertical. These protocols are complex and practically very difficult to implement if processed manually. With a variety of EDI formats (as shown in Figure 6.4) available for different regions and industries, designing an automated tool itself becomes quite complex.

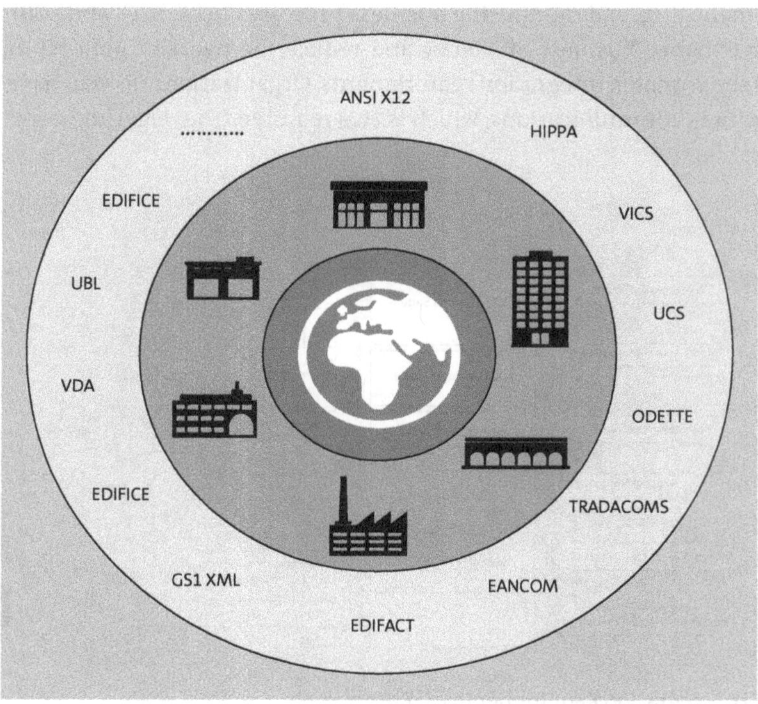

Figure 6.4 Popular EDI Formats in Various Industries and Regions

Solutions are not limited to EDI and require a good platform for catering to a company's end-to-end needs and adhering to both syntax and semantics, thereby harmonizing the complete business process.

Let's quickly cover some of the popular EDI formats:

- **ANSI X12**
  American National Standards Institute (ANSI) X12 is widely popular in the United States for exchanging business documents among companies. It includes many different versions for various transaction types, and documents are represented by numeric identities, such as 850 for orders and 810 for invoices. It was developed in 1979 by the Accredited Standards Committee (ASC) ANSI subsidiary.
- **EDIFACT**
  United Nations/Electronic Data Interchange for Administration, Commerce, and Transport (UN/EDIFACT) was developed by the United Nations and is widely used across Europe. It's maintained and further developed by United Nations Center for Trade Facilitation and Electronic Business (UN/CEFACT) under the UN Economic Commission for Europe.
- **EANCOM**
  EANCOM is a subset of the EDIFACT standard. It's most popular in the European consumer goods industry and is maintained by GS1 organization.
- **EDIFICE**
  The European Network for B2B Integration in High-Tech Industries (EDIFICE) is a subset of EDIFACT and is mainly used in European industrial consortiums for B2B standards in high-tech industries.
- **VDA**
  The Verband der Automobilindustrie (VDA; the German automotive industry association) maintains a series of fixed-format messages that describe business documents typically exchanged between automotive manufacturers and suppliers.
- **ODETTE**
  ODETTE is responsible for developing and maintaining communications standards such as OFTP and OFTP2.0, including EDI formats for the automotive industry in Europe.
- **VICS**
  The Voluntary Interindustry Commerce Standard (VICS) is used mainly by the general merchandise retail industry across North America. It's a subset of the ANSI X12 national standard.
- **Tradacoms**
  Tradacoms is an EDI standard based on the EDI UN/EDIFACT guidelines for trade data interchange primarily used in the UK retail sector. It was introduced in 1982 and was maintained and extended by the UK Article Numbering Association (now called GS1 UK).

- **UCS**
  Uniform Communication Standard (UCS) is mainly used in grocery markets. It's a subset of the ANSI X12 standards.
- **HIPPA**
  The Health Insurance Portability and Accountability Act (HIPPA) defines the national standard for electronic health care transactions. HIPPA is based on ANSI X12 specifications and provides transaction sets to exchange information per its guidelines. It's mainly popular in the healthcare sector.
- **APIs**
  Although legacy EDI structures are sticky and not easy for organizations to change, future B2B communication approaches are moving toward well-governed and secure APIs and standard XML-based formats.

B2B integration platforms should have enterprise-grade qualities required for secure and reliable transmission and processing for B2B documents. This includes secure connectivity, message transfer reliability, avoiding duplicates, EDI-specific design tools, monitoring, and archiving and should cover industry-specific EDI standards. Raw EDI incoming transactions are not human-readable and cannot be processed manually. Automatic tooling is needed to extract, map, and process the right business data. Let's look at some of the important B2B/EDI integration principles supported by SAP Integration Suite, as shown in Table 6.1:

- **Secure and reliable connectivity**
  Business transactions need to be exchanged between trading partners in a secure, encrypted, and reliable way. This is one of the critical requirements for any organization. SAP Integration Suite ensures this fundamental need of integration by providing encryption, signing, and security at a transport level and message level, such as HTTPS, SSH, AUTACK, PGP, PKCS 7, and so on. While standard protocols like SOAP, OData, FTP, and so on are available to transfer messages, SAP Integration Suite supports B2B-specific connectors such as FTPS, SFTP, AS2, AS4, and others.
- **Message and mapping guidelines and maps**
  EDI formats and communication protocols are diversified according to regions and industries. SAP Integration Suite provides support for various diverse EDI standards, such as ANSI X12, EDIFACT, VDA, EANCOM, and so on. It provides an intelligent tool for designing industry-standard message guidelines with collaboration and machine learning-based mapping guidelines for connecting EDI and business systems to reduce the design and mapping efforts.
- **EDI framework**
  This is the backbone of the overall B2B/EDI message processing platform. Normally, EDI documents are exchanged in bulk/batches to optimize the communication bandwidth and send business transactions together for a specific process. Being either on the sender or receiver side of EDI documents, inbound or outbound EDI documents need to be transformed into the right structures to map with your core business backends. Based on the trading partner agreements, there's a need to send

functional acknowledgements back to trading partners in an automated fashion to avoid duplicate transactions and acknowledge the acceptance/rejection of critical business documents.

SAP Integration Suite provides various EDI converters that covert EDI formats (such as X12, EDIFACT, VDA, etc.) into XML and vice versa and provide automatic functional acknowledgment generation. Interfacing with an ecosystem or business network is quite challenging as interactions happen outside the organizational boundaries. There should be effective tracking and monitoring mechanisms available to adhere to legal and compliance requirements and to service-level agreements (SLAs) of the trading partner. Operations staff need to have efficient monitoring tools to search erroneous EDI transactions based on document numbers, such as order ID, invoice ID, EDI interchange ID, partner IDs, and so on. All B2B tools and content are extended with the business expertise of the SAP business solutions (SAP ERP, SAP S/4HANA, SAP Company Relationship Management, SAP Supply Chain Management, etc.) to provide excellent business value and comply with the need for EDI business processes, acknowledgements, auditing, and monitoring. From a legal and compliance standpoint, there is a requirement to store incoming/outgoing business documents, and archiving provides a reliable way to store and retry messages. The number range objects (NRO) functionality of SAP Integration Suite further helps to generate automatic interchange IDs as agreed upon with business partners and to ensure sequential processing of business transactions.

- **Trading partner management**
  The expansion and growth of a business increase interactions with the ecosystem or business network, which results in increased business transactions and connections with business partners. It becomes challenging and complex when enterprises have to maintain huge point-to-point connections with different EDI formats (EDIFACT, ANSI, VDA, etc.), business transactions (orders, invoices, ASN, forecast, payments, etc.), and communication protocols (AS2, AS4, SFTP, etc.). Managing changing specifications and coordinates from business partners becomes a pain point due to increasing costs, timeframes, and requirements for complex monitoring capabilities. Companies have a difficult time finding staff to keep up with the requests from the various LoBs. When an organization has to connect to hundreds or thousands of trading partners, it becomes cumbersome and requires a central repository of trading partner profiles, which includes partner IDs, agreements (such business transactions to be exchanged), acknowledgment requirements, SLAs, technical parameters such as URLs, security certificates, and other partner metadata. Trading partner management (TPM) helps by creating a central dashboard for dynamic deployment of integration scenarios without a need for configuring, deploying, and managing multiple integration scenarios for each partner. This really becomes important when you have connections to value-added network (VAN) providers and via the same technical communication, different partner data is received that needs to be processed separately according to respective partner agreements. SAP Integration Suite provides well-defined APIs for storing and retrieving trading partner–specific data and using it in integration scenarios.

Security	Connectors	Message Guidelines and Maps	EDI Framework	Trading and Partner Management
▪ PGP ▪ HTTPS ▪ Digital Signature ▪ PKCS7 ▪ AUTACK	▪ SFTP ▪ AS2 ▪ AS4 ▪ SOAP ▪ FTPS	▪ ANSI X12 ▪ EdiFACT ▪ EANCOM ▪ Codelist ▪ IDocs ▪ Custom Interfaces	▪ Bulk messages ▪ Splitter ▪ Acknowledgments ▪ EDI converters ▪ EDI monitoring ▪ Archiving	▪ Partner profiles ▪ Partner agreements ▪ Service level agreements ▪ Dynamic deployments ▪ Partner onboarding ▪ Self-service

Table 6.1 Examples of B2B Integration Capabilities Supported by SAP Integration Suite

6.2 Integration Advisor

Now that we've laid the foundation for establishing automated B2B integrations by discussing various components, tools, and concepts, in this section we'll see how the Integration Advisor plays an important role in defining the business interfaces and mapping relationships in an easy and intuitive way for business domain experts. We'll start by exploring the surrounding integration project and then move on to working with the standard libraries, message implementation guidelines (MIGs), mapping guidelines (MAGs), and runtime artifacts.

6.2.1 Overview and Getting Started

An integration project has many phases and follows an implementation methodology, as shown in Figure 6.5. It starts with the requirements gathering phase, including discovery of business processes, gap analysis, and identifying landscapes and associated required business and application systems. Once the business systems and participating stakeholders are finalized, the next task is to develop and design functional specifications for business interfaces that reflect the data carrier structures to be used while performing a business transaction with trading partners. This includes interface definitions and associated mappings between source and target business applications (MIGs and MAGs). As connecting applications use different interfaces (or rather different structures to store/process/transfer the same business document), required fields and message structures must be identified and documented. This task is performed by business domain experts or functional consultants who have expert knowledge of business processes but are not technical experts. Most of the time spent on an implementation project goes into designing the correct message structures, and these definitions are documented mainly in Microsoft Word and Microsoft Excel formats. Once the functional specifications are finalized, they're handed over manually to technical integration experts to be translated into the right designs and tools required for runtime execution. Technical experts take the functional design document as input and implement the business mappings in technical systems. Once this phase is over, both technical and business experts test the results before deploying the final product to production systems. This whole phase is cumbersome and takes up around half of the effort of an integration project.

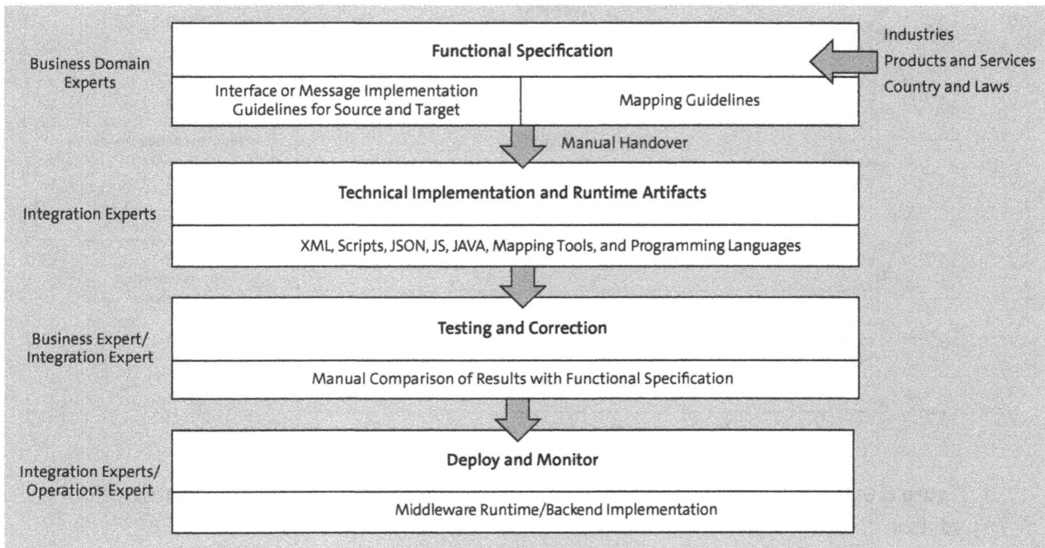

Figure 6.5 Activities and Personas Involved in an Integration Project

The Integration Advisor capability of SAP Integration Suite is an intelligent, cloud-based design-time tool for creating and maintaining integration content via MIGs and MAGs. Using the Integration Advisor, you can simplify and improve the whole end-to-end process and methodology for creating and maintaining integration specifications. Building integration specifications is primarily focused on the business domain experts, as shown in Figure 6.6, who understand the business requirements and have the domain-specific knowledge. As we discussed already, the cost drivers of today's integration project are identifying the gaps between specifications and creating the right business formats and structures to enable process integrations with various internal and external distinct applications. The Integration Advisor provides an innovative platform for business domain experts to close these gaps by unifying all the required tasks for creating integration content based on a comprehensive knowledge base and machine learning technology. It enables the common understanding between all parties and stakeholders involved in the integration project. It provides intelligent tools to facilitate creation of interfaces by providing machine learning-enabled proposals and analytics features for leveraging a streamlined process from specification to testing of interfaces and mappings required for an integration process. Business domain experts and their business partners can collaborate to describe, map, and test the business requirements of interface guidelines. The end-to-end process of designing based on machine learning techniques and automatic technical design save approximately 60% of the effort and time that would go into integration projects.

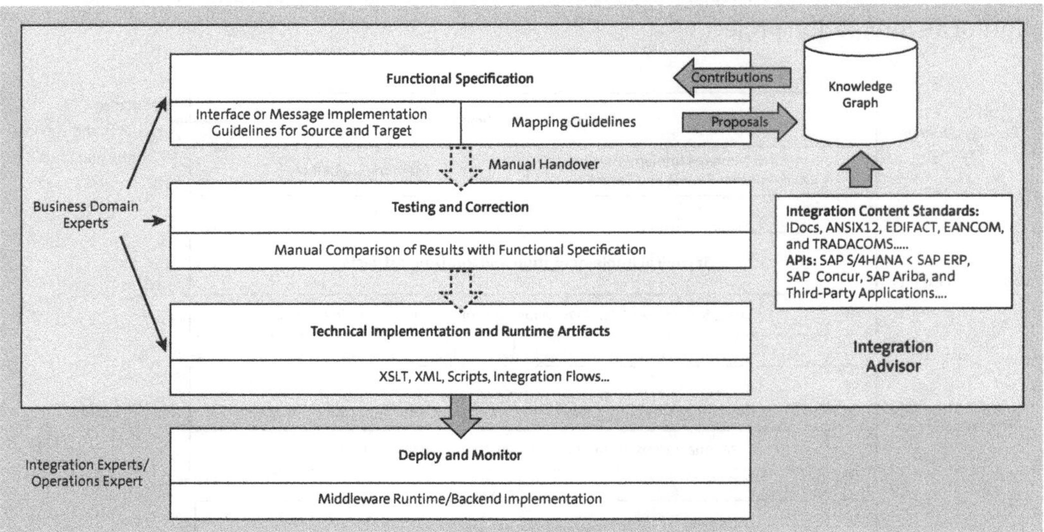

Figure 6.6 Business Domain Expert–Focused Interface Development Using Integration Advisor

The Integration Advisor learns from the content being created in multiple implementations and creates a knowledge graph anonymously that is used to provide proposals and suggestions for MIGs and MAGs. There is multidimensional learning in knowledge graphs, including about the industry, interface types, complexity, versions, fields, business semantics, and much more. There is no manual handover required as the intuitive graphical mapping editor can be used directly by business domain experts to create and test maps. Once the process is done, it automatically creates the required runtime artifacts that can be deployed in SAP Integration Suite.

A *knowledge graph*, as shown in Figure 6.7, is a key backend component of the Integration Advisor. It doesn't store any company business data but rather learns from the created MIGs and MAGs. During creation of MIGs and MAGs, it automatically provides the proposals based on the selected message types and associated business context on each individual node level, with a confidence percentage that reflects the probability of the suggested proposal providing the right inputs for specifications designed by business domain experts. Once you finish designing the new definitions and activate them, this new information is added anonymously to the knowledge graph of the Integration Advisor, along with message definitions, mappings, and associated business contexts for existing learning maps.

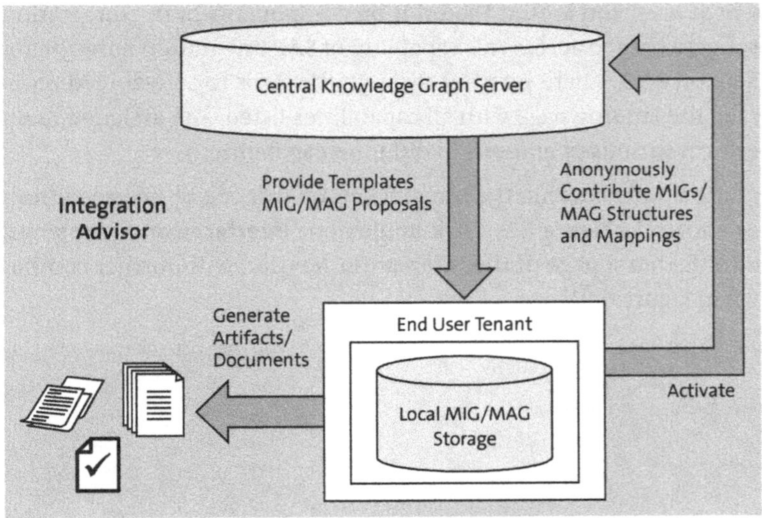

Figure 6.7 Knowledge Graph Server in Integration Advisor

The Integration Advisor has four main components, as shown Figure 6.8: a central repository of standard libraries or support for custom interfaces, finalizing inbound and outbound business message structures, associated mapping required for integration scenarios, and generation of corresponding runtime artifacts that can be used in the design time/runtime of an integration flow. MIG and MAG editors provide business-focused, intuitive, and easy-to-use tools for business domain experts to define business interfaces and transformation logic that are critical for successful execution of any business integration scenario. These are decoupled from the technical design and architecture. These interfaces may be custom-created or standard and are driven by pure business needs for a specific process. They include requirements for many different industry verticals, market segments, regions, and business processes, such as order-to-cash, procure-to-pay, design-to-operate, and so on. Proposal services further make the tasks easy and save a lot of effort and time during an implementation project. In the next sections, we'll go through each of these components in detail.

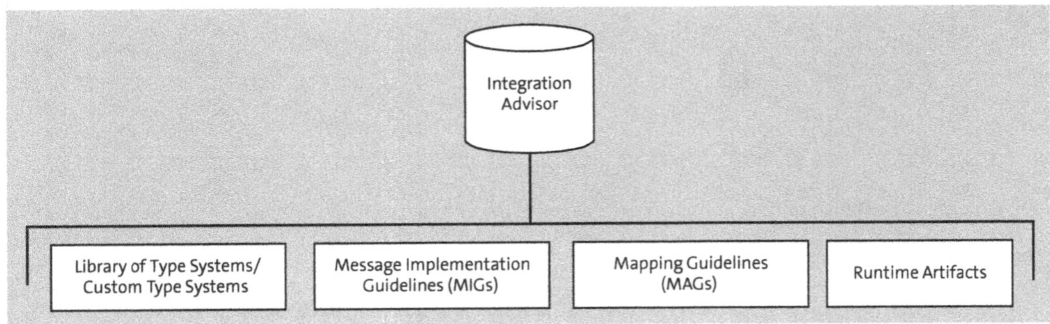

Figure 6.8 Components of Integration Advisor

Before we go to the next level and look at the solution components of the Integration Advisor, let's go through where to access this capability of SAP Integration Suite. (Refer back to Chapter 2, Section 2.3.3, where we went through the steps to activate SAP Integration Suite, accessed the landing page with all capabilities listed, and assigned main roles required to perform various operations in different capabilities.)

From the SAP Integration Suite cockpit, the Integration Advisor can be accessed from the landing page, as shown in Figure 6.9. Click **Implement Interfaces and Mappings**, which will take you to the homepage of the Integration Advisor, with further components listed as shown in Figure 6.10.

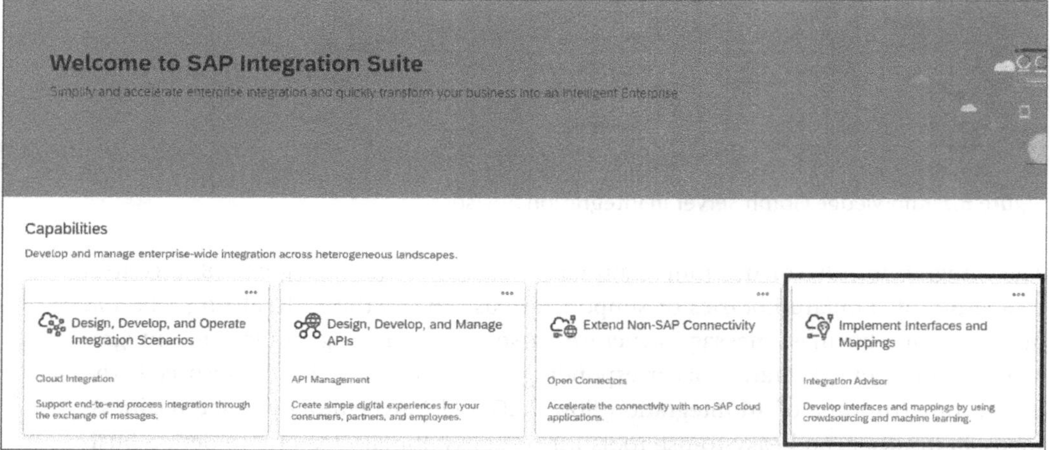

Figure 6.9 Access Point of Integration Advisor from Landing Page of SAP Integration Suite

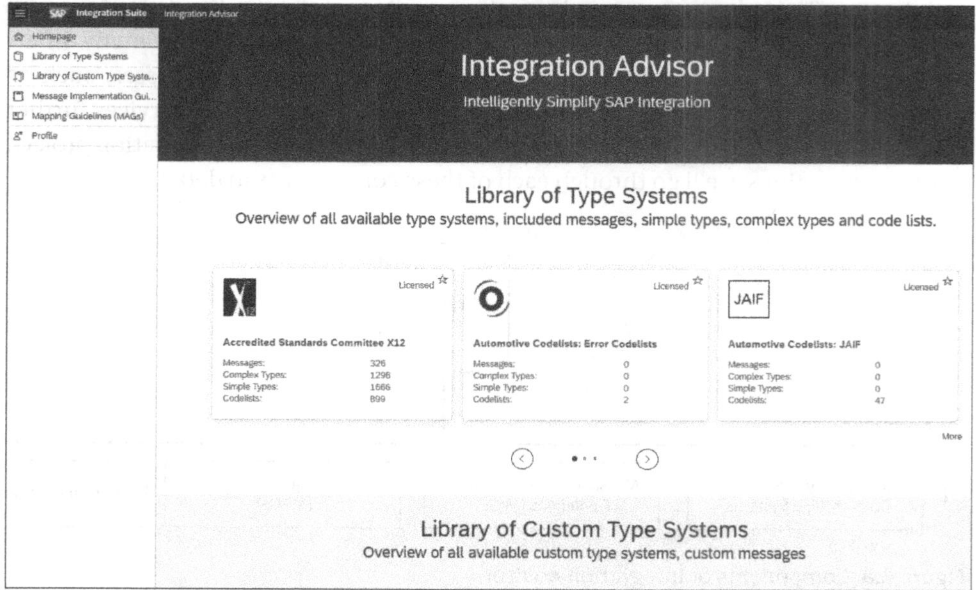

Figure 6.10 Homepage View of Integration Advisor

6.2.2 Library of Type Systems/Custom Type Systems

You can access the repository (or library) by clicking **Library of Type Systems** in the left-hand panel, as shown in Figure 6.10. The Integration Advisor provides a central repository for standard libraries and custom messages, as shown in Figure 6.11. Standard libraries are provided out of the box from SAP and are based on specifications from agencies such as ASC (ANSI X12), United Nations (EDIFACT, EANCOM), SAP (IDOCs, cXML, SAP S/4HANA APIs), Odette, VDA, Tradacoms, and so on.

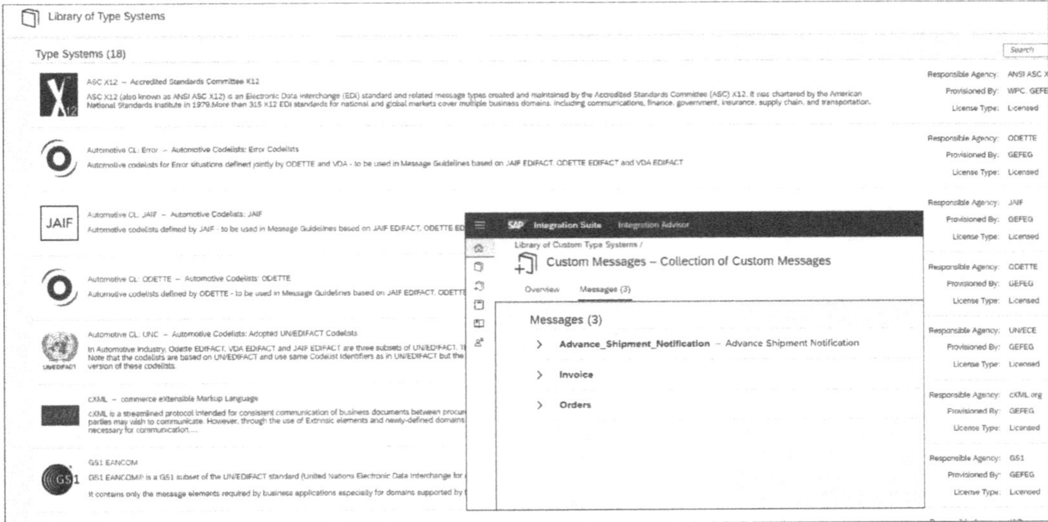

Figure 6.11 View from Library of Type Systems and Custom Type Systems

An example view of the EDIFACT library is shown in Figure 6.11. Each type system included in the standard list consists of the following elements, as shown in Figure 6.12:

- **Versions**
  Different EDI libraries may have many versions that responsible agencies keep on revising based on market requirements.
- **Messages**
  Actual structural representations of business transactions included in the library.
- **Codelists**
  Fixed and predefined collections of values that can be assigned to a field. (e.g., country codes or order types).
- **Simple Types**
  Provides information on properties of individual nodes in a message structure, such as type (float, numeric, string etc.), length, cardinality, and if any codelist is associated.
- **Complex Types**
  Provides a definition for an aggregated message on a parent level and includes further child elements inside it. For example, address can be a complex type that

includes further fields such as street number, city, state, and country as individual nodes.

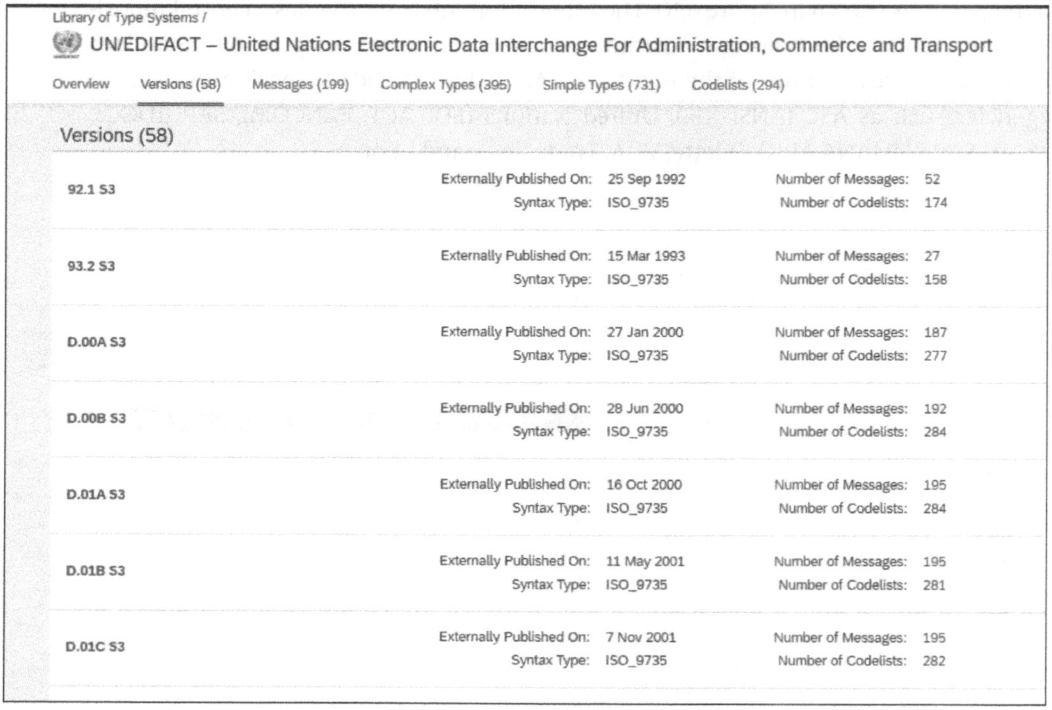

Library of Type Systems /

UN/EDIFACT – United Nations Electronic Data Interchange For Administration, Commerce and Transport

Overview | Versions (58) | Messages (199) | Complex Types (395) | Simple Types (731) | Codelists (294)

Versions (58)

Version				
92.1 S3	Externally Published On:	25 Sep 1992	Number of Messages:	52
	Syntax Type:	ISO_9735	Number of Codelists:	174
93.2 S3	Externally Published On:	15 Mar 1993	Number of Messages:	27
	Syntax Type:	ISO_9735	Number of Codelists:	158
D.00A S3	Externally Published On:	27 Jan 2000	Number of Messages:	187
	Syntax Type:	ISO_9735	Number of Codelists:	277
D.00B S3	Externally Published On:	28 Jun 2000	Number of Messages:	192
	Syntax Type:	ISO_9735	Number of Codelists:	284
D.01A S3	Externally Published On:	16 Oct 2000	Number of Messages:	195
	Syntax Type:	ISO_9735	Number of Codelists:	284
D.01B S3	Externally Published On:	11 May 2001	Number of Messages:	195
	Syntax Type:	ISO_9735	Number of Codelists:	281
D.01C S3	Externally Published On:	7 Nov 2001	Number of Messages:	195
	Syntax Type:	ISO_9735	Number of Codelists:	282

Figure 6.12 Example View of EDIFACT Library Elements

Moreover, you can also access and create custom messages based on your integration requirements by clicking on **Library of Custom Type Systems** from the left-hand menu, as shown previously in Figure 6.10. For a custom library, any external definition can be imported in XSD format and can be included in the custom library list. You can add, delete, view, and compare custom libraries from this repository.

6.2.3 Message Implementation Guidelines

As you learned in previous sections, MIGs define the main business interface used in integration scenarios for transporting data from one business application to another. The MIG designer tool provides an easy and intuitive wizard-based approach, as shown in Figure 6.13, for MIG creation. It can be accessed from the left-side panel by clicking **Message Implementation Guidelines**, as shown previously in Figure 6.10. It provides the option to create a MIG from (1) a custom interface uploaded in the **Library of Custom Type Systems** provided by your application provider or (2) the standard structures available out of the box in **Library of Type Systems**.

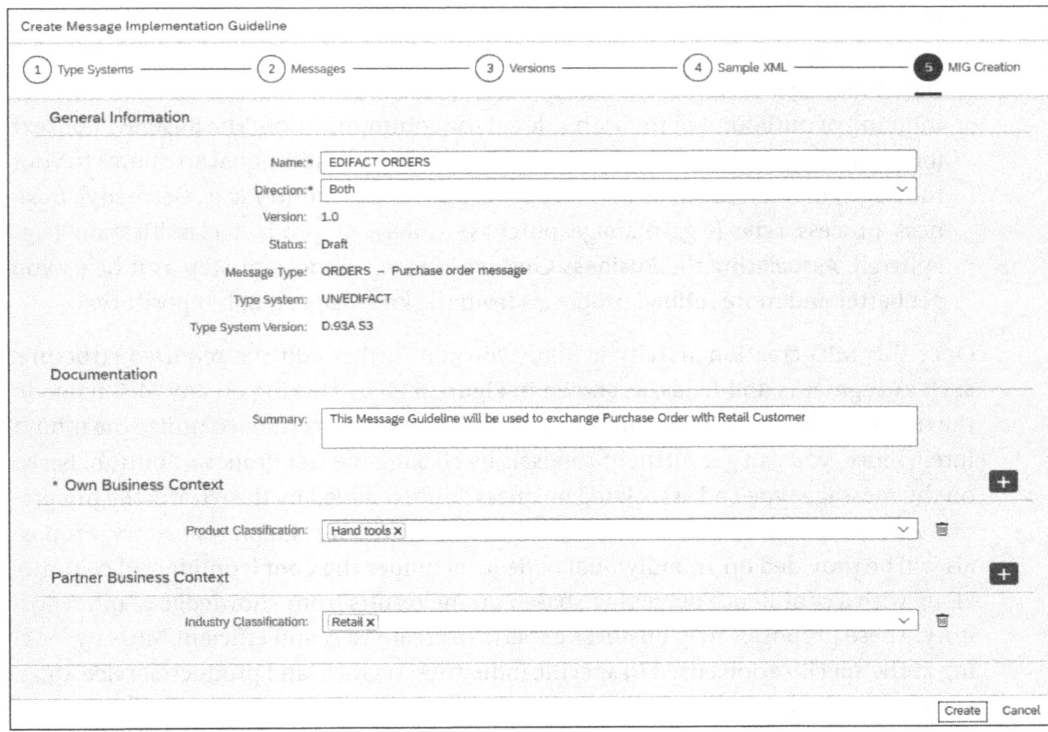

Figure 6.13 MIG Creation Wizard

Let's discuss the different steps involved when creating a MIG from standard libraries via the creation wizard:

1. **Type Systems**
   You need to choose the respective **Type System** provider to be used (such as **UN/EDIFACT**).
2. **Messages**
   Once you've selected the type system provider, you need to select an actual message type to use (such as **Orders**).
3. **Versions**
   Now choose the right version of the message, as required for your integration scenario. There are many different versions available released over many years (such as **D.93A** for **EDIFACT** orders).
4. **Sample XML**
   You can optionally choose to upload a sample XML file for the selected messages. Based on the uploaded sample payload, the Integration Advisor intelligently determines elements of the message structure, such as optional segments and fields. However, this step is optional.

5. **MIG Creation**
   The last step is to define metadata related to this MIG, such as **Name** and **Version. Direction** helps determine if this specific message guideline will be used only for inbound or outbound or for both sides of the communication. The **Business Context** (both partners' and your own) parameters help you add additional attributes to your message guideline, such as industry type (e.g., retail), country (e.g., Germany), business process type (e.g., manage purchase order), or product classification (e.g., apparel). Associating the **Business Context** is a very important step as it helps you get better and more refined proposals from the knowledge graph repository.

Once this MIG creation activity is done, you can further edit the required structure, such as segments and fields, as shown in Figure 6.14, by clicking on any MIG name in the main view of the **Message Implementation Guideline** section. To utilize the inbuilt intelligence, you can get further proposals by clicking the **Get Proposals** button. Based on the message type and associated business context added by the wizard, the Integration Advisor will provide the proposals from the knowledge graph repository. Proposals will be provided on an individual node level (under the **Conf** [confidence] column), along with a confidence percentage based on the results from knowledge graph repository. These proposals help business experts to create easy and efficient MIGs by looking at the specifications used in specific industries, regions, and product/service areas. Node attributes (cardinality, type, length, codelist, etc.) in the editor can be edited, and documentation notes can be added for further information. The MIG editor also supports lifecycle management of a MIG with various status levels (**Create, Review, Accept, Reject**, and so on) if you switch to the **Status** tab.

In addition, you can also click the **Simulate** button to validate the MIG with the uploaded XML payload (optionally created during MIG creation phase). This will add test values in the MIG and will automatically validate against the codelist and qualifiers.

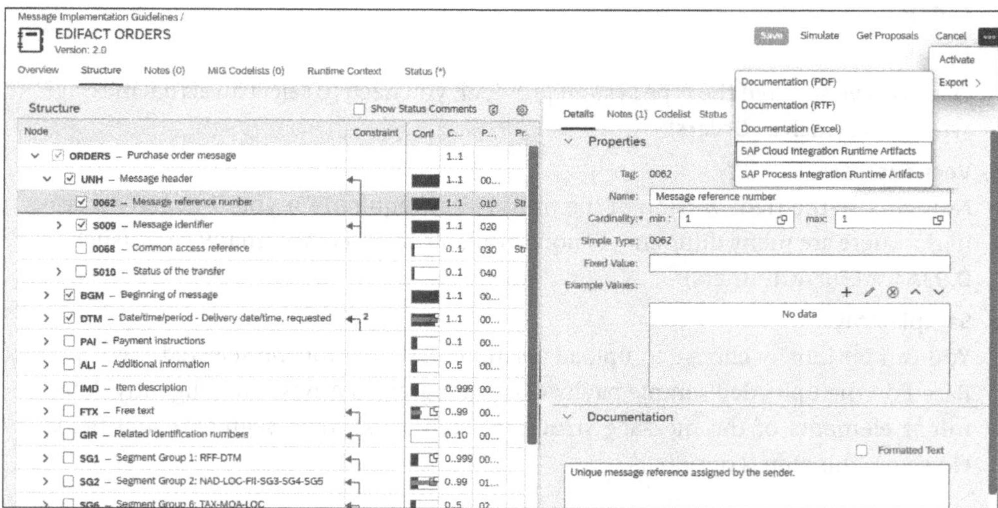

Figure 6.14 View from Message Implementation Guide Editor

Once you have created the MIG, you can save it to be used as a source or target structure in mapping guidelines by clicking the **Save** button. Activating it will push the final interface structure and associate business context values with the knowledge graph repository in an anonymized manner. You may want to provide this interface definition to your trading partners for them to incorporate it in their tooling for sending/receiving business messages. There is an option to download the message definition (by clicking the **Export** link and choosing an option) in PDF, RTF, or XLS format, along with complete documentation that can be sent to your business partners and that can used for your project documentation as well. Further, you can download the interface definition artifacts that can be readily used in SAP Integration Suite's Cloud Integration capability.

6.2.4 Mapping Guidelines

As message guidelines (or interfaces) are different semantically, there needs to be a mapping between these two message guidelines to be used by an integration flow during runtime execution of scenarios. A MAG is based on a source and a target MIG.

To create or access MAGs, you can access the MAG designer by from the left-side panel by clicking **Mapping Guidelines (MAGs)**, as shown previously in Figure 6.10. You can create a new MAG or open any existing MAG by clicking it. While creating a new MAG, you need to select **Source** and **Target** MIGs, as shown in Figure 6.15.

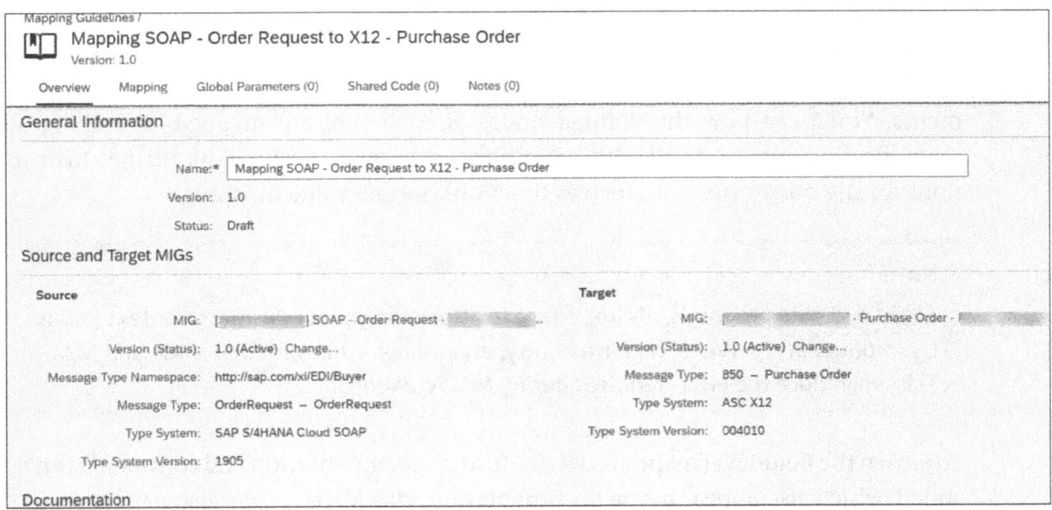

Figure 6.15 Selecting Source and Target MIGs in Mapping Editor

Handover from the MIG to MAG phase is seamless and doesn't have any media breaks. Business domain experts can continue to create functional mappings with domain expertise without involving technical experts at this stage. This helps make the whole mapping guideline journey easier for business experts by providing proposals derived from the knowledge collected from various industry and business scenarios. With the traditional manual process, this activity takes most of the time spent on an integration project implementation.

Example

You're getting a purchase order from you trading partner and you need to post that order into your backend SAP S/4HANA system. You are receiving the purchase order in EDIFACT format from a partner, but you need to convert it per the API specification supported by SAP S/4HANA. You need two different message guidelines to be created in the Integration Advisor from the provided repository content for the incoming EDIFACT orders message (MIG1) and the SAP S/4HANA API for posting the order (MIG2). After creating the respective MIGs, you need to define the mapping between these structures (MIG1 and MIG2) using MAGs.

You can use a graphical message editor by switching to the **Mapping** tab, as shown in Figure 6.16, to map nodes between the source and target. This is supported by out-of-the-box functions like string and date conversion, and additional custom transformation logic can be defined using the Extensible Stylesheet Language Transformations (XSLT) programming language. It also provides a detailed view with field cardinality, field type, and documentation. The **Proposal** button provides proposals for mapping nodes with a confidence level based on the selected message guidelines and associated business context information. The **Confidence** bar is based on the intelligence derived from the integration content present in the knowledge graph. You can accept the proposals suggested by the Integration Advisor or customize them based on your requirements. You'll see how the defined nodes at each side are mapped, describing all mapping elements in detail with definitions or notes and providing further instructions for the transformation, such as functions or code value mappings.

Note

To get accurate proposals, it's important to attach the correct **Business Context** (industry, product and services, region/country, etc.) values while creating MIGs and MAGs. This will reduce the effort required during MAG creation.

To enrich the field-level mapping details, further documentation and comments can be added, which also appear in the documentation. Like MIGs, MAGs also have complete lifecycle governance processes and can be created in a collaborative manner, with status options in the **Status** tab such as **In Progress**, **In Review**, **Accepted**, **Rejected**, and so on.

Once MAGs are created, it's important to ensure they're mapped correctly and update the mapping if required. This can be done using a sample payload. The **Simulate** button provides a test conversion from the source to the target format without deploying the mapping in runtime. A business domain expert can check the functional correctness of the target results to ensure that the right fields are mapped. Simulation can be done with the payload provided with MIG (optionally during creation), or a sample payload can be uploaded from your file system.

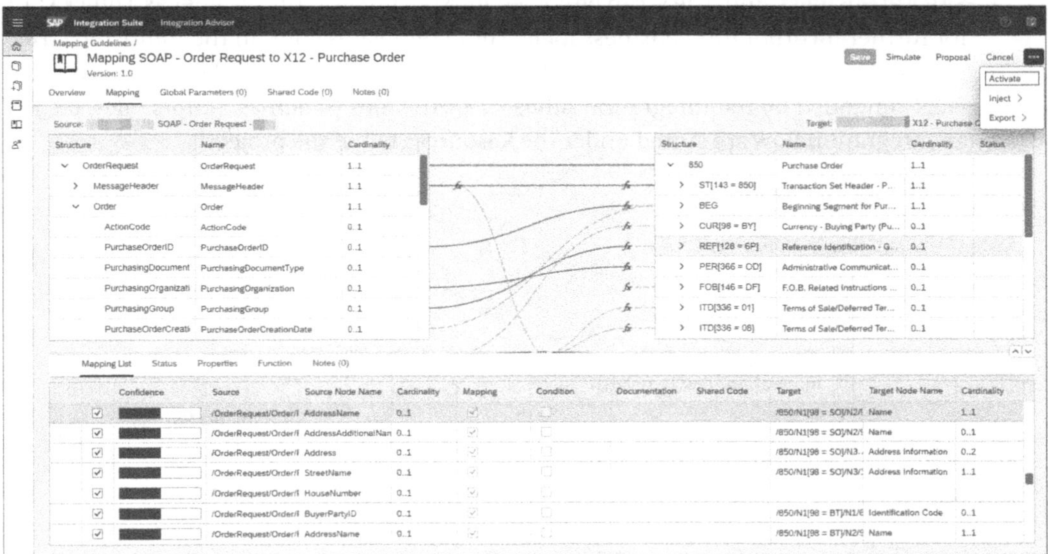

Figure 6.16 Detailed View of Mapping Guidelines Editor

Once you **Save** the mapping, you have the following options available:

- Use the **Export** link to download the mapping guidelines in the format of XSL and XSD files that can be imported into SAP Integration Suite's Cloud Integration capability while creating integration flows. You can also download the documentation of MAGs in XLS, PDF, and RTF formats. This documentation can be used as a standard asset required for project documents.
- Use the **Inject** link to inject the MAGs directly by selecting the available integration packages and integration flow.
- Use the **Activate** link to trigger sending the scenario knowledge anonymously back to the knowledge graph repository of the Integration Advisor.

6.2.5 Runtime Artifacts Generation

Once MIGs and MAGs are created, corresponding artifacts need to be incorporated into integration flows in the Cloud Integration capability. Both MIG and MAG editors provide an option to automatically generate the runtime artifacts (XSLs, XSD, etc.), as explained in previous sections. These artifacts can be manually uploaded in integration scenarios under the **Resources** tab (see Chapter 5, Figure 5.40) and further associated with integration flow steps. In addition, MAGs (along with their respective MIGs) can be injected directly into specific integration scenarios from the Integration Advisor. The **Inject** button provides direct selection of integration packages and corresponding integration flows into which these artifacts can be imported.

In addition, there are integration flow templates, as shown in Figure 6.17, available in SAP API Business Hub. These can be copied into the integration designer as a quick start for further incorporating business logic. These templates contain the multistep structure for implementing various B2B integration scenarios in accordance with the artifacts generated by the Integration Advisor. All the files (schemas, scripts, files, etc.) of an integration flow are stored under the **Resources** tab of the project.

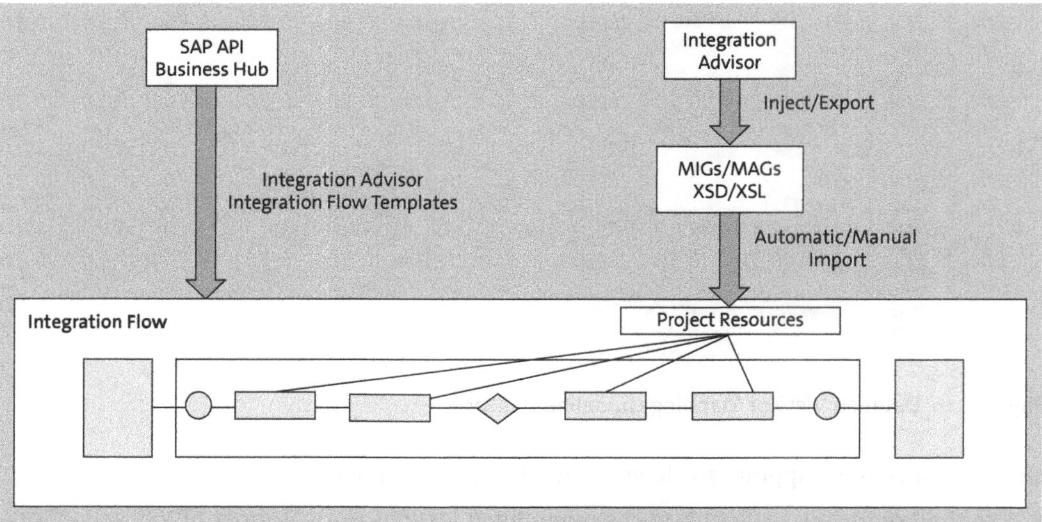

Figure 6.17 Association between Integration Advisor and Cloud Integration Artifacts

6.3 Summary

In this chapter, we went through the requirements of and need for B2B and EDI integrations and how organizations are exchanging business documents with trading partners. In addition, there is a complete end-to-end framework required for B2B integration, including security, protocols, functional acknowledgements, and specifications. SAP Integration Suite can be a central platform for carrying out B2B integrations in addition to A2A integration scenarios.

The Integration Advisor capability of SAP Integration Suite provides an intelligent machine learning–enabled proposal service to design interfaces. It reduces the implementation effort significantly by providing more power to business domain experts when designing the business interfaces and mappings (MIGs and MAGs), including testing and automatic generation of runtime artifacts.

In the next chapter, we'll move on to third-party integration with SAP Integration Suite.

Chapter 7
Third-Party Integration

In this chapter, we'll take our first look at the Open Connectors capability of SAP Integration Suite and learn how it helps to simplify and accelerate connectivity to third-party cloud applications. We'll walk you through managing the Open Connectors instance using the API Management capability and will design a scenario to replicate product details from the web application to Slack using the Open Connectors adapter in the Cloud Integration capability of SAP Integration Suite.

Over the past few years, it became extremely important for a company to modernize its IT infrastructure and take steps to make its business truly digital. Back then, the need for integration was different. It mainly focused on how the physical system comes together and works cohesively in a coordinated and unified manner. In the SAP world, the way to integrate to and from SAP was through the on-premise middleware known as SAP Process Integration, which was a combined ABAP and Java solution that then further evolved into SAP Process Orchestration, which included an option to run on Java only. When moving to the cloud, SAP revamped the solution so it could handle heterogeneous integration needs and offer a holistic integration approach. This is known today as SAP Integration Suite.

SAP customers have diverse integration requirements that not only include SAP to SAP but also SAP to non-SAP and non-SAP to non-SAP scenarios. SAP's strategy is to have a holistic view and offer you out-of-the-box integrations, application programming interfaces (APIs), and connectors to cater to all your integration needs, be they on-premise to cloud, cloud to on-premise, cloud to cloud, on-premise to on-premise, SAP to SAP, SAP to non-SAP, or non-SAP to non-SAP.

In this chapter, you'll see how SAP simplifies and accelerates non-SAP integrations with its cloud-based offering, Open Connectors, which provides robust, feature-rich, pre-built connectors for an extensible library of over 160 connectors for the most popular third-party cloud applications. SAP's central catalog, SAP API Business Hub, also offers prepackaged third-party packs and integrations to help organizations quickly become intelligent enterprises. You'll see how to seamlessly connect to the non-SAP universe with the Open Connectors capability of SAP Integration Suite through hub-based, one-to-many integration hubs such as customer relationship management (CRM), enterprise resource planning (ERP), cloud storage, social media, and more.

7.1 Connecting to Third-Party Applications

In addition to an extensive library of over 160 connectors, there are APIs and integration packs available on SAP API Business Hub from third-party providers. Here you can experience a full collection of APIs, integrations, and Open Connectors from third-party (non-SAP) providers that will reduce your overall costs, effort, risk, and time until go-live. This helps make the configuration activity for some common use cases simple.

Let's begin our journey by enabling the Open Connectors capability as part of onboarding for the SAP Integration Suite trial. We'll then explore the Open Connectors views and walk through an example connection to Slack.

7.1.1 Navigating to Open Connectors

To begin, you need to enable your SAP Integration Suite trial. We discussed these steps in Chapter 2, Section 2.3.3. Once you've provisioned the trial account for SAP Integration Suite, follow these steps to add the Open Connectors capability:

1. Click **Go to Application** to navigate to SAP Integration Suite from the SAP BTP cockpit.
2. Click the **Add Capabilities** button, as shown in Figure 7.1.

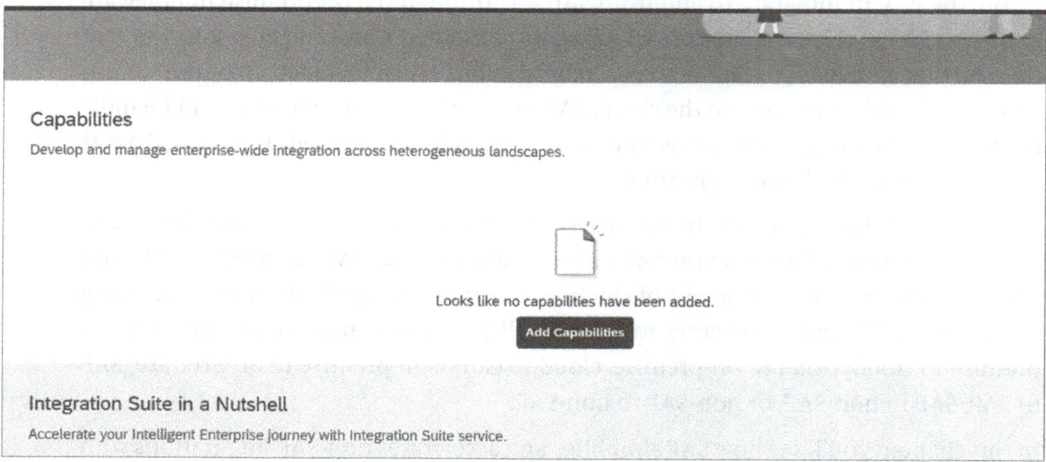

Figure 7.1 Add SAP Integration Suite Capabilities

3. Select **Extend Non-SAP Connectivity** to add the Open Connectors capability and select **Next** as shown in Figure 7.2.

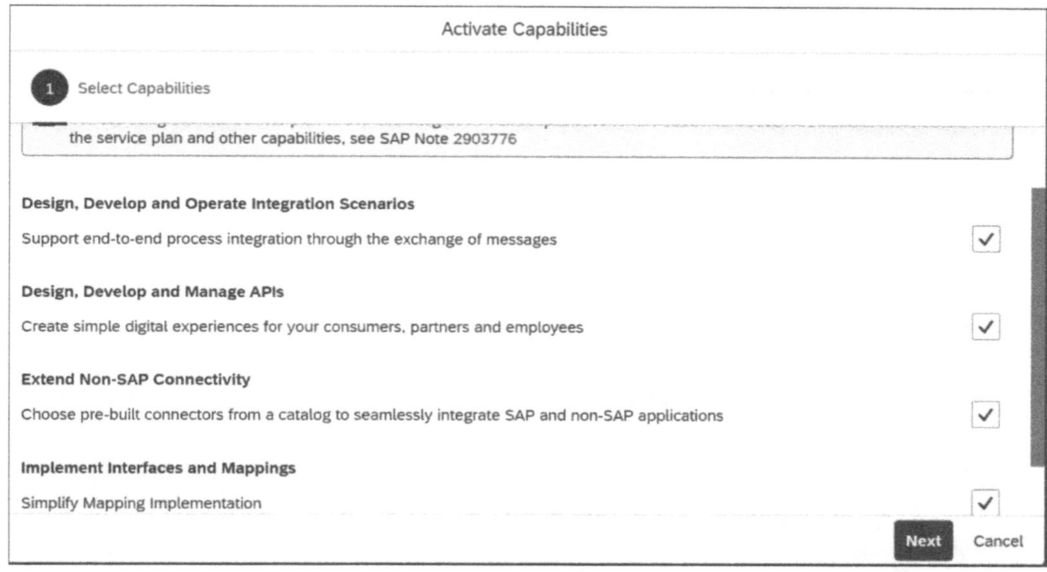

Figure 7.2 Capabilities of SAP Integration Suite

4. Select the **Message Queues** checkbox for **Cloud Integration** and **Enable Developer Portal** for **API Management**. On the **Summary** screen, click **Activate**, as shown in Figure 7.3.

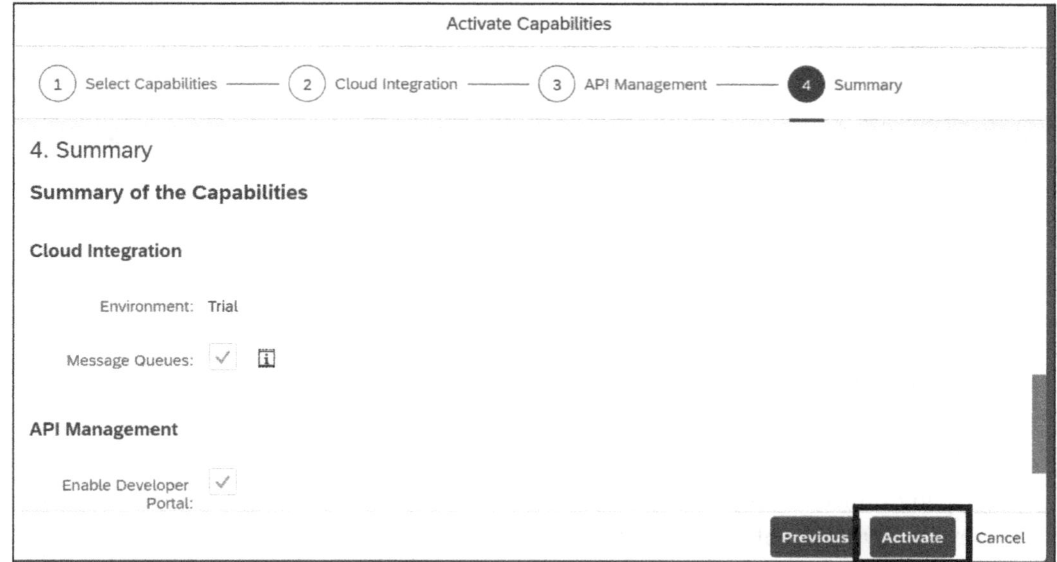

Figure 7.3 Additional Message Queues and API Business Hub Enterprise

5. Once the services are activated, navigate to the **Open Connectors** tab and click the **URL** shown in Figure 7.4.

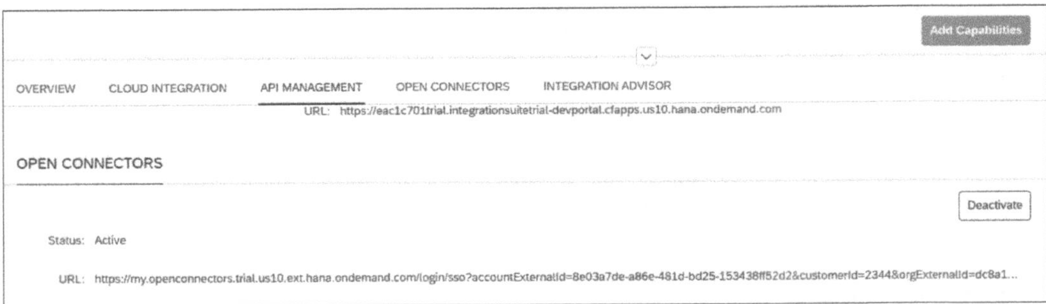

Figure 7.4 Open Connectors URL

6. You'll be taken to the Open Connectors home or landing page, as shown in Figure 7.5.

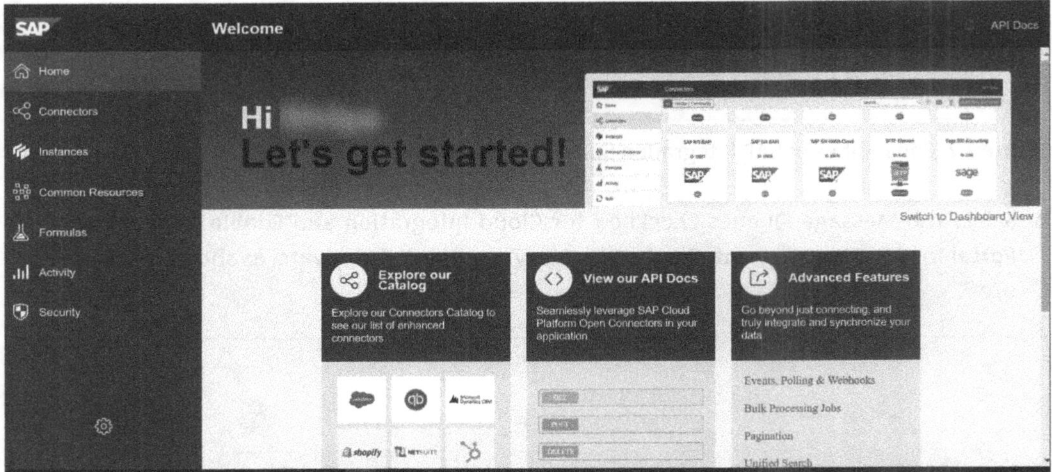

Figure 7.5 Open Connectors Home Page

7.1.2 Simplify Connectivity to Third-Party Cloud Applications

Now that you've accessed Open Connectors, let's explore how to use it in this section. We'll walk through the views offered with the Open Connectors capability and then connect to an example third-party application (Slack).

Open Connectors Views

Open Connectors create a unified API layer to unify the developer experience across every environment. Let's look at the value adding that SAP Integration Suite provides for non-SAP integration with its Open Connectors capability. The following views are available:

- **Connectors**
  You can check out the prebuilt connectors catalog in the **Connectors** view, as shown

in Figure 7.6. This is the where an integration developer would go to use the prebuilt connectors, which provide normalized REST interfaces to 160-plus API providers. All connectors have normalized authentication, content, error handling, pagination, bulk, and HTTP structures. You can check out the interactive API documentation that comes with each API to try out the connectors before you deploy them.

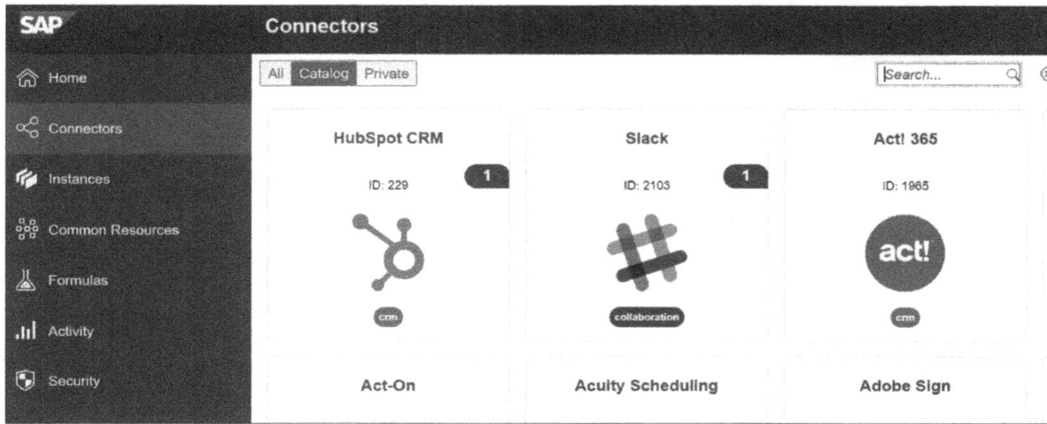

Figure 7.6 Connectors Tab

If the connector you need doesn't exist in the list, you can navigate to the connector builder by clicking the **Build New Connector** button, as shown in Figure 7.7.

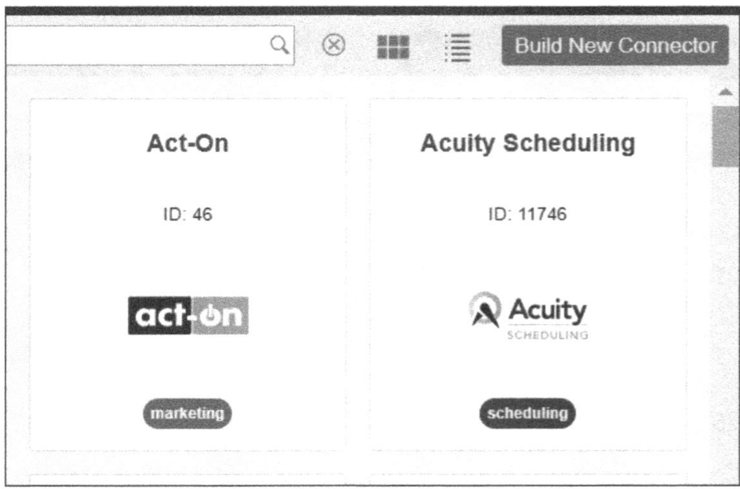

Figure 7.7 Build New Connector

Custom connectors can be built for REST, SOAP, and OData APIs, and even database applications under the **Builder** option, as shown in Figure 7.8.

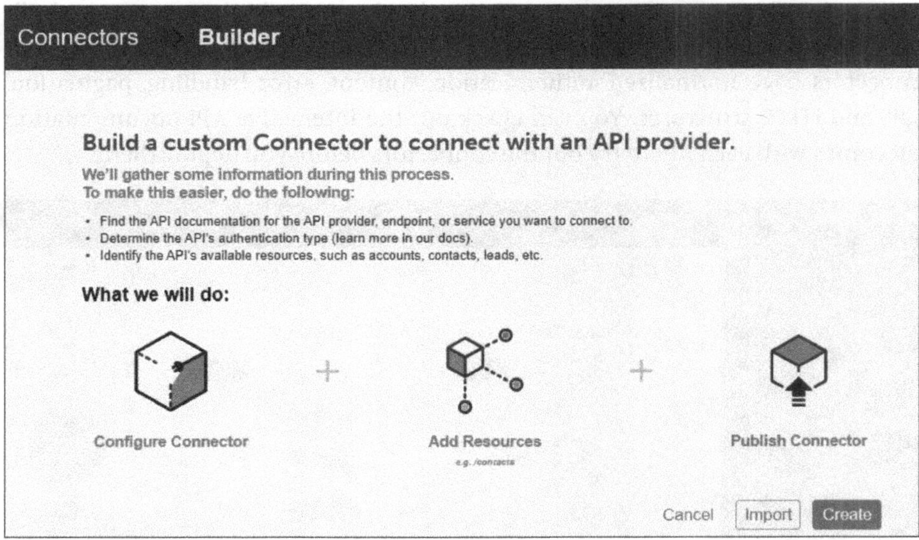

Figure 7.8 Custom Connectors

- **Instances**
  The **Instances** view gives you a way to manage **Connector** and **Formula** instances, as shown in Figure 7.9. The connector instance is your authenticated connection to API providers that you have configured. The formula instance is the view of the formulas you have created.

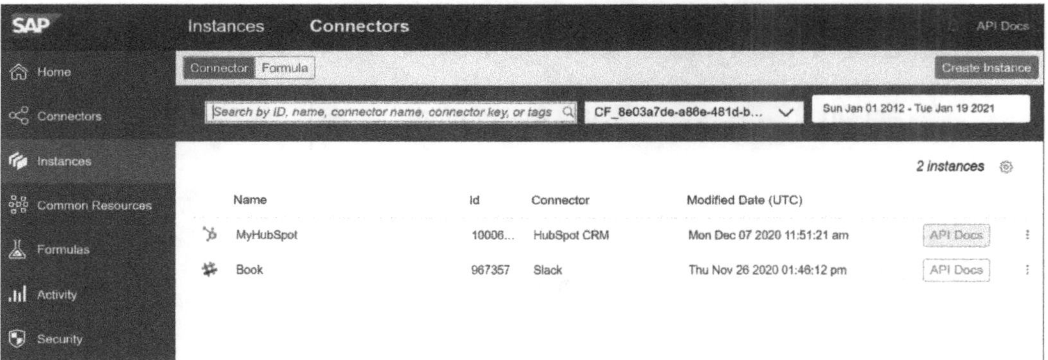

Figure 7.9 Instances Tab

- **Common Resources**
  The **Common Resources** view provides a uniform API wrapper to transform fields provided by your service providers into a single, normalized resource, as shown in Figure 7.10. This allows you to take advantage of routing, mapping, and transforming these requests and responses to third-party providers, resulting in a one-to-many

integration approach. You write a single common object to integrate the transformed resources from your various service providers.

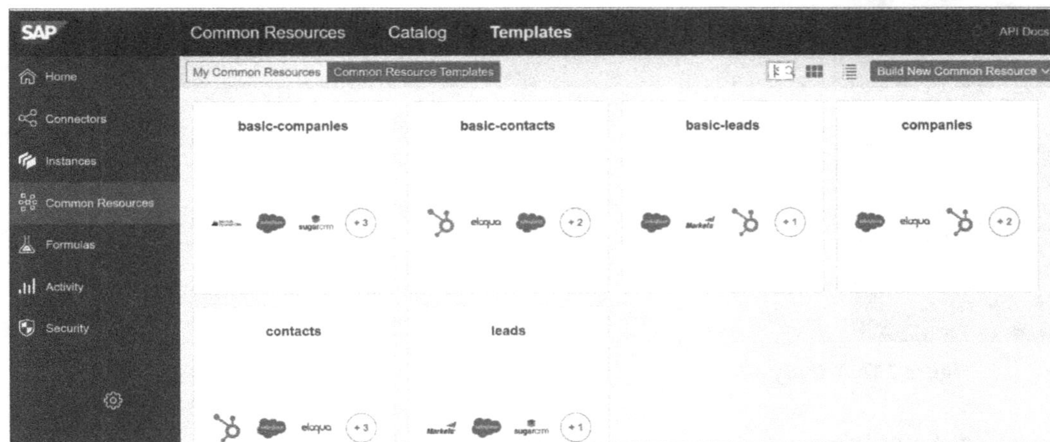

Figure 7.10 Common Resources

- **Formulas**
  The **Formulas** view shows all the formula templates created by you and enables you to create a new template, as shown in Figure 7.11. You can also leverage the **Formulas** tab to set up a new formula instance.

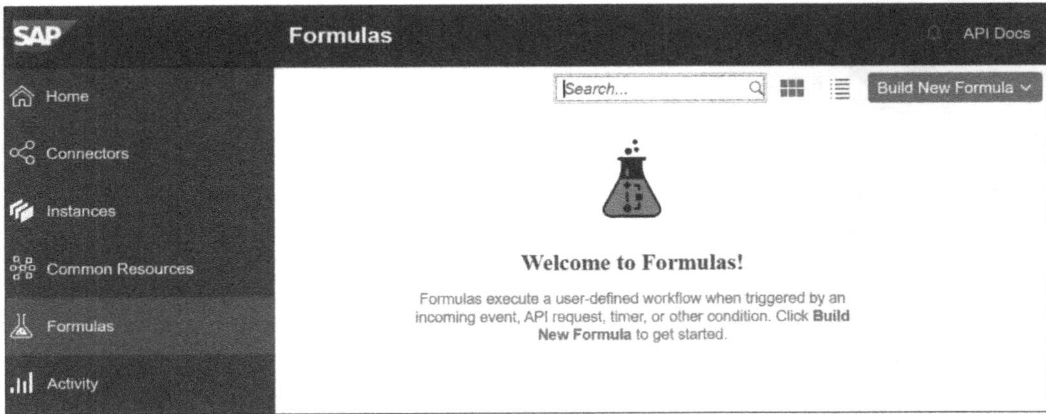

Figure 7.11 Formulas

- **Activity**
  This is the central view to manage your logs and metrics, as shown in Figure 7.12. You can search for a specific log entry and filter based on the date and time. An enhanced filter displays results based on user accounts, connectors, authenticated connector instances, and instance tags.

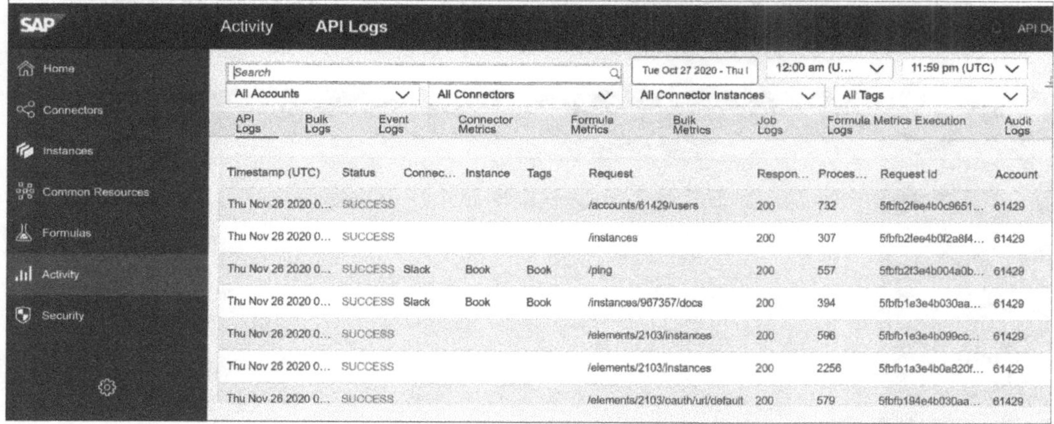

Figure 7.12 Activity

- **Security**
  The **Security** view, as shown in Figure 7.13, gives account details like the **ID**, **Created** date, and whether the instance is **Active** or not.

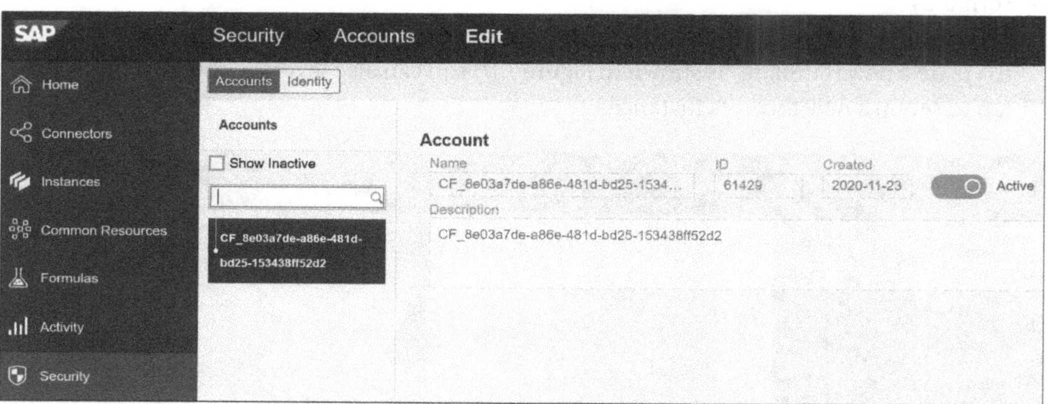

Figure 7.13 Security

Connect to Slack Using Open Connectors

In this section, we'll connect to a non-SAP system using the Open Connectors capability of SAP Integration Suite. We'll use the Slack collaboration hub as an example non-SAP system.

> **Prerequisite**
>
> You can create a Slack account and workspace at *https://slack.com/get-started#/create*.

Follow these steps:

1. Log into Slack and navigate to your workspace, as shown in Figure 7.14.

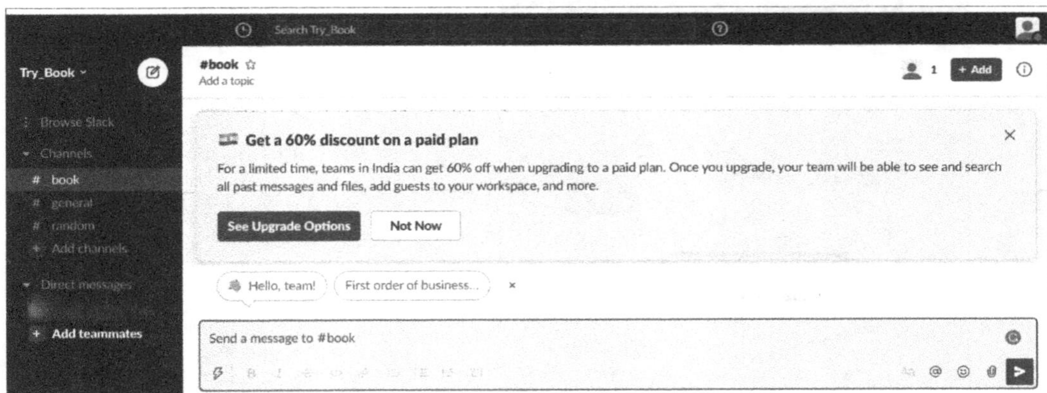

Figure 7.14 Slack Workspace

> **Note**
> Make sure that you've logged into your Slack workspace using the same browser window in which you're logged into Open Connectors.

2. Select the **Connectors** tab in your Open Connectors application and use the search bar to find the **Slack** connector, as shown in Figure 7.15.

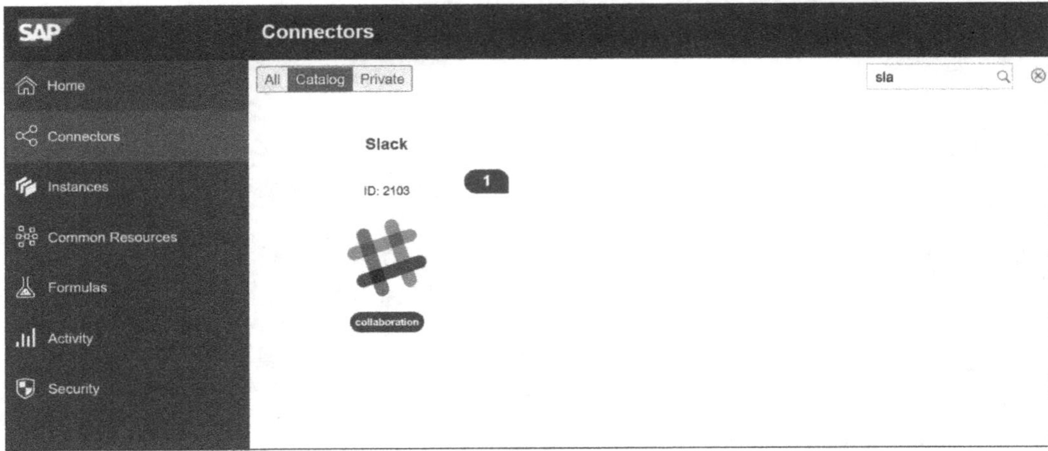

Figure 7.15 Slack Connector

3. Hover over the **Slack** connector and click **Authenticate**, as shown in Figure 7.16.

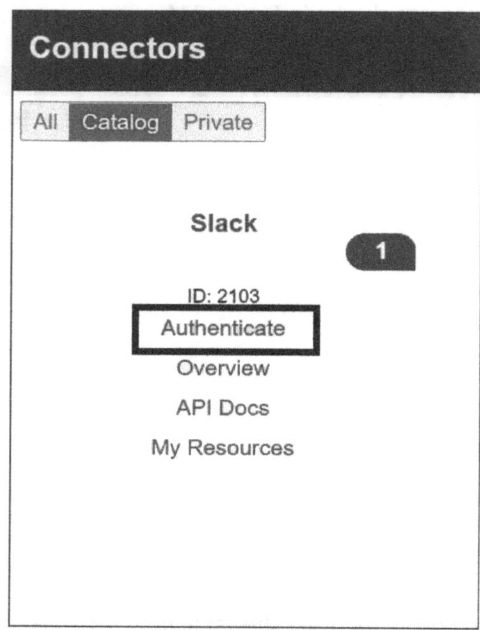

Figure 7.16 Authenticate Slack Connector

4. Enter a **Name** for your Slack connector instance (the **OAuth Scope** is filled in automatically) and click **Create Instance**, as shown in Figure 7.17.

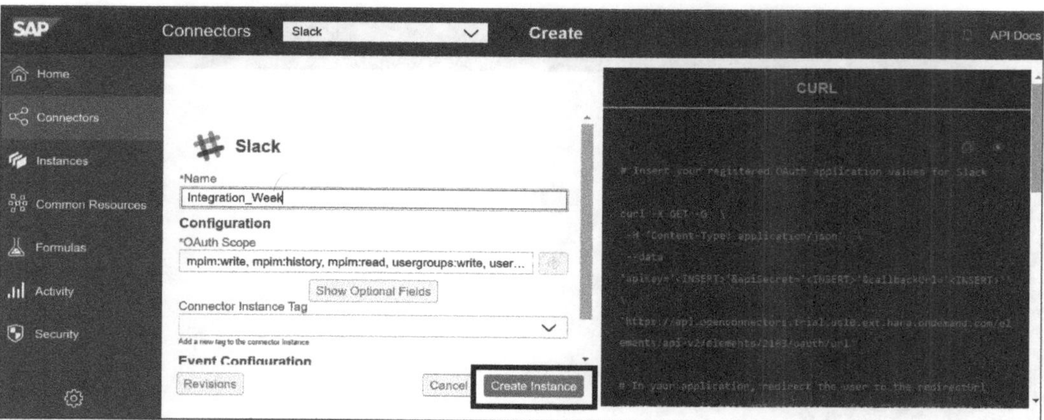

Figure 7.17 Create Slack Instance

5. You'll be redirected to a new browser tab in which your Open Connectors Slack instance will request permission to access your Slack workspace. Click **Allow**, as shown in Figure 7.18.

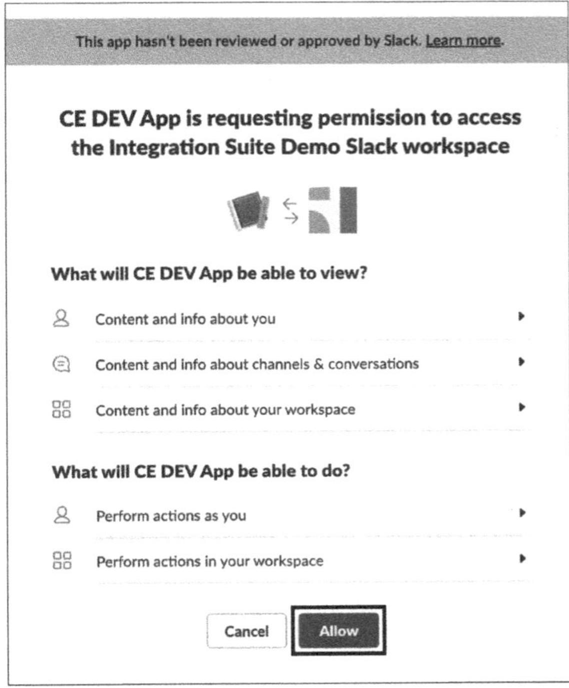

Figure 7.18 Access Slack Workspace

6. The browser will automatically close, and you will be redirected to your Open Connectors application, where you will see a success message indicating that the connector instance has been authenticated. Click the **Test in the API docs** link to do a quick ping test to verify the connectivity, as shown in Figure 7.19.

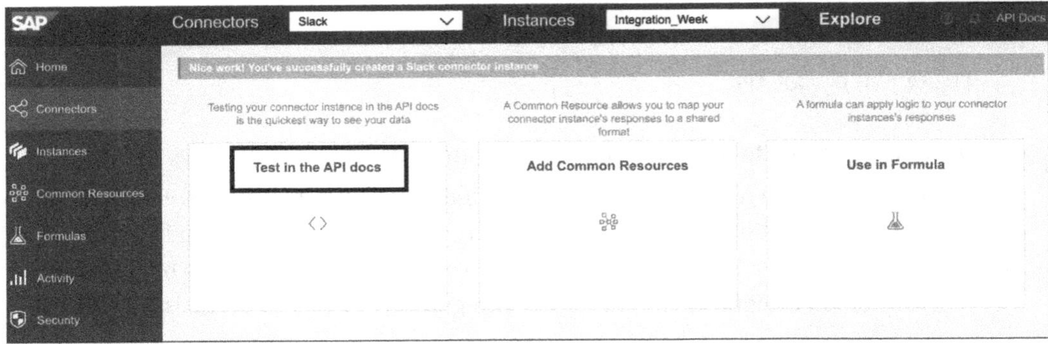

Figure 7.19 Test API Doc

7. In the API docs, scroll down to find the **ping** API, as shown in Figure 7.20.

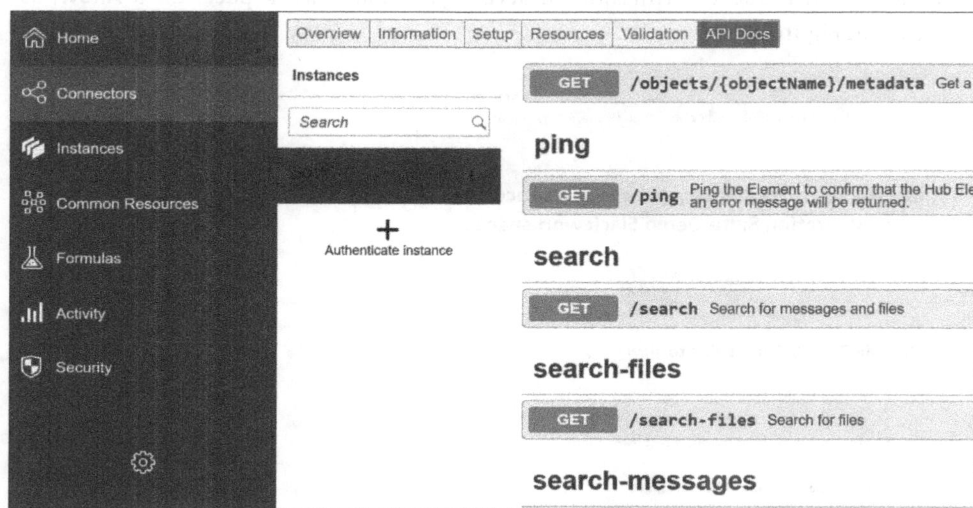

Figure 7.20 Ping API

8. Click the **ping** API and then click the **Try it out** button, as shown in Figure 7.21.

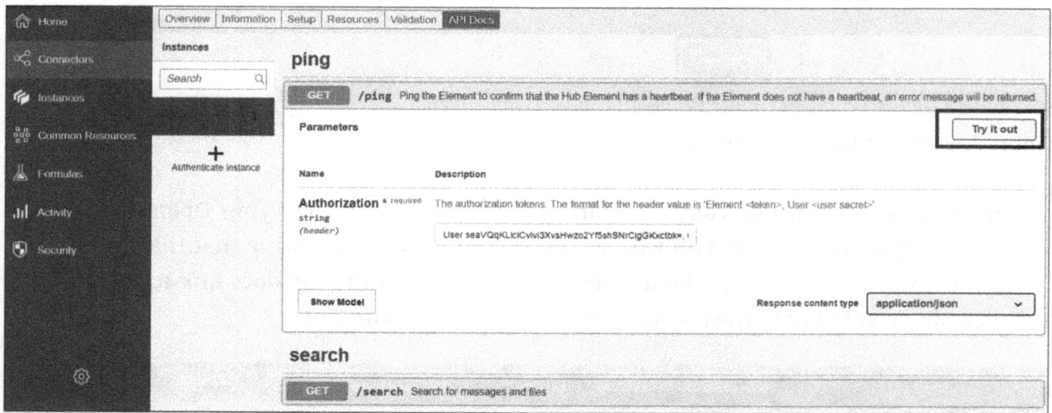

Figure 7.21 Try Out API

9. Click the **Execute** button, as shown in Figure 7.22.
10. You should see response code **200**. This indicates a successful response from your Slack account. Copy the `User`, `Organization`, and `Element` from the `curl` codeblock, as shown in Figure 7.23. Also copy the URL. These parameters have to be deployed in your Cloud Integration tenant of SAP Integration Suite as credentials.

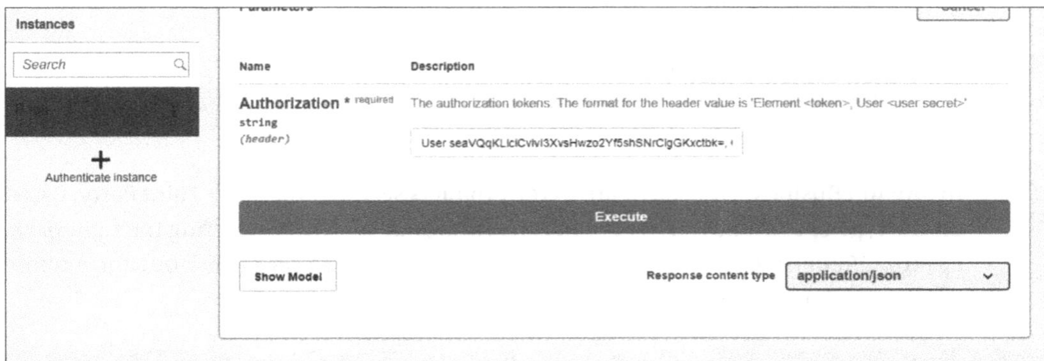

Figure 7.22 Execute API

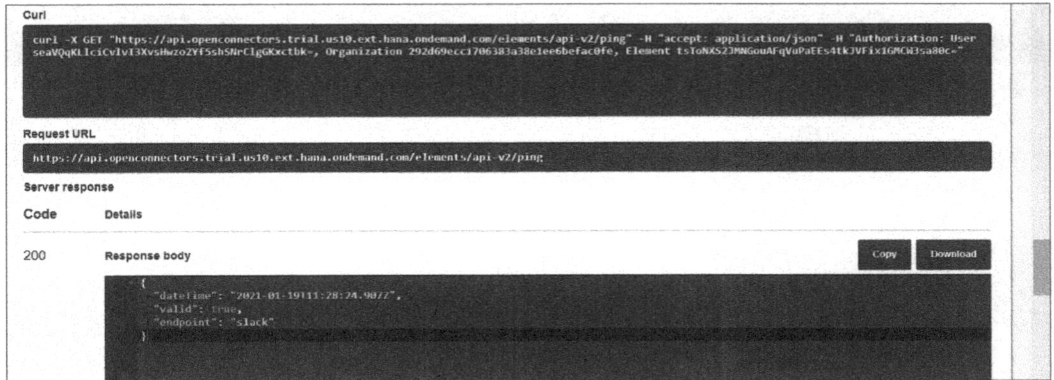

Figure 7.23 Test Successful

7.2 Exploring Third-Party SaaS Digital Content

The complete collection of third-party digital content, including APIs, integrations, and Open Connectors, can be experienced from the central catalog available on SAP API Business Hub. To strengthen and simplify non-SAP integration further, SAP's partner ecosystem has contributed the following adapters and integration packs:

- **Adapters**
  - Salesforce
  - Amazon Web Services
  - Microsoft Dynamics CRM
  - Sugar CRM
  - ServiceNow
  - Workday

- **Integration packs**
  - Integration between SAP S/4HANA and Salesforce
  - Integration between SAP S/4HANA Cloud and Salesforce
  - Integration between SAP ERP and Salesforce

In SAP API Business Hub, navigate to the **Products** section and select **Third Party**. Based on the type of content you need, you can refine your search by selecting the right package from **Refine By**, as shown in Figure 7.24. For an example, let's check out and explore some social media APIs.

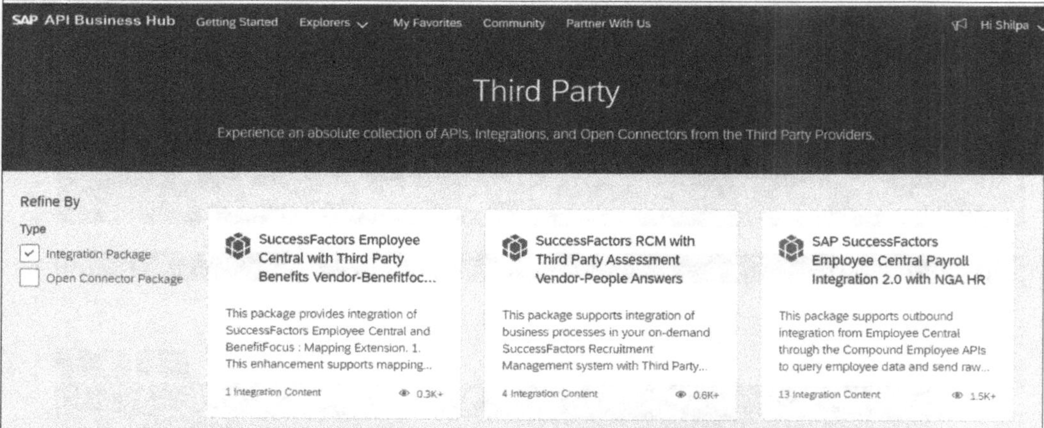

Figure 7.24 Third-Party Products

Choose the **Open Connector Package** and click the **Social—Open Connectors** package, as shown in Figure 7.25.

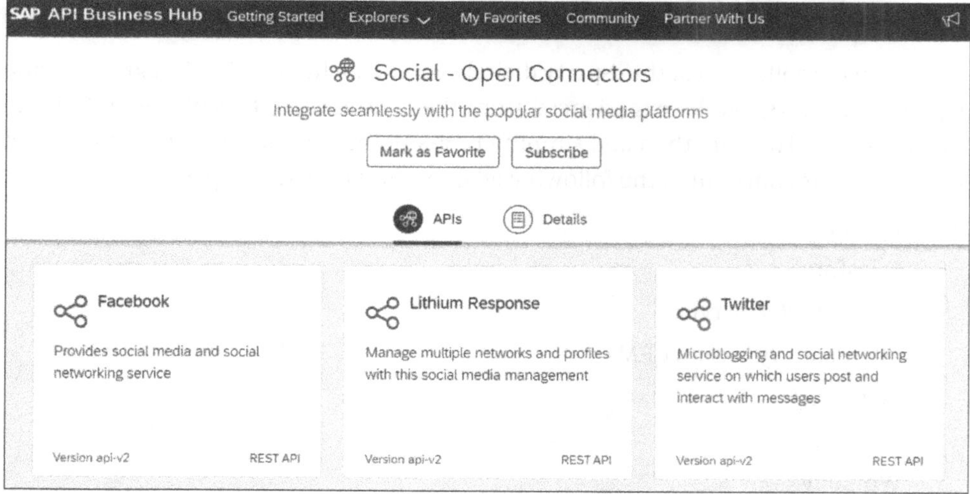

Figure 7.25 Third-Party APIs

You can navigate to the **Details** section and check out the documentation and other details, as shown in Figure 7.26.

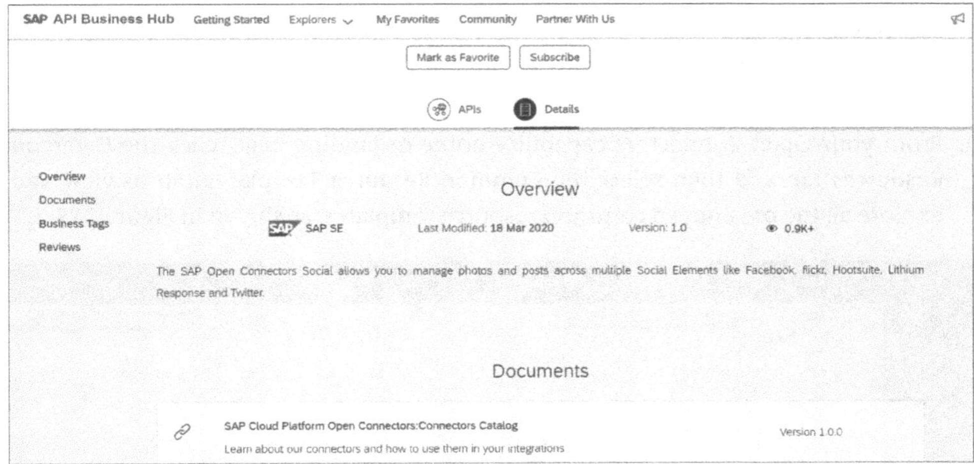

Figure 7.26 Details of APIs

7.3 Managing a Third-Party CRM Instance

To streamline integration in hybrid and multicloud environments, SAP Integration Suite can be leveraged as a single platform for managing your APIs and connector instances. You can discover your Open Connectors instances in the API portal and then create API proxies for these instances, which you can later strengthen using the rich set of policies available as part of the API Management capability. In the following sections, we'll use the **Common Resources** templates to map necessary fields in Open Connectors, create a provider in API Managment, discover the Open Connectors instance, and finally manage it using out-of-the-box policies.

7.3.1 Creating a Common Resources Template

In Section 7.1.2, we connected to a Slack instance; similarly, you can connect to a HubSpot CRM instance, as shown in Figure 7.27.

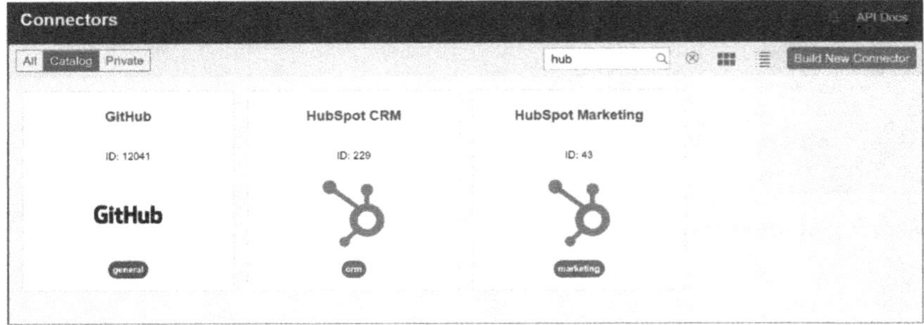

Figure 7.27 HubSpot CRM

In this section, we'll show how to create and test a template for a third-party HubSpot instance via the **Common Resources** view. In our example, we'll use the **basic-companies** common resource template that is premapped to multiple non-SAP connectors from the CRM hub via a normalized interface for the accounts or companies entity, providing basic information like name, description, phone, website, shipping address, and billing address. Follow these steps to clone the template:

1. From your Open Connectors capability home or landing page, click the **Common Resources** tab and then select the **Common Resource Templates** tab to view and explore all the preshipped common resource templates, as shown in Figure 7.28.

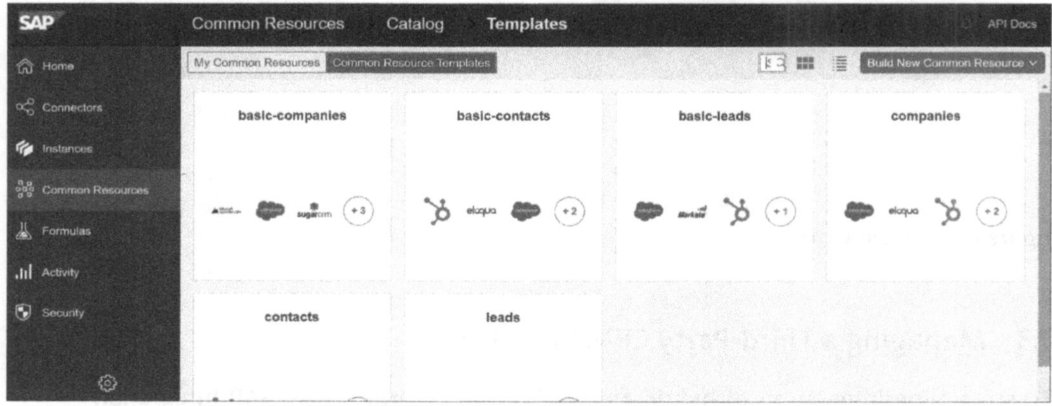

Figure 7.28 Common Resources

2. Select the **companies** tile and then select **Mappings**, as shown in Figure 7.29.

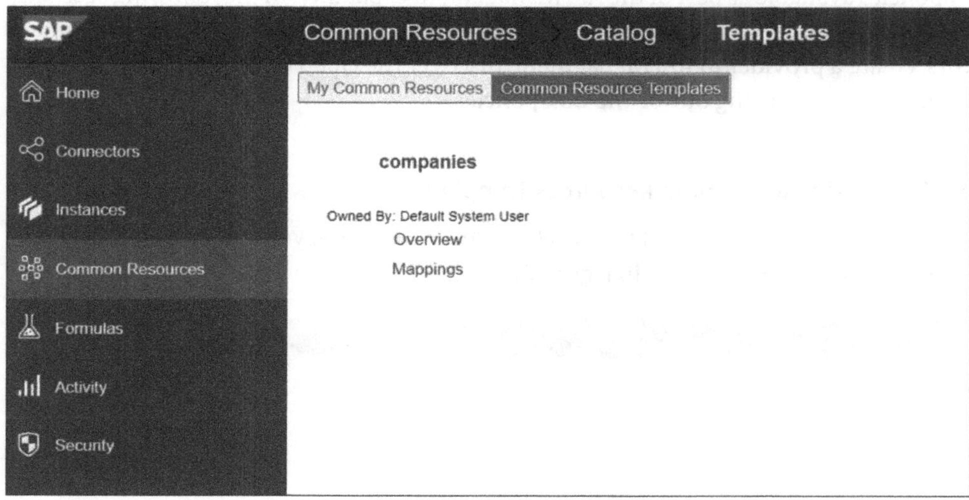

Figure 7.29 Mapping Companies

3. Select **Clone** to open the **Clone Into** popup. Here you can copy or clone the selected common resource template, as shown in Figure 7.30, to your Open Connectors tenant of SAP Integration Suite. Select the checkbox next to **Connector Name** to select all the connectors, then click **Save & Next**.

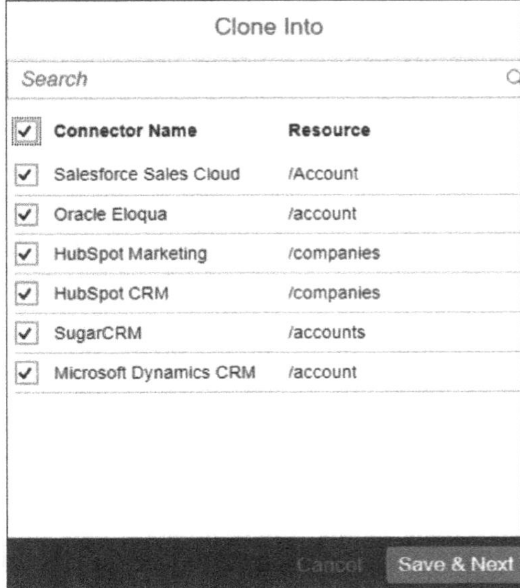

Figure 7.30 Cloning Common Resource Template

4. Select the level (**account**) and click **Save & Next**, as shown in Figure 7.31.

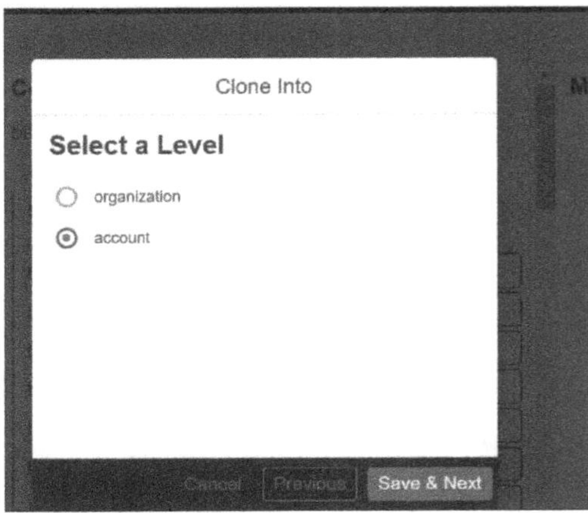

Figure 7.31 Clone Info

5. Provide a unique name for the cloned common resource and click **Save & Next** to complete the setup.

Now your common resource template is created. To test it, follow these steps:

1. Select the connector instance for HubSpot that you have in your Open Connectors tenant, as shown in Figure 7.32.

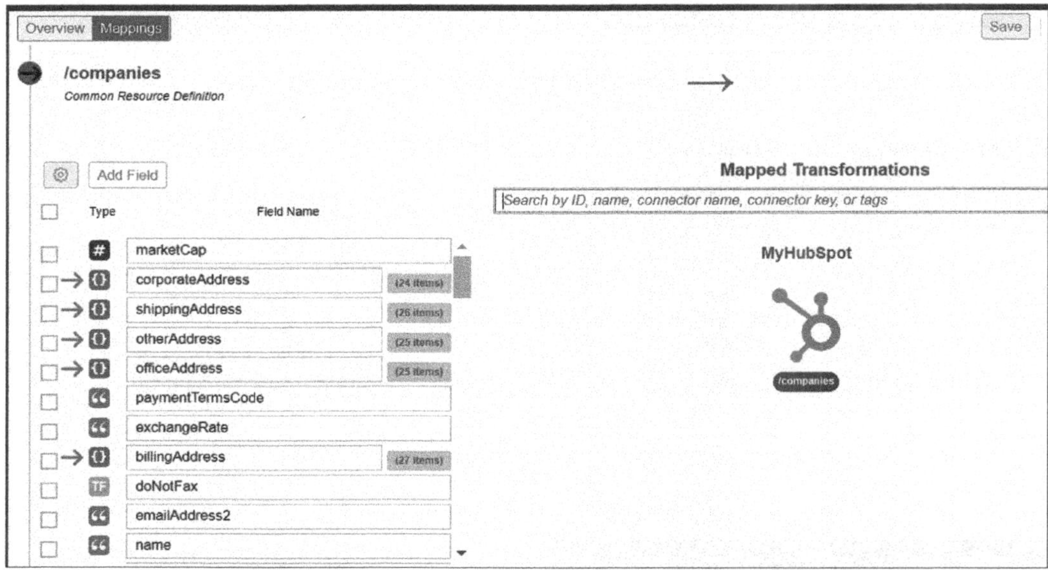

Figure 7.32 Test Common Resource Template

2. Select a non-SAP CRM connector instance, like **myhubspotdemo**, and select **Try it Out** to test the common resources, as shown in Figure 7.33.

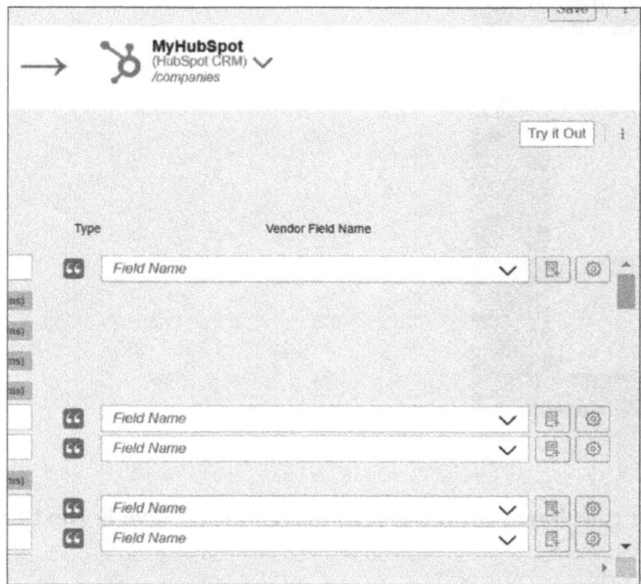

Figure 7.33 HubSpot CRM Instance

3. All the necessary fields are automatically filled in. Click **Execute** to test the common resources template, as shown in Figure 7.34.

Figure 7.34 Execute Common Resources Template

7.3.2 Managing the HubSpot Instance Using API Management

You've now set up your common resources template and created the HubSpot instance. To manage your HubSpot instance using the API Management capability, follow these steps:

1. Navigate to your API portal (refer back to Chapter 4, Section 4.3.1).
2. Navigate to **Configure** and click the **Create** button, as shown in Figure 7.35.

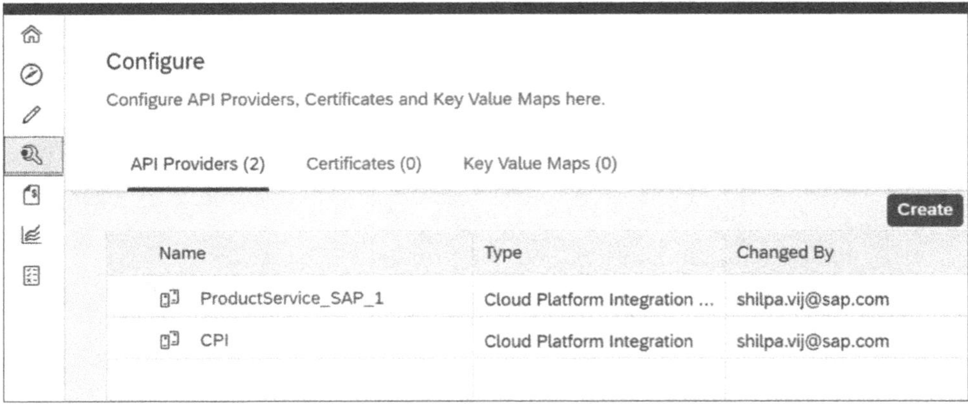

Figure 7.35 Create Provider

3. In the **Overview** section, give a name for a provider (see Chapter 4, Section 4.3.1).
4. Go to the Open Connectors capability and copy the **Organization Secret** and **User Secret** for the HubSpot instance you created in the previous section by clicking **Settings** (the gear icon), as shown in Figure 7.36.

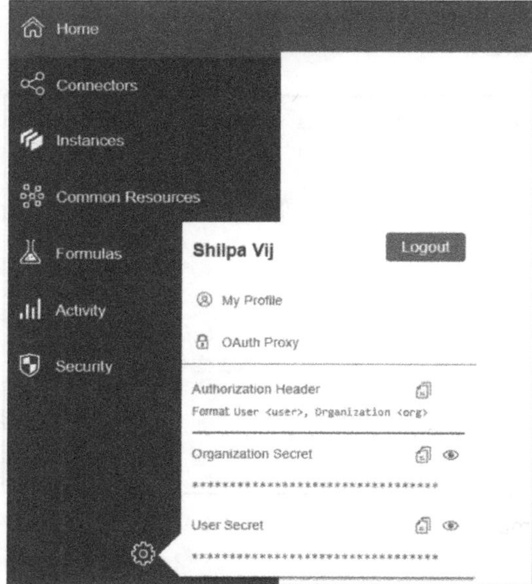

Figure 7.36 Open Connectors Credentials

5. In API Management, switch to the **Connection** tab, as shown in Figure 7.37. Select **Open Connectors** from the **Type** dropdown, choose your **Region**, enter your **Host** (your Open Connectors host), enter your **Port** (**443**), and paste the **Organization Secret** and **User Secret**. Click **Save**.

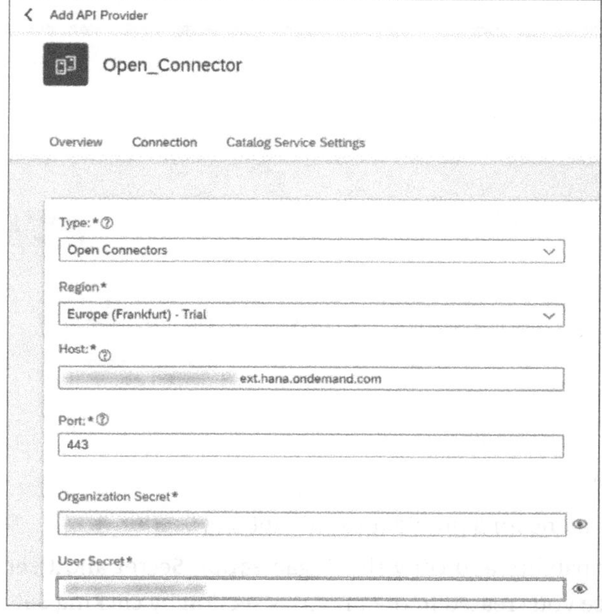

Figure 7.37 Connectors Details in API Management

6. You can test the connection for this Open Connectors instance of SAP Integration Suite by clicking the **Test Connection** button, as shown in Figure 7.38.

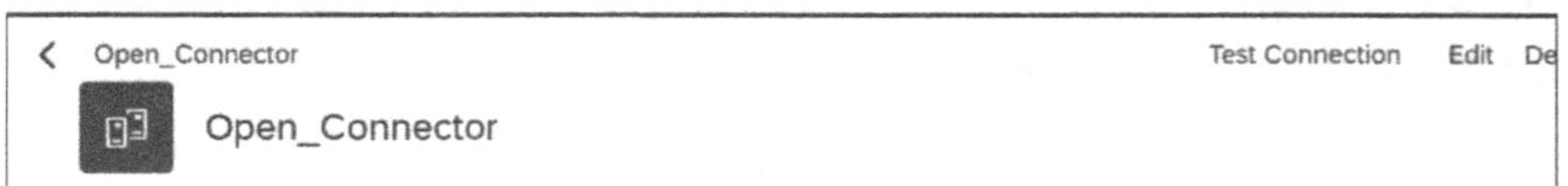

Figure 7.38 Testing Connection

7.3.3 Discovering the Open Connectors Instance

Next, we'll discover the Open Connectors instance that we've created. Follow these steps:

1. Navigate to **Develop** and click the **Create** button, as shown in Figure 7.39.

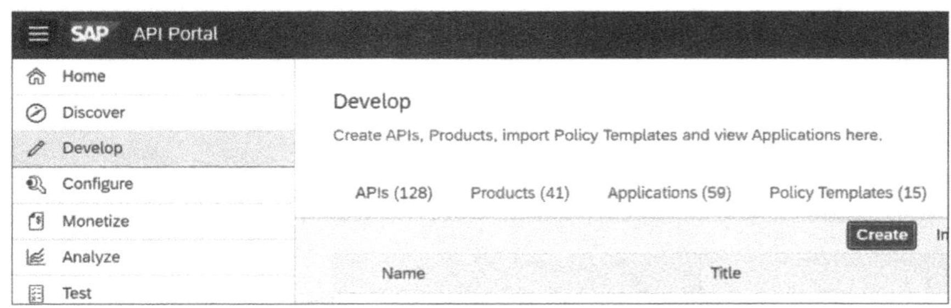

Figure 7.39 Create API Proxy Pointing to Open Connectors Instance

2. Select the API provider pointing to the Open Connectors instance that you created in the previous section. Click **Discover** to discover the associated APIs, as shown in Figure 7.40.

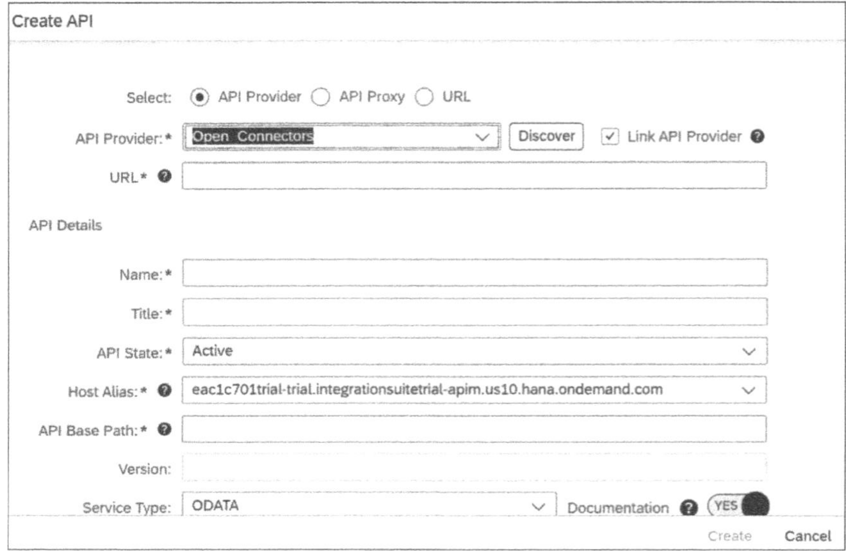

Figure 7.40 Discover Open Connectors Instance

3. Create an API proxy by selecting the HubSpot CRM connector (**MyHubSpot**, shown at the top of Figure 7.41).
4. Select the instance you created in Section 7.3.1 from the list, as shown in the lower portion of Figure 7.41, and click **OK**.

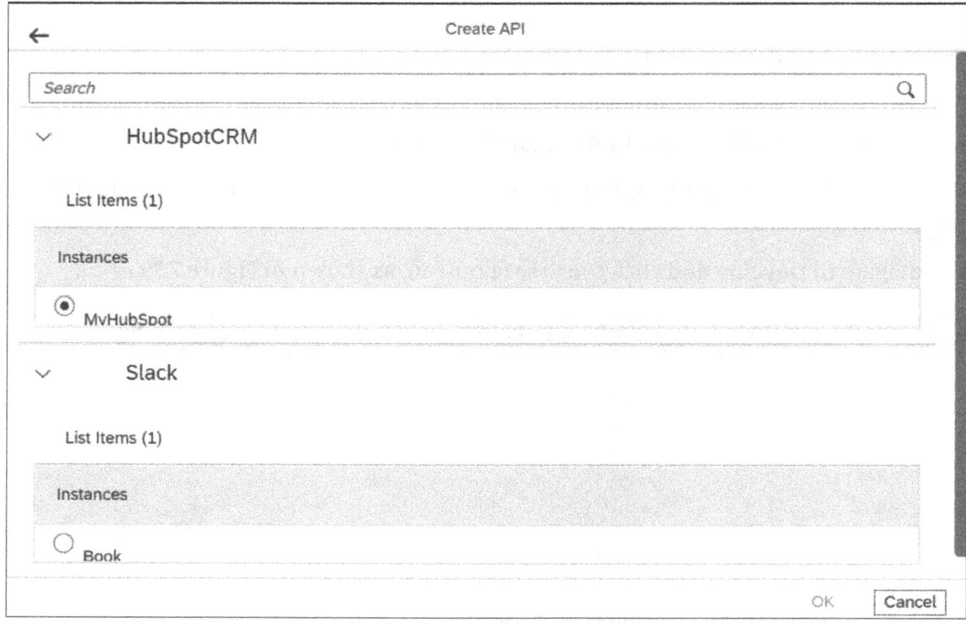

Figure 7.41 Select Instance

5. All the necessary details will be prefilled. Click **Create** and then click **OK**, as shown in Figure 7.42.

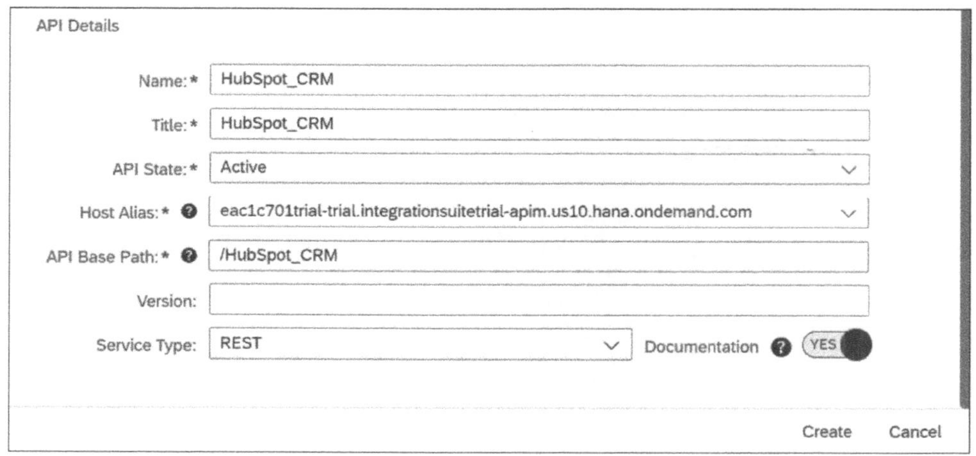

Figure 7.42 Deploy Proxy

6. Once you've deployed the proxy, you can try out the URL of the proxy by clicking the **Try It Out** button to get the details of the accounts from the HubSpot instances in the browser, as shown in Figure 7.43.

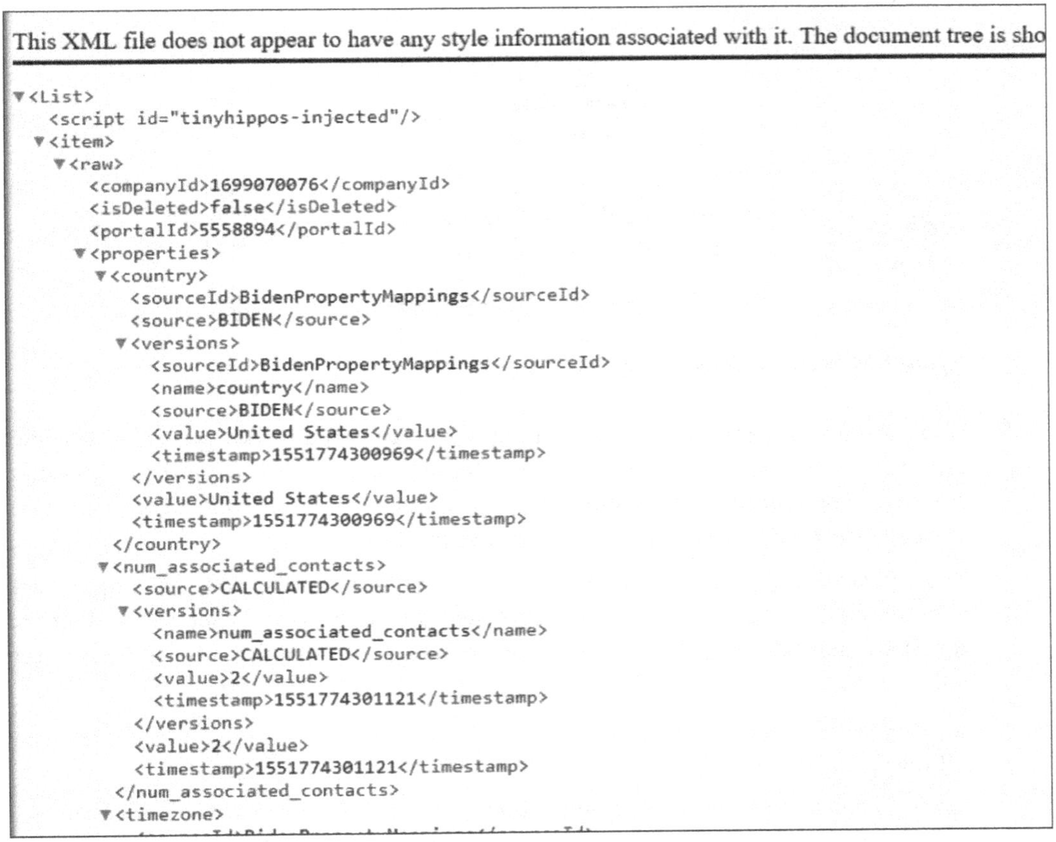

This XML file does not appear to have any style information associated with it. The document tree is sho


```
▼<List>
   <script id="tinyhippos-injected"/>
  ▼<item>
    ▼<raw>
       <companyId>1699070076</companyId>
       <isDeleted>false</isDeleted>
       <portalId>5558894</portalId>
      ▼<properties>
        ▼<country>
           <sourceId>BidenPropertyMappings</sourceId>
           <source>BIDEN</source>
          ▼<versions>
             <sourceId>BidenPropertyMappings</sourceId>
             <name>country</name>
             <source>BIDEN</source>
             <value>United States</value>
             <timestamp>1551774300969</timestamp>
           </versions>
           <value>United States</value>
           <timestamp>1551774300969</timestamp>
         </country>
        ▼<num_associated_contacts>
           <source>CALCULATED</source>
          ▼<versions>
             <name>num_associated_contacts</name>
             <source>CALCULATED</source>
             <value>2</value>
             <timestamp>1551774301121</timestamp>
           </versions>
           <value>2</value>
           <timestamp>1551774301121</timestamp>
         </num_associated_contacts>
        ▼<timezone>
```


Figure 7.43 Browser Test

Now you're all set to apply policies to manage your instance, as described in Chapter 4, Section 4.3.3.

7.4 Building an Integration Scenario

In this section, you'll learn how to use SAP Integration Suite to implement an integration scenario involving non-SAP applications. You'll implement a product catalog scenario, a use case involving webshop applications and collaboration tools like Slack, as shown in Figure 7.44.

You'll work with Slack, a cloud-based non-SAP application that provides communication services. You can use this application's trial offering to create a new channel.

Note

You could also use Gmail as a target system to send the catalog via email, but we won't use Gmail in this section.

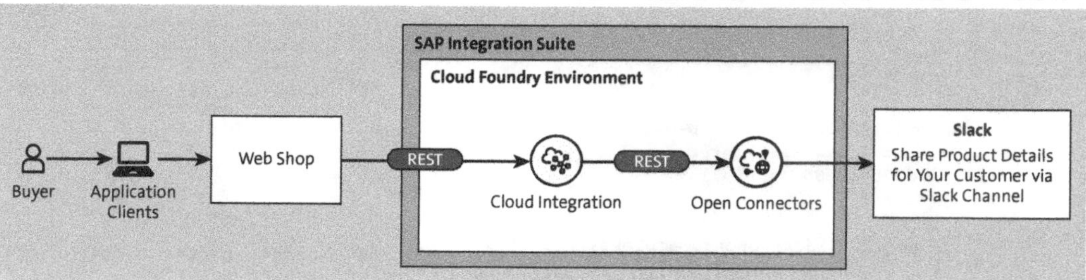

Figure 7.44 Scenario Walkthrough

Let's consider an example: you as an integration expert are working closely with your sales team to implement a scenario in which the product details from the online website should be sent to your consumer base via a Slack channel so that subsequent tasks can be performed by the team for further distribution.

You will send a request to an HTTP endpoint, similar to a REST API. This request will contain the ID of the product, and the corresponding details can be fetched and sent to the Slack channel using the Open Connectors adaptor capability of SAP Integration Suite.

Prerequisites

You'll need to provision a Cloud Integration tenant of SAP Integration Suite (see Chapter 2, Section 2.3.3). Then you'll need to set up the connectivity of the Open Connectors capability with Slack (see Section 7.1.2).

In the following sections, we'll walk through the steps to create your service instance and key, deploy your credentials, model an integration flow with the Cloud Integration capability, and test your scenario.

7.4.1 Creating a Service Instance and Service Key

You'll need to create a service instance and a key to authenticate your integration flow endpoint using Postman to test the scenario you designed. This is also needed when you deploy an integration flow with an HTTP endpoint. In the next section, you'll use the client ID and client secret from this service key as authentication details.

Follow these steps to create a service instance and service key with the Cloud Integration capability:

1. From your trial account cockpit, access the space created for your account, as shown in Figure 7.45.

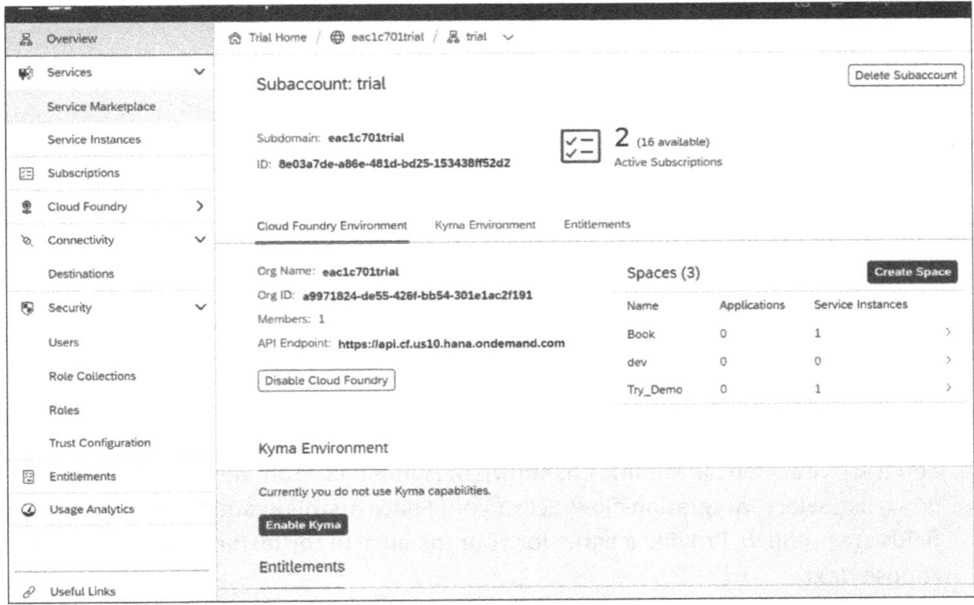

Figure 7.45 Trial Account

2. Select the **Service Marketplace** tab. Filter for "process integration" and select the **Process Integration Runtime** tile, as shown in Figure 7.46.

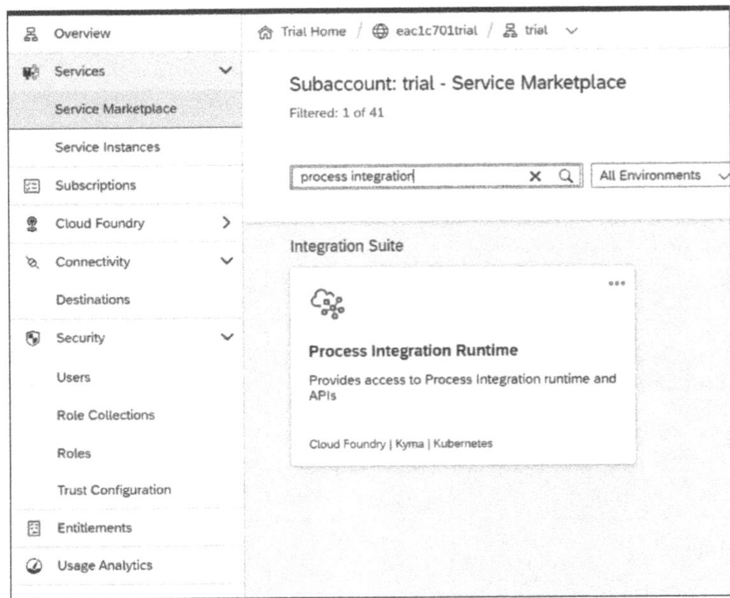

Figure 7.46 Service Marketplace

3. Click the **Create Instance** button, as shown in Figure 7.47.

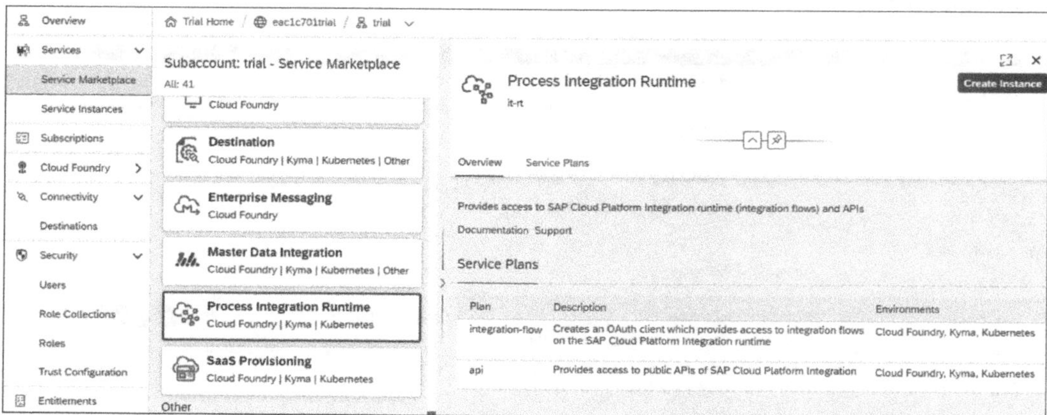

Figure 7.47 Process Integration Runtime

4. You'll arrive at step 1 (**Basic Info**), as shown in Figure 7.48. From the **Service Plan** dropdown list, select **integration-flow**. Select your **Space** (in this example, **dev**). The other fields are prefilled. Provide a name for your instance in the **Instance Name** field and choose **Next**.

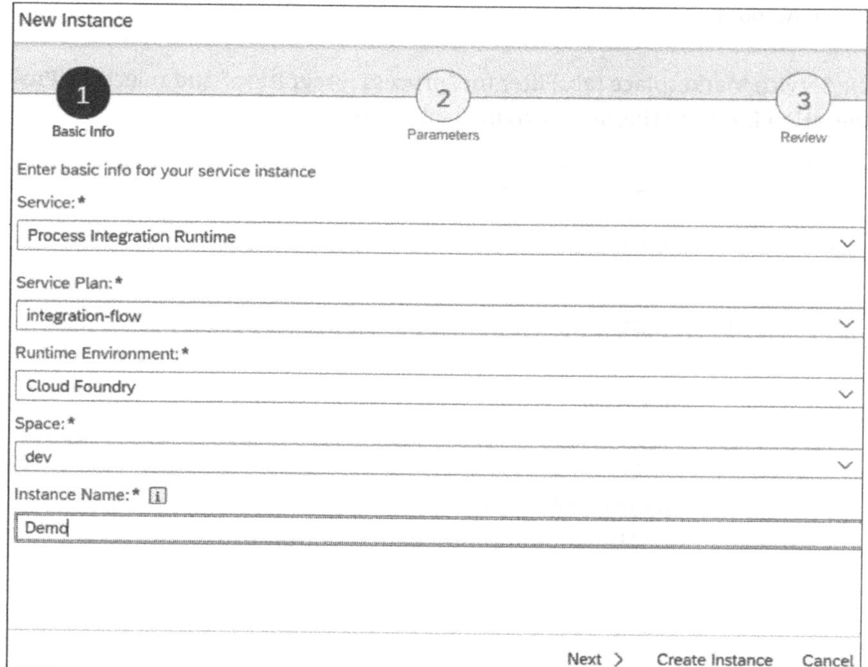

Figure 7.48 Instance Details

5. You'll arrive at step 2 (**Parameters**), as shown in Figure 7.49. Enter the parameters for the `ESBMessaging.send` role, as shown Listing 7.1, and click the **Next** button.

```
{
"roles":["ESBMessaging.send"]
}
```

Listing 7.1 Message Send Role

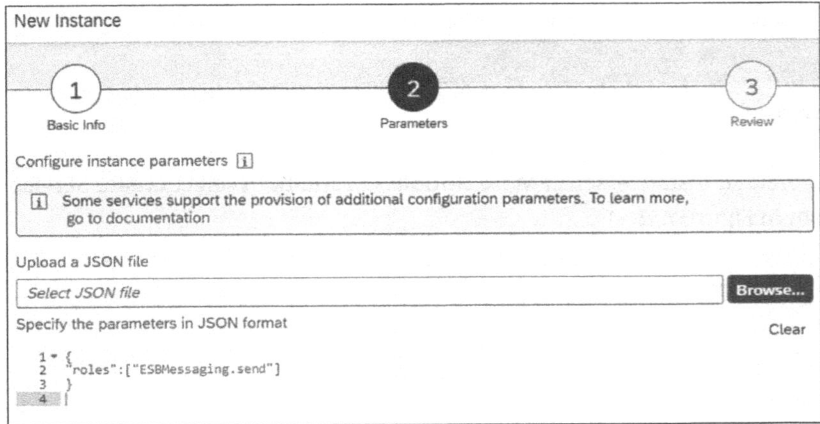

Figure 7.49 Instance Parameter

6. You'll arrive at step 3 (**Review**). Click the **Create Instance** button, as shown in Figure 7.50.

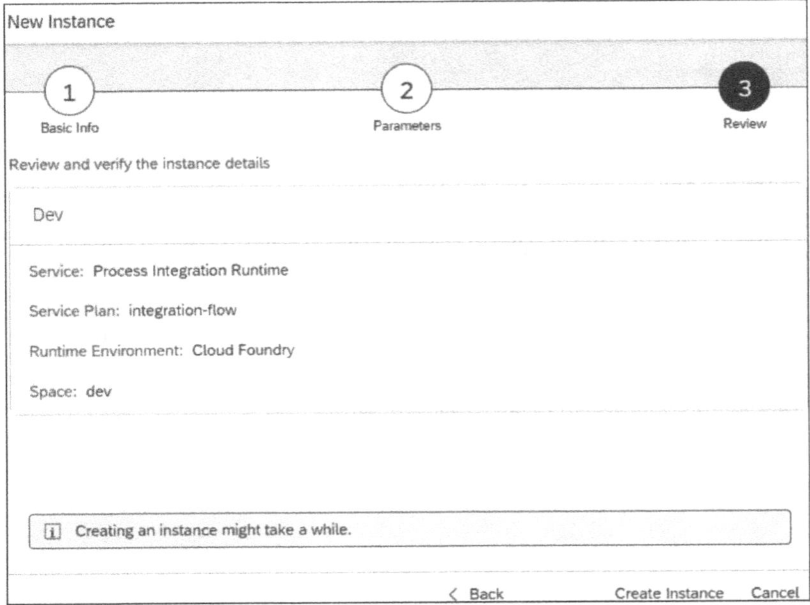

Figure 7.50 Create New Instance

7. In the **Success** popup, select **View Instance**, as shown in Figure 7.51.

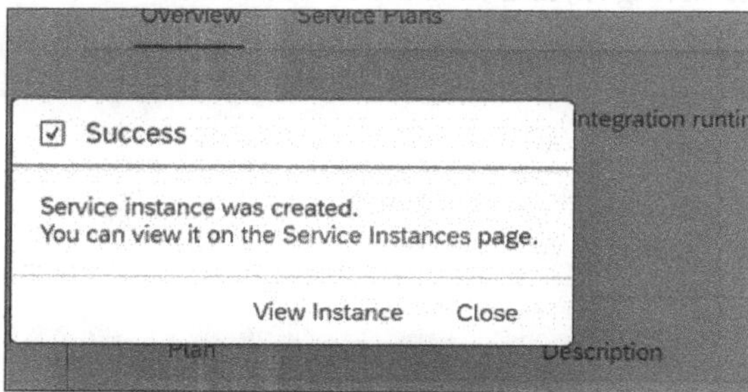

Figure 7.51 View Instance

8. In the newly created instance, select **More Options** (...) and then select **Create Service Key**, as shown in Figure 7.52.

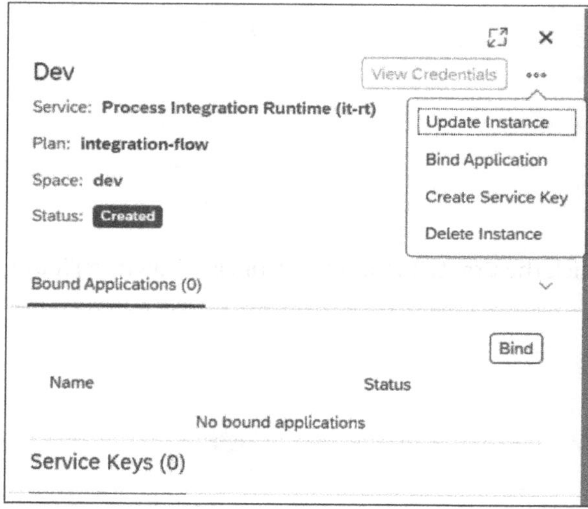

Figure 7.52 Create Service Key

9. Enter a unique name for the service key, enter the parameters for the role as shown in Listing 7.1, and click the **Create** button.
10. Once the instance is created, click the **View Credentials** button, as shown in Figure 7.53.
11. Copy the `clientid` and `clientsecret`, as shown in Figure 7.54.

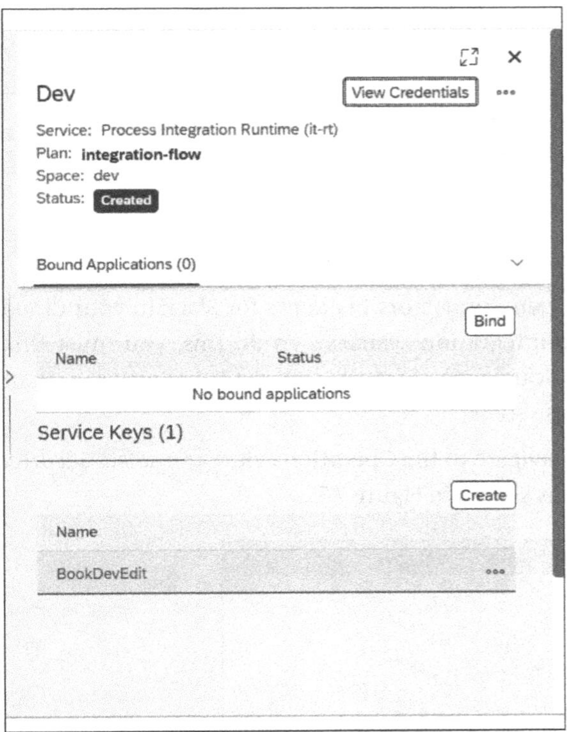

Figure 7.53 View Credentials

Credentials

BookDevEdit


```
{
    "oauth": {
        "clientid": "s",
        "clientsecret":
            -82lNoUuQI=',
        "url": "https://eac1c701trial.it-cpitrial-rt.cfapps.us10.hana.ondemand.com",
        "authorizationurl": "https://eac1c701trial.authentication.us10.hana.ondemand.com/oauth
            /authorize",
        "grant-types": [
            "refresh_token",
            "urn:ietf:params:oauth:grant-type:saml2-bearer",
            "client_credentials",
            "password",
            "authorization_code",
            "user_token",
            "urn:ietf:params:oauth:grant-type:jwt-bearer"
        ],
        "roles": [
            "ESBMessaging.send"
        ],
        "tokenurl": "https://eac1c701trial.authentication.us10.hana.ondemand.com/oauth/token"
```


Copy JSON Download Close

Figure 7.54 Client ID and Client Secret

Note

Be sure to jot down the values for the client ID and client secret parameters. You'll use these values of the user ID and password respectively when you're making a call to the integration flow endpoint using the Postman client.

7.4.2 Deploying Your Credentials on Your Cloud Integration Tenant

In this section, you'll be using the Open Connectors instances for Slack in your Cloud Integration tenant for building the integration scenario. To do this, you must first deploy the credentials, or authorization details, for your connector instances on your Cloud Integration tenant. Follow these steps:

1. In the Cloud Integration tenant, navigate to the **Operations** view and select **Security Material** under **Manage Security**, as shown in Figure 7.55.

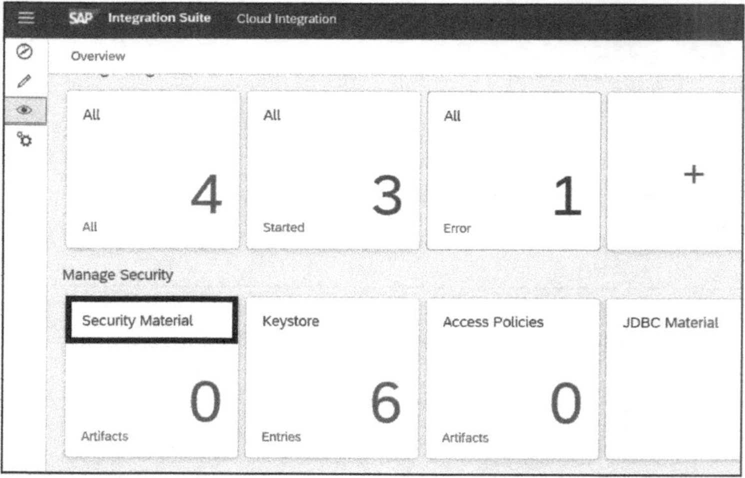

Figure 7.55 Select Security Material

2. Create user credentials by choosing **Create • User Credentials**, as shown in Figure 7.56.

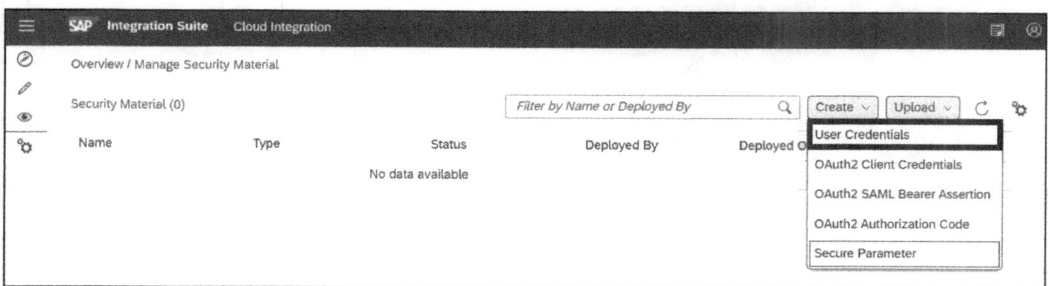

Figure 7.56 Create User Credentials

3. Enter a **Name** and **Description** that identifies this credential as the one specific to your Slack Open Connectors instance, as shown in Figure 7.57. In the **Type** dropdown list, select **OpenConnectors**. In the **User** and **Organization** fields, enter the user secret and organization secret from your Open Connectors application, as explained in Section 7.3.2. In the Open Connectors application, you can click the **Copy** icon next to each of the fields (refer back to Figure 7.36), then paste each into the Cloud Integration application.

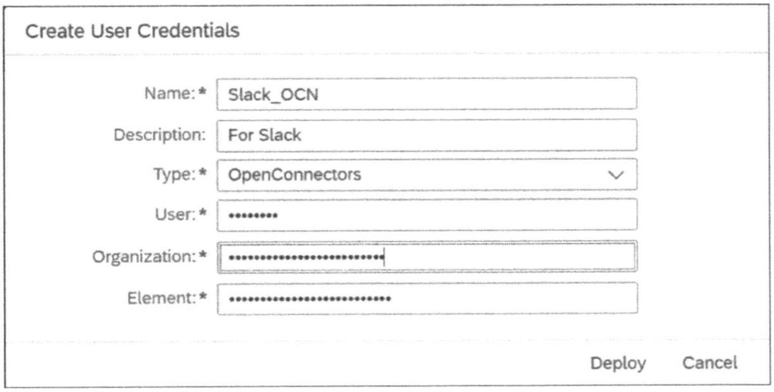

Create User Credentials

Name: * Slack_OCN
Description: For Slack
Type: * OpenConnectors
User: * ••••••••
Organization: * ••••••••••••••••••••••••
Element: * ••••••••••••••••••••••••••

Deploy Cancel

Figure 7.57 Create User Credentials

4. After entering all the details, select **Deploy** to deploy or activate the credentials.

7.4.3 Modeling an Integration Flow

This is the heart of this section. You've just deployed the credential for connectivity purposes; now you'll use the Cloud Integration capability of SAP Integration Suite to model the integration flow. Follow these steps:

1. Navigate back to the Cloud Integration capability of SAP Integration Suite.
2. Choose the **Design** tab and click the **Create** button to create a new integration package, as shown in Figure 7.58.

Figure 7.58 Create Integration Package

3. Enter a **Name** and **Short Description** for your integration package, as shown in Figure 7.59. Optionally, you can also specify a **Version** and **Vendor**, but we won't use those for this exercise. When you're finished, click **Save**.

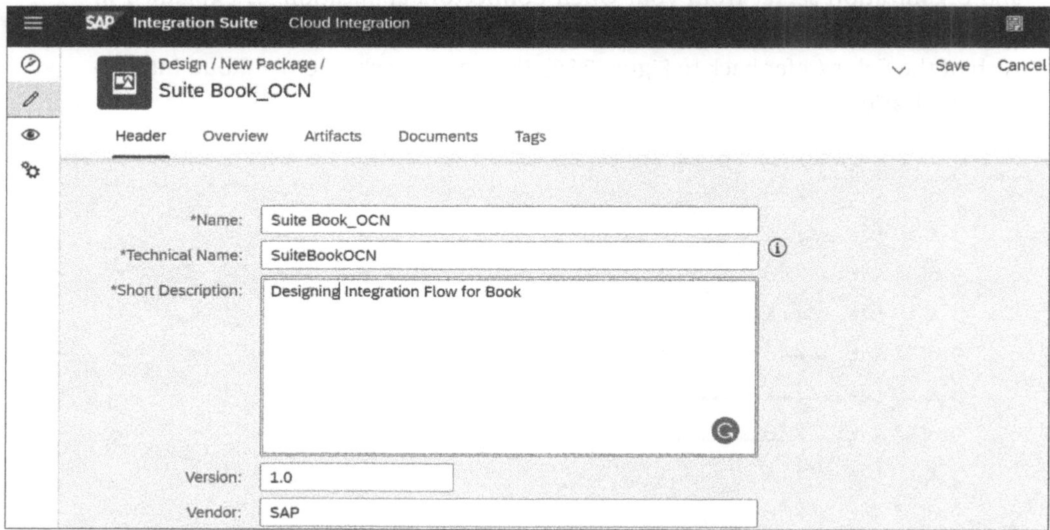

Figure 7.59 Package Details

4. Navigate to the **Artifacts** tab. In the **Add** dropdown, select **Integration Flow**, as shown in Figure 7.60. This action will add a new integration flow to the integration package you just created.

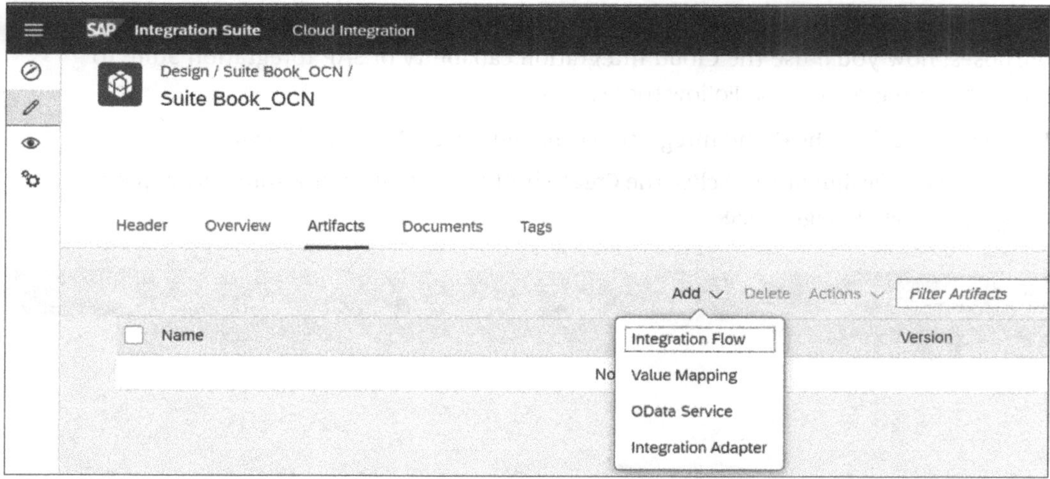

Figure 7.60 Add Integration Flow

5. Provide a **Name** (**OCN**, in this example) and **Description** (optional) for your integration flow, as shown in Figure 7.61, and click **OK**.

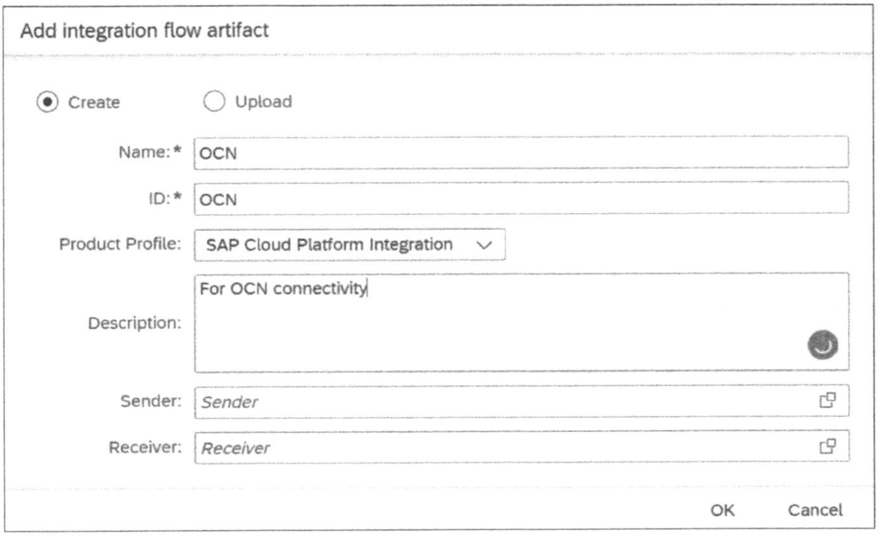

Add integration flow artifact

(●) Create ( ) Upload

Name:* OCN

ID:* OCN

Product Profile: SAP Cloud Platform Integration

Description: For OCN connectivity

Sender: Sender

Receiver: Receiver

OK Cancel

Figure 7.61 Integration Flow Details

6. Select the integration flow and click **Edit** to start editing it, as shown in Figure 7.62.

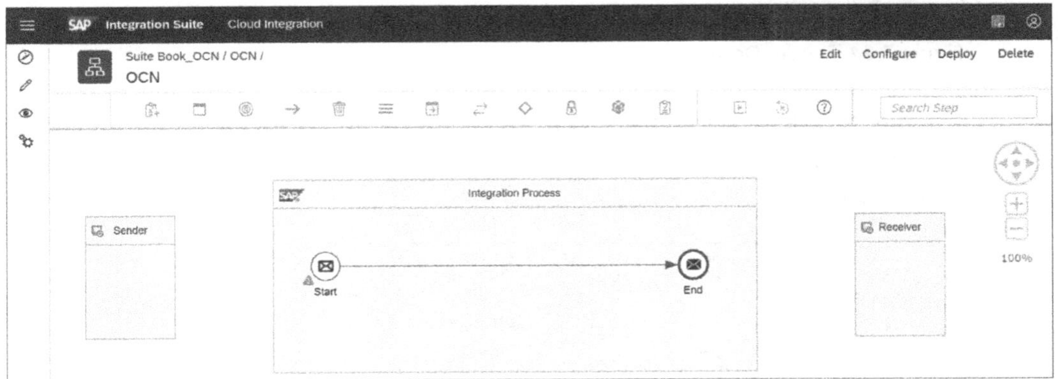

Figure 7.62 Edit Integration Flow

7. Using the **Connector** (arrow) icon, connect the **Sender** to the **Start Message** step, as shown in Figure 7.63.
8. In the adapter properties sheet, made visible by clicking the adapter, choose the **Connection** tab, as shown in Figure 7.64. In the **Address** field, enter a "/" followed by a value of your choice to give a unique identifier to the HTTP endpoint. This will be the endpoint you will call using Postman. Also, disable the **CSRF Protected** checkbox.

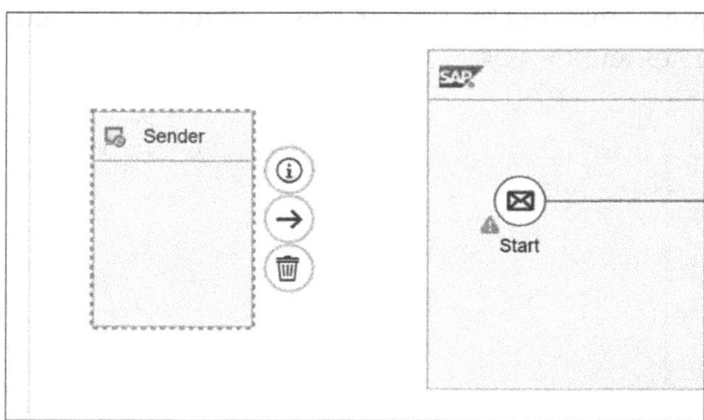

Figure 7.63 Connecting Sender

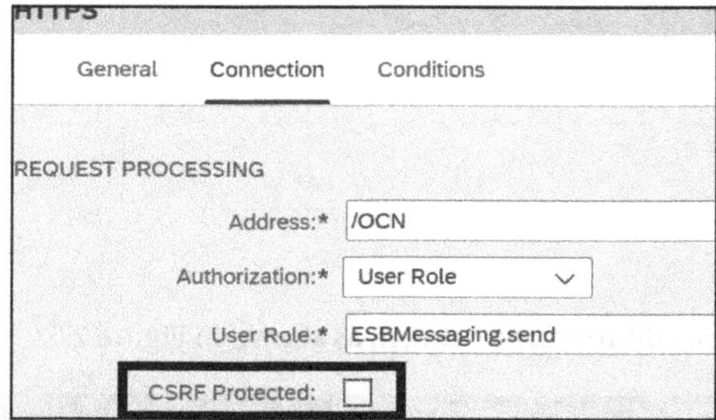

Figure 7.64 Sender Details

9. From the palette, select the **JSON to XML** converter under **Message Transformers** by selecting the seventh icon from the left and add it to the integration flow, as shown in Figure 7.65.

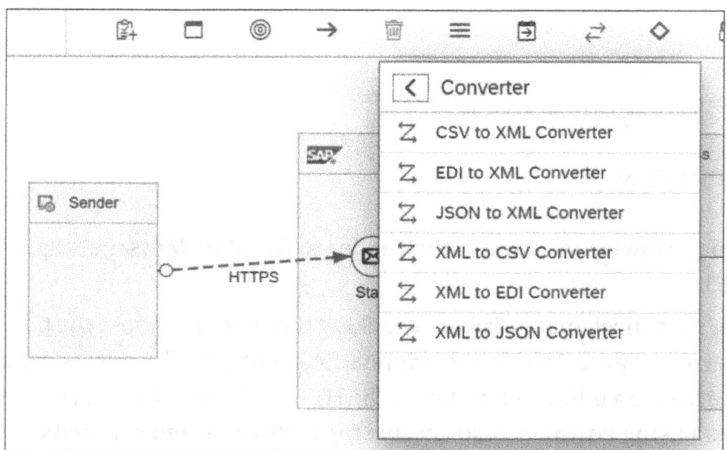

Figure 7.65 JSON to XML Converter

10. Again, click the seventh icon from left, which is **Message Transformers**, and add the **Content Modifier** to the integration flow, as shown in Figure 7.66. You use this step to read the product ID you provided as input and set it as a header value.

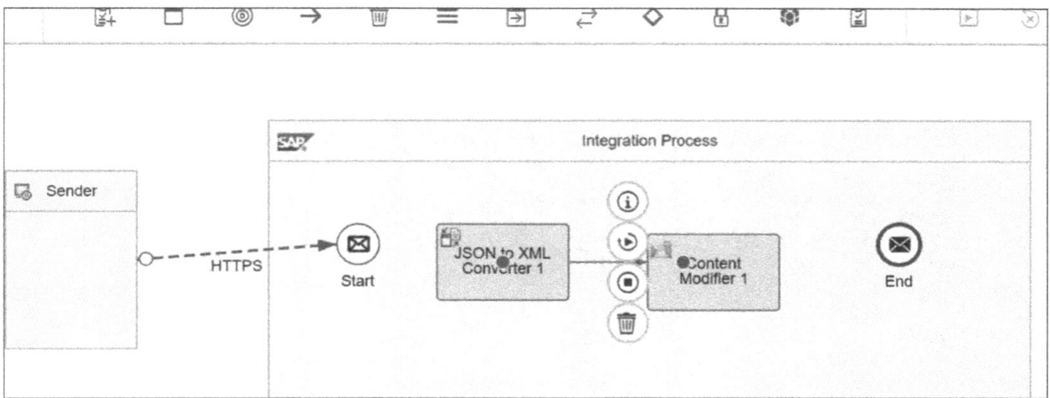

Figure 7.66 Add Content Modifier

11. Select the **Content Modifier** and navigate to its property sheet. In the properties sheet, select the **Message Header** tab and create a new header by choosing **Add**. Enter "productid" for the **Name**, select **XPath** in the **Type** dropdown, and enter "java.lang.String" for the **Data Type** and "//productIdentifier" for the **Value**.
12. From the palette, choose the icon showing two arrows (**Call**), then **External Call • Request Reply**, and add it to the integration flow as shown in Figure 7.67. You will use this step to call the webshop and request product details.

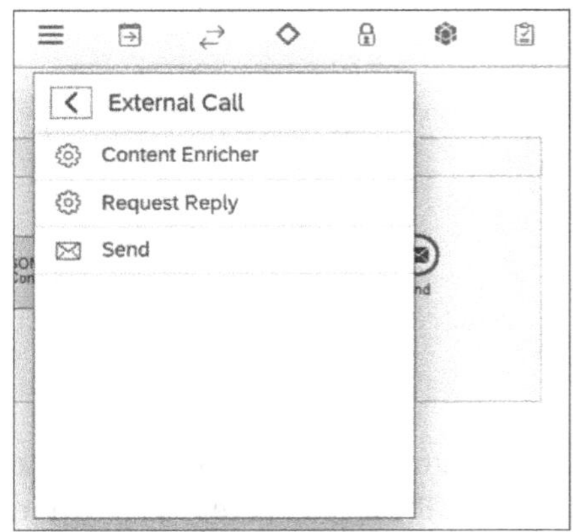

Figure 7.67 Request Reply

13. Connect the **Request Reply** step to the receiver. In the **Adapter Type** prompt, select **OData** as the adapter and **OData V2** as the message protocol, as shown in Figure 7.68.

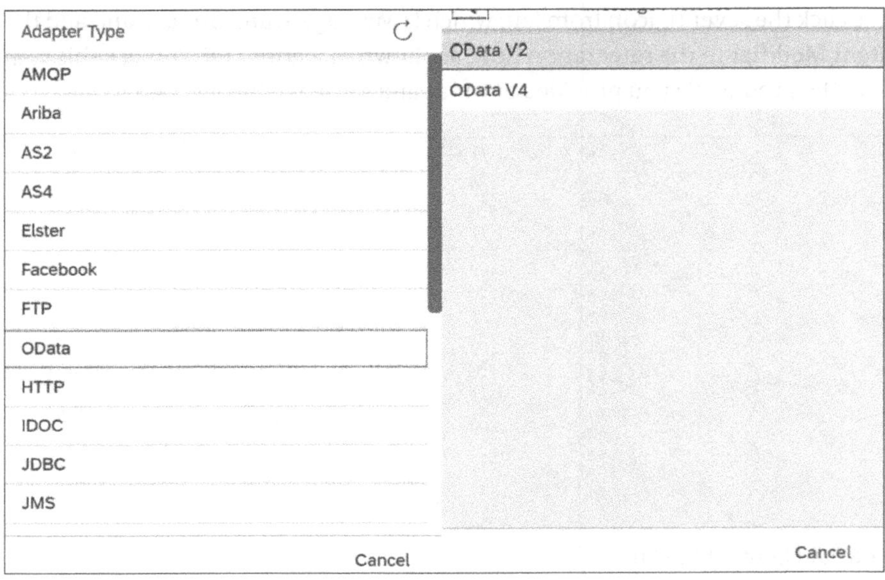

Figure 7.68 Add OData Reciever

14. In the **Connection** tab, enter *https://refapp-espm-ui-cf.cfapps.eu10.hana.ondemand.com/espm-cloud-web/espm.svc* as the **Address**. This is the URL of the OData service of the webshop, as shown in Figure 7.69.

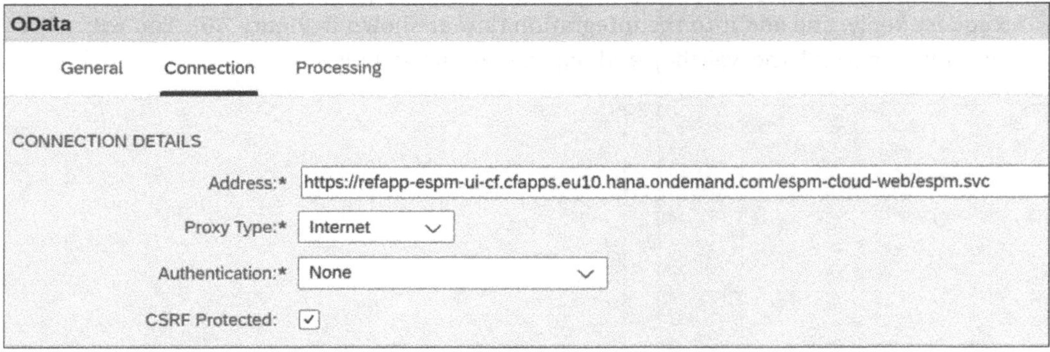

Figure 7.69 Webshop Address

15. Select the **Processing** tab and click **Select** in the **Resource Path** field to arrive at the **Connect to System** step, for which everything is prefilled. Select **Step 2**, shown in Figure 7.70.

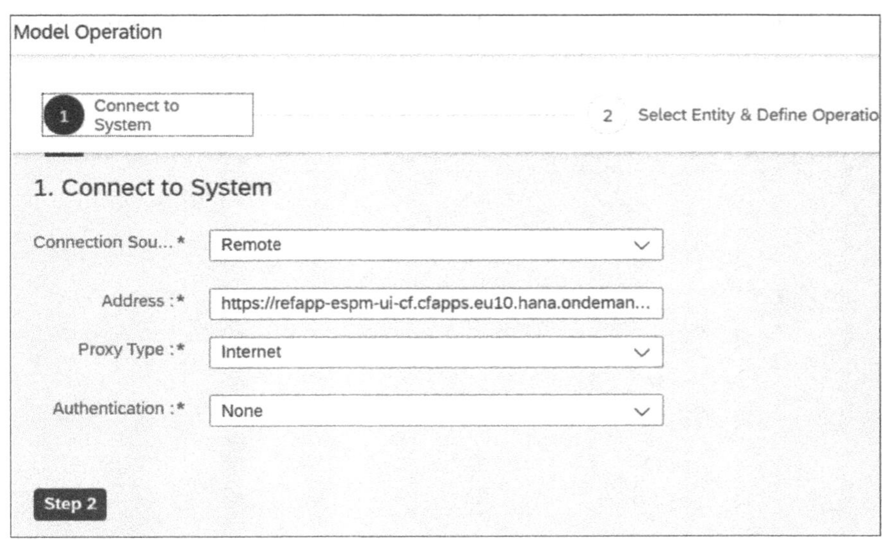

Figure 7.70 System Details

16. Click **Select Entity** and choose **Products** from the dropdown. Here you're choosing to fetch the **Products** entity, as shown in Figure 7.71.

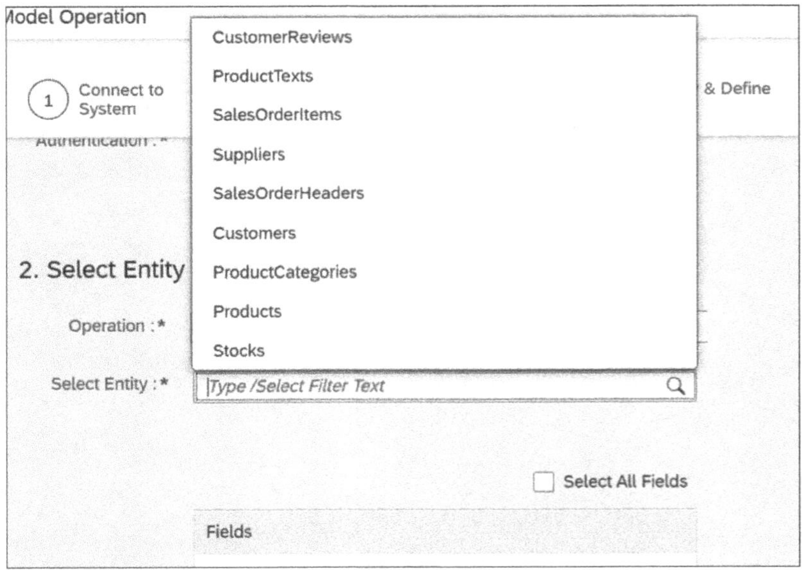

Figure 7.71 Select Entity

17. Enable the **Select All Fields** checkbox and choose **Step 3**, as shown in Figure 7.72.

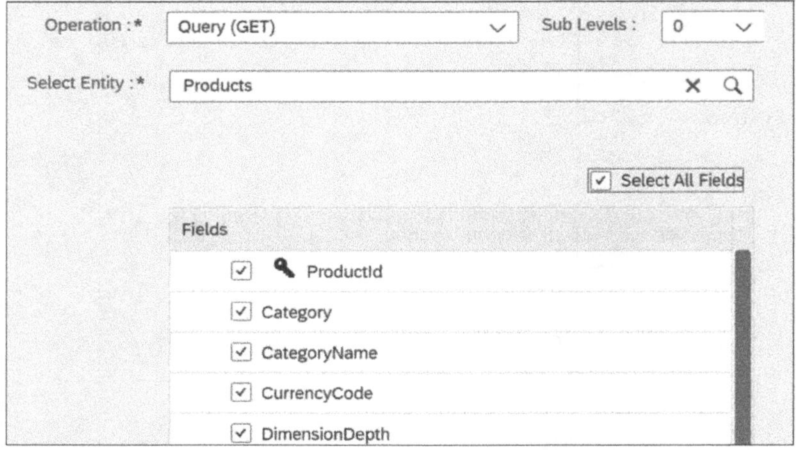

Figure 7.72 Select Entity Fields

18. Choose the **Select Field** icon and select **ProductId**. Click the **OK** button, shown in Figure 7.73.

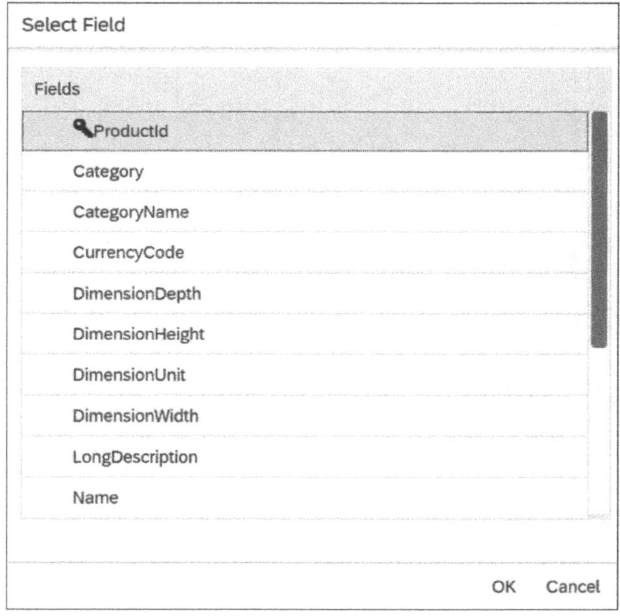

Figure 7.73 Select Product ID

19. You'll arrive at the product ID details screen. In the dropdown list, choose **Equal**. In the value field, enter "${header.productid}", as shown in Figure 7.74. Here, choose **ProductId** from the header.

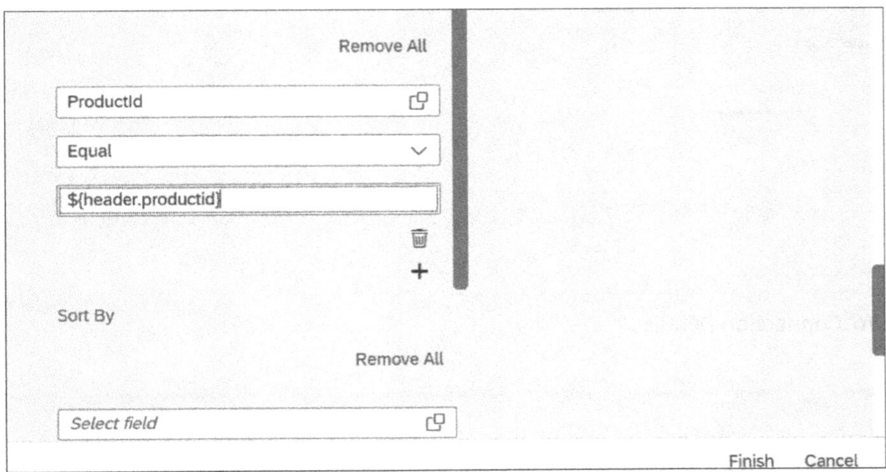

Figure 7.74 Product ID Details

20. Click **Finish**. Navigate back to the integration flow and connect the end event to the **Receiver**. In the **Adapter Type** popup, choose **OpenConnectors**, as shown in Figure 7.75.

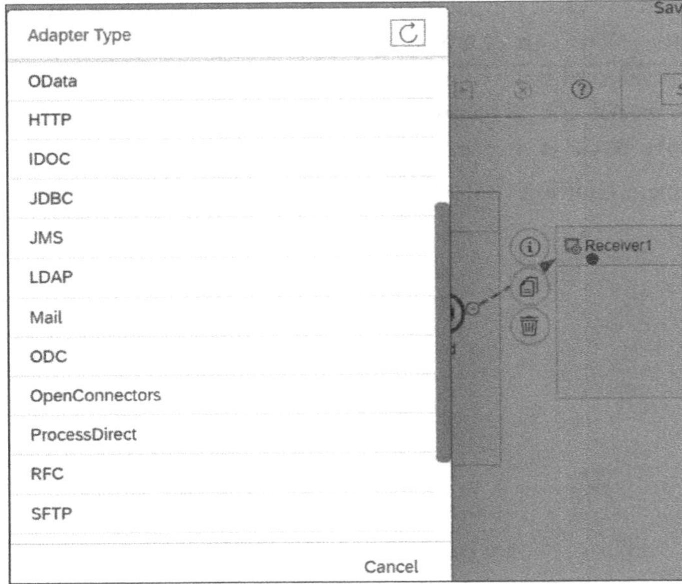

Figure 7.75 Open Connectors Adaptor

21. Select the receiver channel and click the **Connection** tab. In the **Base URI** field, enter the base URI you copied while setting up your Open Connectors instance in Section 7.1.2, as shown in Figure 7.76.

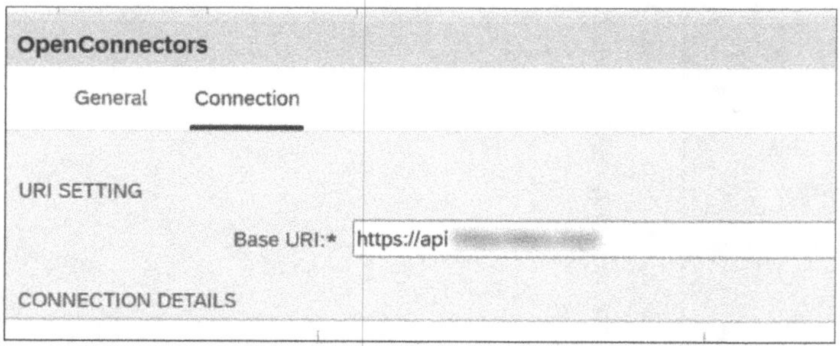

Figure 7.76 Connection Details

> **Note**
> If you missed copying these details, you can access your Open Connectors Slack instance and execute a ping test (see Section 7.1.2). You'll see the base URI in the results.

22. In the **Credential Name** field, click **Select**. In the dialog, select the credential name that you deployed before you started designing the integration flow (**Slack_OCN** in this example; refer to Section 7.4.2).
23. Click **Select** in the **Resource** field. In the list of resources, select **/channels/{channelId}/messages**. This is the API resource for posting a message in your Slack channel. Replace the `{channelId}` value with the channel ID of your slack channel. For example, if your Slack workspace URL is *https://app.slack.com/client/ACBDEFGHIJK/12345678901*, then "12345678901" is your channel ID.
24. In the **Method** dropdown, choose **POST** because you're posting the message, as shown in Figure 7.77.

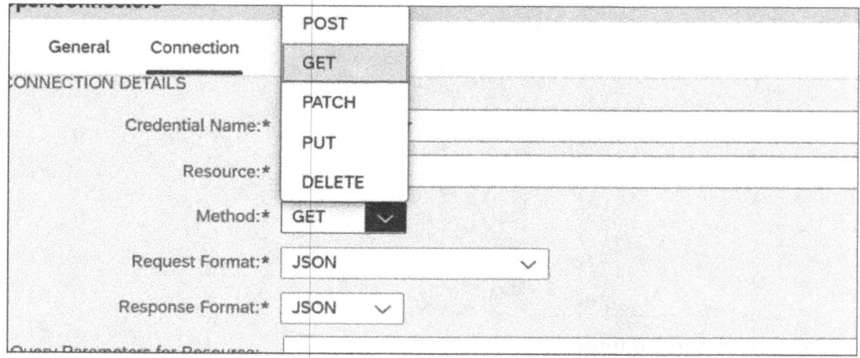

Figure 7.77 Post Method

25. Click **Save** and then click **Deploy**, as shown in Figure 7.78.

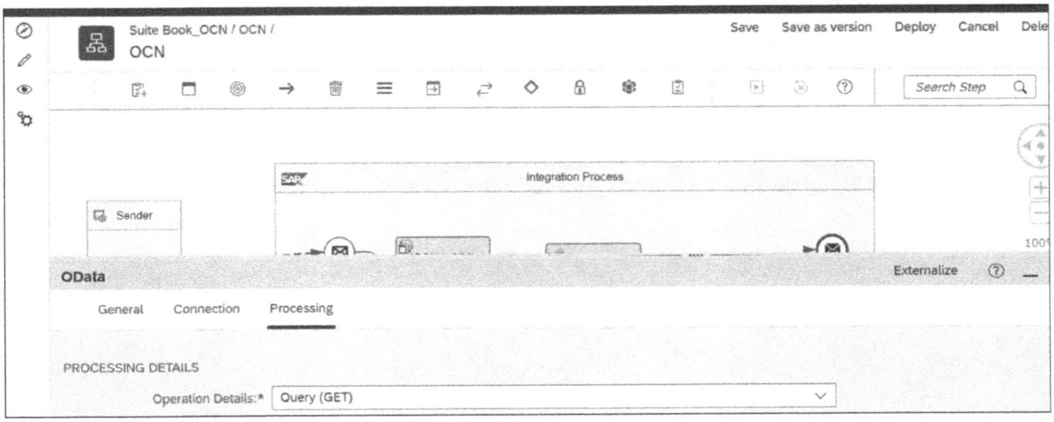

Figure 7.78 Deploy Integration Flow

26. Navigate to the **Operations** view to check the status. It should be **Started**, as shown in Figure 7.79. Copy the endpoint shown in the **Endpoints** tab for testing in the next section.

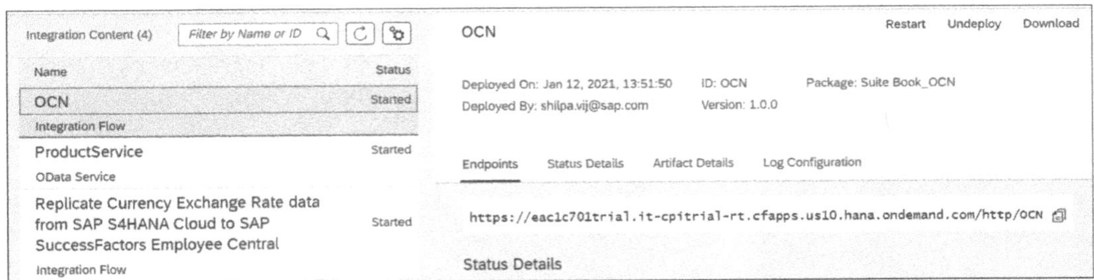

Figure 7.79 Operations View

7.4.4 Testing the Scenario

In this section, we'll test the scenario by creating a new product from Postman and monitoring how the message is processed in the Cloud Integration **Monitor** tab.

> **Note**
>
> Postman is a collaboration platform that simplifies API development and eliminates dependencies by reducing the time to production. It helps to simulate the expected behavior of API endpoints and their response. You can download the Postman REST client at *https://www.postman.com/product/rest-client/*.

Follow these steps to replicate product details from webshop to your Slack channel using Postman:

1. Open Postman and click the **+** icon.
2. Enter the copied integration flow endpoint in the top bar and choose the **POST** request, as shown in Figure 7.80.

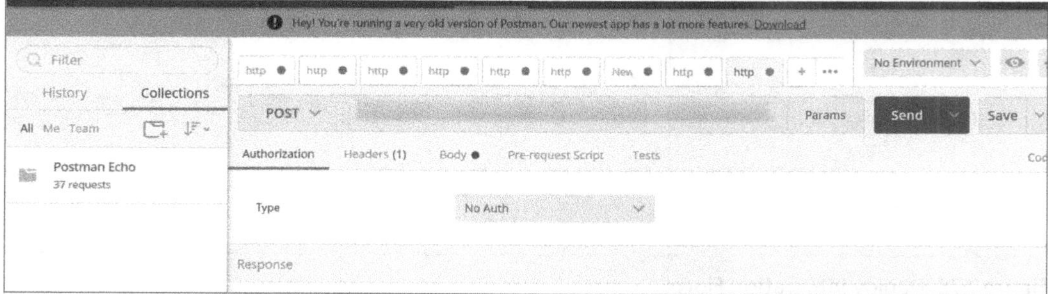

Figure 7.80 Test in Postman

3. Select the **Authorization** tab, as shown in Figure 7.81, and choose **Basic Auth** from the **Type** dropdown list. In the **Username** and **Password** fields, enter the client ID and client secret from your service key.

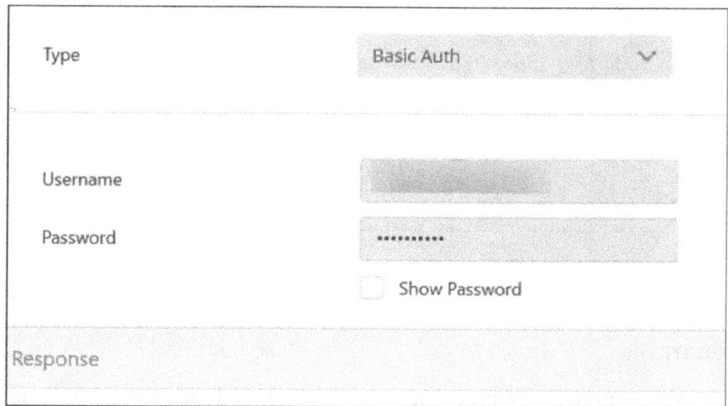

Figure 7.81 Authentication in Postman

4. Select **Body • Raw • JSON.** Enter the input as shown in Listing 7.2. You should see the details of your product posted in your slack channel.

```
{
"productIdentifier": "HT-2000"
}
```

Listing 7.2 JSON Input

> **Note**
> Feel free to change the product identifier based on the products available in the webshop.

7.5 Summary

In this chapter, you learned to simplify and accelerate connectivity to third-party cloud applications using the 160-plus open connectors offered by the Open Connectors capability of SAP Integration Suite. This capability focuses on harmonizing access to third-party cloud backends—REST, SOAP, proprietary SDKs, databases, and so on—by bringing API integration and management together with a data-centric approach.

You learned how to use formulas and resources as templates to unify the developer experience. We also focused on the significant value companies get from reimagining their businesses in the digital age via the out-of-the-box predelivered third-party content available on SAP API Business Hub. You learned to connect to the HubSpot CRM and a Slack instance and manage a HubSpot instance using the API Management capability of SAP Integration Suite. You also saw the tight integration between Cloud Integration and Open Connectors while sketching and developing an integration scenario in which you replicated product details from webshop to your Slack channel.

In the next chapter, we'll move on to event-based integrations using SAP Integration Suite and SAP Event Mesh. We'll also explain how SAP Integration Suite can act as both a publisher and consumer of events.

Chapter 8
Event-Driven Integration

Event-driven integration is becoming more popular as business use cases require real-time performance and more enterprise systems are supporting event-based concepts, facilitating a more asynchronous approach to system integration architectures. In this chapter, we'll cover a few key event-driven concepts.

In this chapter, we'll review SAP Event Mesh as it relates to event-driven integration frameworks. We'll start by reviewing SAP Event Mesh in relation to the event hub concept, followed by how to decouple integration scenarios using event patterns and event-driven extensions. We'll then move on to how SAP Event Mesh can be leveraged as part of SAP Integration Suite. Finally, we'll review how event-driven integration concepts can be leveraged with SAP S/4HANA and third-party applications.

8.1 SAP Event Mesh as an Event Hub

In this section, we'll begin our discussion of SAP Event Mesh with the asynchronous integration concept and how asynchronous communication via messaging compares to synchronous integration with application programming interfaces (APIs). We'll then provide an overview of event messaging scenarios, messaging protocols, and the centralized event bus before describing the general setup of SAP Event Mesh.

8.1.1 Asynchronous Integration

Traditionally, application-to-application (A2A) communications have been conducted in a synchronous manner. As a result of synchronous communications, when a significant change occurs in one system, a resulting change is made in a corresponding application. At scale, synchronous communication between two applications can represent quite a burden, resulting in performance issues and service degradation. For SAP users, SAP Gateway is a classic example of synchronous communication: connected applications needed to wait on each other to process updates, which had the effect of slowing down business processes.

Consequently, there has been consistent interest in asynchronous communication technology for the past two decades. Asynchronous integration is built around the concept that source systems can publish events to a central service bus, with an event being a significant change to a record within the source application. In turn, subscribing applications can be alerted to events taking place on a particular topic. The published event is typically extremely light in terms of information, providing only summary data that the subscribing systems can assess in order to make a decision about whether access to more information relating to this event is desired, at which point an API call can be made. Conversely, if the subscribing application isn't interested in a particular event, it can be ignored. Ultimately, this more efficient approach optimizes network traffic and lowers storage costs.

Figure 8.1 illustrates how an event message in the source system is sent to the event hub in response to a change in the source system. The event is subsequently consumed by the extension application, which can place an API call if the extension application is interested in the change event.

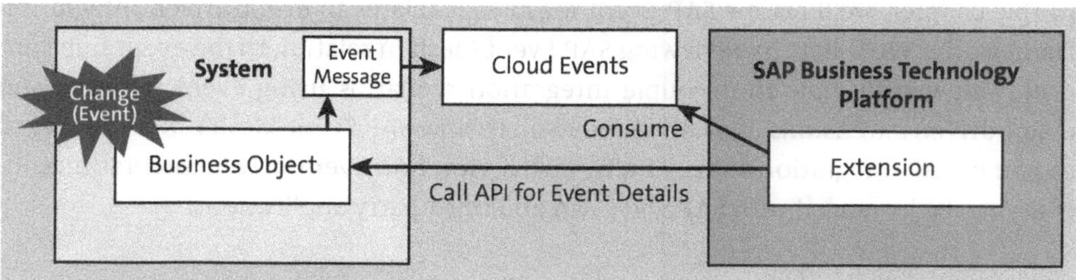

Figure 8.1 Event Messaging

Another potential benefit of publishing the event to a central event hub is that the event does not have to be responded to immediately (if the event is stored in a queue). Subscribing applications or system elements can access the event hub on a periodic basis to identify what events may have occurred for a given topic since the last check, and those events can be evaluated by the subscribing system to identify if they are of consequence. Subscribing applications or system elements can be located across global regions, cloud and on-premise landscapes, and internal and external systems.

Applications can vary significantly in their transactional frequency, with some applications being "chattier" than others. For example, a customer-facing application may provide updates on a continual basis, whereas backend applications may only have state changes on an ad hoc basis. Subscribing applications can therefore receive different numbers of events depending on how often they check in.

Asynchronous approaches to integration can also be beneficial for core application extensions. Extensions are typically leveraged by developers to provide custom functionality to a group of users. These extension applications can be a subset of functionality for one core application, or they may be presenting data to a user from multiple applications. The extension approach has the advantage of keeping the core application standard or "clean," which facilitates straightforward updates. With technologies such as the API Management capability, it also offloads some of the core applications' functional responsibilities. With the use of asynchronous updates, the subscribing extension can process which updates are relevant and thereby reduce the number of requests made by the extension to the underlying core application.

Now, let's consider an alternative: synchronous communication with APIs. APIs are extremely popular in today's enterprise IT landscapes. They are popular due to their relative simplicity and power. However, API popularity can sometimes lead to their overuse. The popularity of APIs in synchronous integration concepts such as database calls and request/response type application queries has trained developers to expect a fast response to their calls. Although this immediate validation can feel good in the long term, this single approach can lead to problems as end points increase and the number of calls being made grow in parallel (e.g., with the growing adoption of microservices in the composable enterprise model).

Synchronous communications require a tight coupling. When endpoints are added to a landscape, the associated code for communicating between systems needs updating. Robustness needs to be factored in; in terms of what happens when a called system is down, the requesting application needs to handle the error and try again. This can add a lot of orchestration and communication management. In terms of network efficiency, the requesting application has to wait on the sending application to send a reply. This ties up the requesting application while it waits for the response and prevents the requesting application from performing other tasks.

Asynchronous communication has the advantage of there being no need for knowledge of the associated applications. A single event can be consumed by multiple applications. There is no blocking for requesting applications while they wait for the sender to provide a response. It's therefore extremely simple to scale your landscape, with microservices growing or shrinking according to the immediate system needs. When outages occur, subscribing applications experience minimal impact as the message-based architecture can guarantee delivery and retry when the source application is back online and ready.

API gateways are very powerful, with capabilities such as load balancing, health checks, versioning and routing, authentication and authorization, data transformation, analytics, and more. API gateways provide a solid layer of defense and provide powerful capabilities such as low complexity and minimal latency. However, they can also be extremely centralized, with a single point of view for registering and changing APIs. This can lead to backlogs, delays, and overreliance. In contrast, message queues can easily support complex communication patterns within an organization with topic-based routing. Message queues provide buffering of task queues that facilitate processing over time. Message queues eliminate the need for health checks, routing, load balancing, and end point discovery. Message queues are ideal for auto-orchestration and scaling with highly efficient leveraging of resources.

8.1.2 Event Messaging Scenarios

In 2020, SAP released event-based technology with SAP Event Mesh that facilitates event-based communications for both SAP S/4HANA and SAP's legacy SAP ERP applications, meaning event-based approaches can be leveraged for customers with legacy SAP ERP and net new SAP S/4HANA implementations. A software development kit (SDK) is leveraged with SAP ERP to enable companies to take advantage of asynchronous event-based architecture for their legacy SAP ERP systems (SAP NetWeaver 7.31 and higher) in a similar fashion to SAP S/4HANA–sourced events. Other SAP applications with supported events include SAP SuccessFactors, SAP Subscription Billing, and SAP Marketing. SAP S/4HANA and SAP Marketing events can be found at *https://api.sap.com* (click **Events** under **CONTENT TYPES**), as shown in Figure 8.2.

The process for enabling applications events will vary. For SAP SuccessFactors, all intelligent services can be enabled as events. For SAP Customer Experience, all business objects can be event enabled.

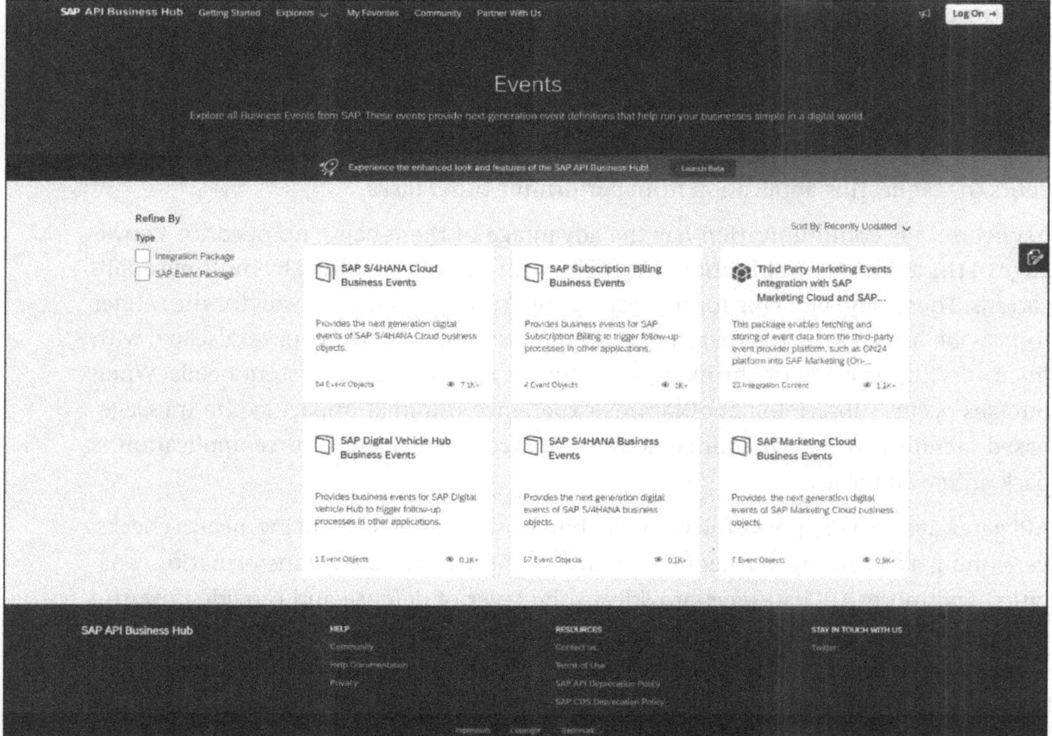

Figure 8.2 SAP API Business Hub, Cataloging Event-Based Integration Details

For these applications, SAP has made available a catalog of events, with on-premise SAP S/4HANA supporting more than sixty, SAP S/4HANA Cloud supporting over fifty, and so on. These SAP S/4HANA events cover changes to the master record, billing, business partners, cost centers, and more, relating to creation, cancellations, and changes.

A lot of internal enterprise applications and business processes can take advantage of events occurring within enterprise resource planning (ERP) and customer relationship management (CRM) systems as they often serve as the systems of record, so embracing an asynchronous approach to integration offers significant benefits for both the publishing and subscribing applications in terms of efficiency, security, and scalability (replacing legacy synchronous-based approaches).

As described previously, in most cases event messages are designed to be very light and standardized. The exceptions are application events that are data events, which include a lot of information—for example, for SAP SuccessFactors. Other applications, like SAP S/4HANA, allow you to choose which sorts of events you want published, including notification-only or data events.

For notification events that are lightweight, it's very easy for the subscribing systems to consume and process the events. This has the advantage of removing the barriers to cross-application integration with seamless data flows and the elimination of data silos. For example, a typical notification event message will only include the context model (ID, source type), a field catalog, and an optional attribute.

Listing 8.1 shows an example of a message payload. The event describes an outbound delivery change, and the document number is 00800005. The payload of the event only includes the type of business object (purchase order, sales order, etc.), key information, and the action (creation, change, or deletion).

```
{"eventType":"BO.OutboundDelivery.Changed",
"cloudEventsVersion":"1.0.1",
"source":"https://sap.corp",
"eventID":"QgEKigAJHuqnOlaStiBciA==",
"eventTime":"2021-01-25T14:19:12Z",
"schemaURL":"https://sap.corp/sap/opu/odata/IWXBE/BROWSER_SRV/",
"contentType":"application/json",
"data":{"KEY":[{"OUTBOUNDDELIVERY":"0080000005"}]}}
```

Listing 8.1 Event's Typical Message Payload

This message payload references cloudEventsVersion 1.0.1, which is the latest specification for the CloudEvents standard that was released in December of 2020. The CloudEvents specification is part of the Cloud Native Computing Foundation, which has been working with industry stakeholders to establish a common approach for describing events. The CloudEvents specification aims to provide a consistent way for developers to write event-handling logic into their code, which in turn will achieve a standard approach and simplify development efforts. By establishing a standard approach to CloudEvents, common libraries, tooling, and infrastructure will be possible. By leveraging a common event standard, it will be easier for developers to deliver data across environments (platforms, services, and beyond).

Finally, the use of a standard event specification will enable portability of information and aid developer productivity through common approaches to the usage of events. Version 1.0.1 is backward-compatible with version 1.0. SAP is part of the Cloud Native Computing Foundation, and as such SAP application teams are seeking to leverage the CloudEvents standard in association with application events.

Figure 8.3 illustrates how multiple elements of SAP Integration Suite (API Management, Open Connectors, and SAP Event Mesh) can be leveraged for bringing together data from SAP and Microsoft applications and technologies. You can see multiple Microsoft and SAP elements illustrated. First, end users leveraging SAP Extension Suite will typically interface with API Management to gain access to source data across SAP, Microsoft, and third-party applications. In this use case, the API Management capability provides an abstraction layer, interacting with other services of SAP Integration Suite, including Cloud Integration, Open Connectors, SAP HANA Cloud, and SAP Event Mesh.

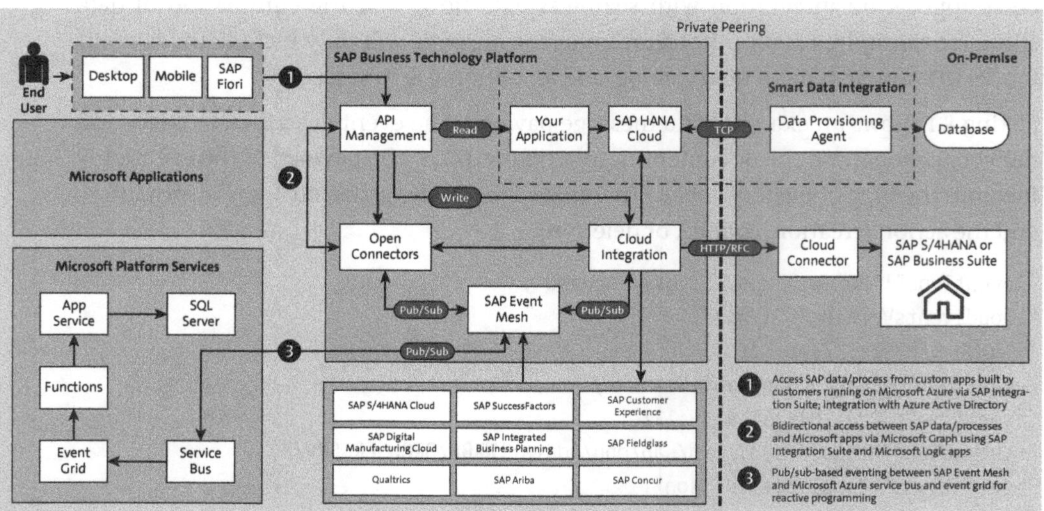

Figure 8.3 Microsoft Azure Platform Accessing SAP Application Data

In this scenario ❶, SAP Event Mesh is shown supporting events for SAP, Microsoft, and third-party applications. The second use case depicted ❷ shows Microsoft Applications interfacing with both API Management and Open Connectors, in turn connecting to SAP Event Mesh, which enables users of Microsoft applications to access data generated by SAP Event Mesh events within their core user experience. The third scenario depicted ❸ is a direct link between Microsoft Azure platform services and SAP Event Mesh, where SAP Event Mesh is able to notify of and receive events from Microsoft

Azure platform services. SAP Event Mesh is therefore extremely versatile and flexible, supporting your event needs across the landscape. As we noted earlier, sometimes asynchronous integration will be preferred over synchronous integration and vice versa depending on the underlying use case.

8.1.3 Messaging Protocols

SAP Event Mesh supports the following standard messaging protocols:

- AMQP v 0.9.1 and AMQP v 1.0.0
- MQTT v 3.1.11
- HTTP 1.1 REST

The Advanced Message Queuing Protocol (AMQP) is an open standard application layer protocol. AMQP provides a powerful, rich set of messaging scenarios, with an explicit design for extensibility (forward-compatible capabilities), but it's more costly in terms of messaging overhead. AMQP supports concepts such as sandboxes, multitenants, and multiple user security schemes. The defining features of AMQP are message orientation, queuing, routing (including point-to-point and publish-and-subscribe), reliability, and security. AMQP is the asynchronous complement to HTTP.

Message Queuing Telemetry Transport (MQTT) is an ISO standard (ISO/IEC PRF 20922), publish-and-subscribe-based messaging protocol. MQTT has low overhead and is a simple way to implement and send data, especially from embedded devices. The MQTT protocol handles retransmission and ensures that an *event* (a specific type of message) is delivered regardless of network reliability. In terms of quality of service tied to delivery, MQTT supports *at most once* and *at least once* scenarios:

- **Quality of service 0 (at most once)**
  Messages are not acknowledged by the receiver or stored and redelivered by the sender. This is often referred to as fire and forget and is used to transmit noncritical data.
- **Quality of service 1 (at least once)**
  Messages are delivered at least once to the receiver. Consequently, a receiver might receive the same message multiple times.

Because MQTT is constrained in supporting basic messaging concepts, it's not a good fit for many enterprise use cases—for example, use cases featuring multitenant servers or for supporting dynamic migration from development to production. MQTT's handling of security and associated users for resource sandboxing is also limited. Table 8.1 highlights the major differences between AMQP and MQTT messaging standards.

Feature	AMQP	MQTT
Scope	Bigger range of scenarios—peer to peer, load balancing, etc.	Simple queuing, no advanced features, typically used in embedded use cases
Background	Originally from the finance community; open source; improves over time with no customization	Vendor-driven (IBM); tends to have high implementation costs
Design	TCP for asynchronous transfer independent of operating system, hardware, and programming language	TCP as well, designed for small devices with small memory and low bandwidth
Control	Varied control of network users and infrastructure resources	Parties are managed by private infrastructure
Framework	Supports buffering, which improves server performance	Supports streaming for devices with minimal memory; bulk transfer is not supported
Message services	Supports message lifetime, remains in queue until received; supports concepts such as round-robin, classic queue, save, and forward	Supports publish/subscribe models; transient, active routing; and limited extended queuing
Transmission of message	Different acknowledgement support; supports expiry concept	No transmission support; only general acknowledgement is supported
Security	TLS/SASL unified to use connection (or can be eliminated)	MQTT doesn't react to security issues
User security	Authentication with user before broadcasting	Needs small username and minimal password
Last value	Not supported in queue	Provides retain command; supports last value
Reliability	Only fire and forget; once received cannot be retrieved	Similar to AMQP

Table 8.1 AMQP and MQTT Comparison

Finally, the HTTP 1.1 REST protocol exposes messaging capabilities using REST APIs, and messages can be published to and consumed from queues and topics.

8.1.4 Centralized Event Bus

The business event bus concept is a key component of SAP's intelligent enterprise suite, providing a mechanism for communicating events between applications, services, and systems. As Figure 8.4 shows, a single event bus can support all cloud and on-premise applications across physical locations and technology providers. Because events are distributed globally but are only consumed locally as needed, this approach brings greater efficiency to the enterprise landscape while optimizing network and storage costs.

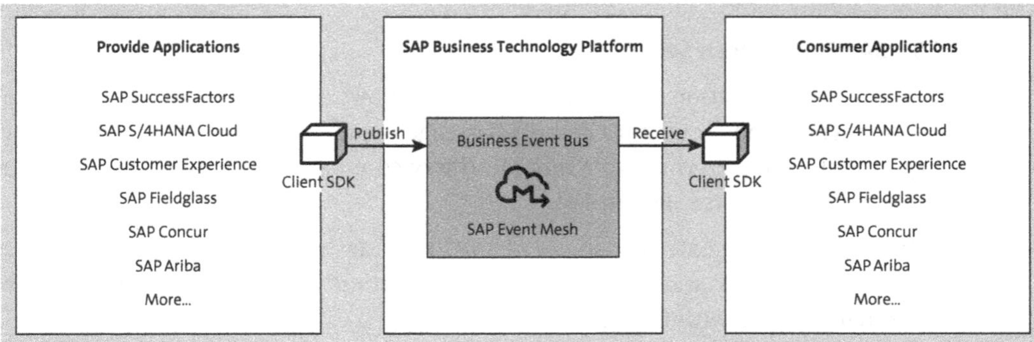

Figure 8.4 Centralized Event Bus

The SAP service that is applicable to an event hub framework is SAP Event Mesh. SAP Event Mesh is a fully managed cloud service that is part of SAP Integration Suite and operates within a Cloud Foundry environment, which facilitates the use of the service within different hyperscaler offerings.

In terms of SAP Integration Suite, asynchronous integration elements can be leveraged in the context of other integration components and flows. Figure 8.5 shows how an event from SAP SuccessFactors (SAP's HR solution) can be published to the event hub and SAP's Cloud Integration process integration tool can initiate associated tasks.

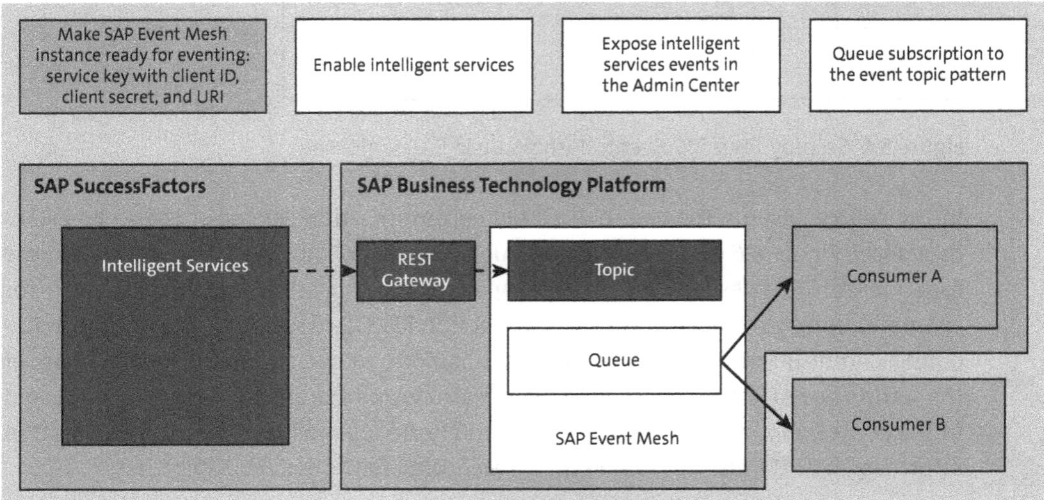

Figure 8.5 Human Resources Event Scenario

A single event (new employee being hired) can trigger multiple related tasks and steps being taken. For example, a welcome email can be sent, an employee can be added to appropriate communication tools, and related tasks can be initiated in additional HR applications. We'll cover this scenario in more detail in Section 8.4.

8.1.5 SAP Event Mesh Setup

As part of SAP Integration Suite, SAP Event Mesh allows applications to communicate through asynchronous events as described in this chapter. So, let's briefly describe how you go about setting up your SAP Event Mesh service as part of SAP Integration Suite to serve as an event bus (see Figure 8.6).

The first step ❶ is to create an instance of SAP Event Mesh, which in turn generates the service key, client ID, and secret, which are important details when setting up the secure channel from within the source application.

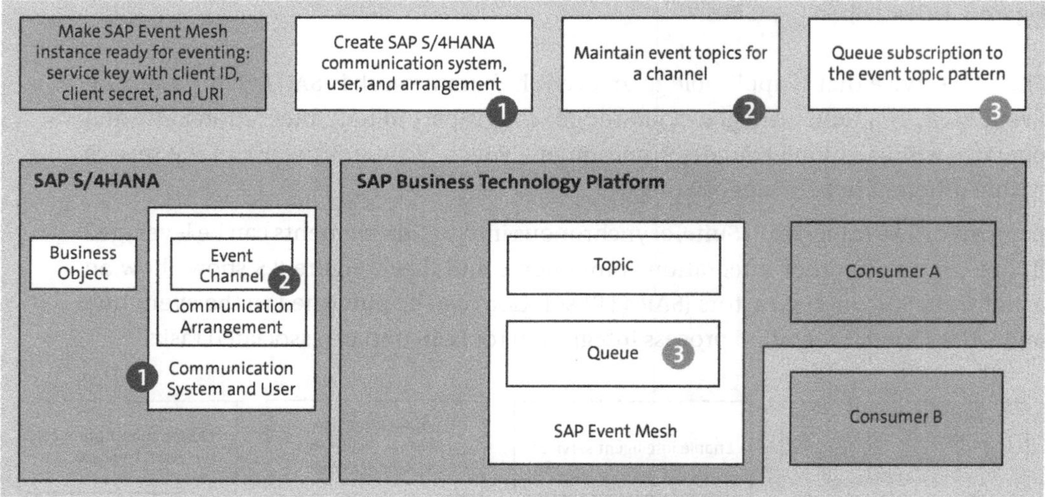

Figure 8.6 Step-by-Step SAP Event Mesh Setup in SAP S/4HANA

In the source system, the channel-related communication settings need to be established (see Figure 8.6 ❷). In the case of SAP S/4HANA, as shown in Figure 8.7, a remote function call (RFC) destination can be setup with authentication credentials (OAuth, for example, along with role definition [Transaction PFCG]) to facilitate inbound and outbound communications, confirming which supported events are enabled. As part of the setup, the SAP Event Mesh service key is also imported. Finally, a channel is created in the source application, which includes the protocol, quality of service, destination information (URL), and associated topics according to the desired topic filters.

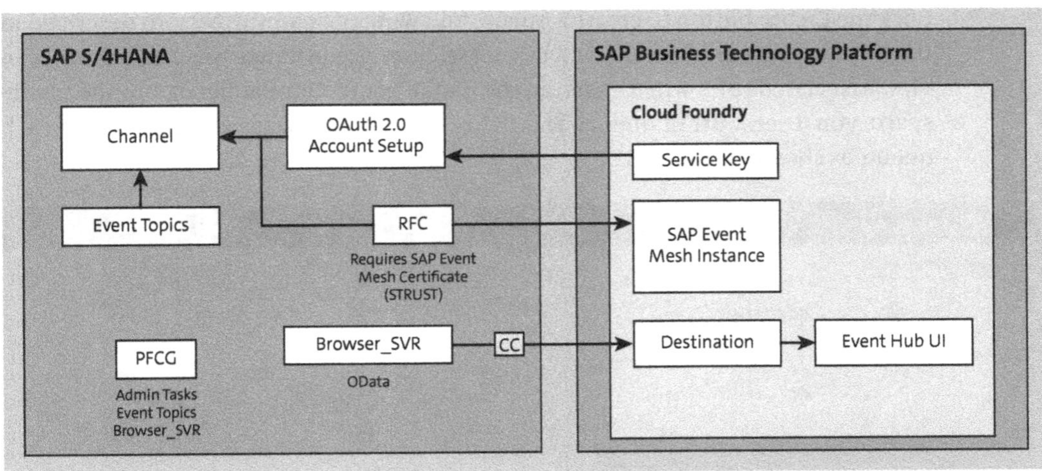

Figure 8.7 SAP S/4HANA On-Premise Setup with SAP Event Mesh

The SAP Event Mesh instance can then subscribe to the topic and associate a queue for subscribing applications to subscribe to (refer back to Figure 8.6 ❸). You can subscribe to a queue in SAP Event Mesh in three easy steps:

1. Open the SAP Business Technology Platform (SAP BTP) cockpit, click **Subscriptions**, and open SAP Event Mesh. Then click the message client, which opens an **Overview** tab, as shown in Figure 8.8.

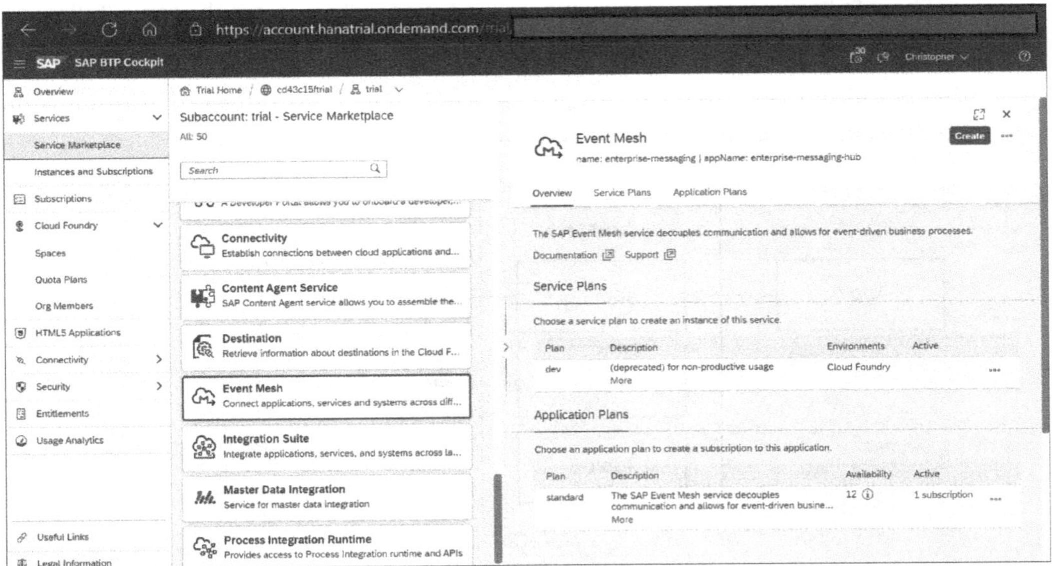

Figure 8.8 SAP Event Mesh Overview

2. Click the **Create** button to create a queue, following the naming pattern described in the JSON descriptor when the SAP Event Mesh service instance was created. You can view associated rules when entering the queue name. Once appended to the namespace, you'll see a list of queues and the number of messages associated with each queue, as shown in Figure 8.9.

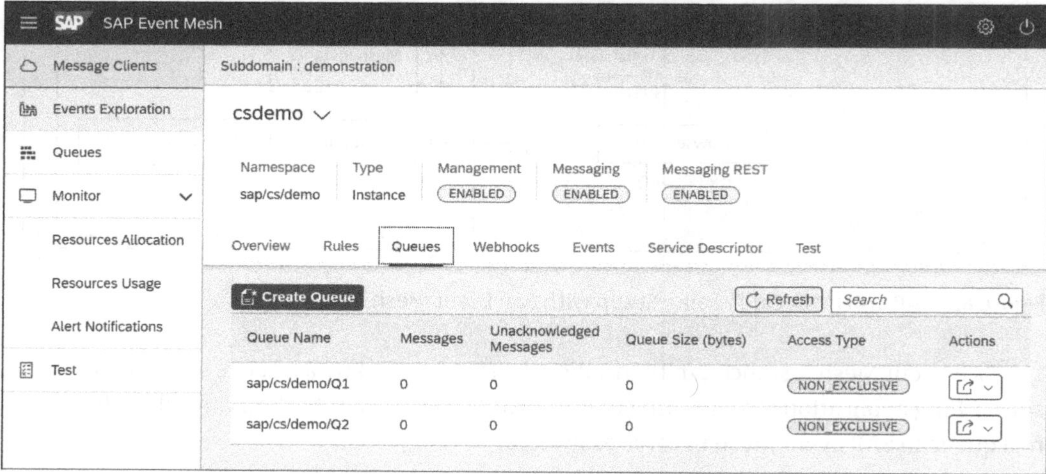

Figure 8.9 Create Queue

3. Create a relevant topic—for example, "SAP S/4HANA Sales Order". To do this, click **Show Subscriptions** for this queue under **Actions**. Then create the topic (following the rules defined in the service descriptor).

Figure 8.10 shows topics working together in SAP Event Mesh, with events flowing from application A to applications B and C via a topic.

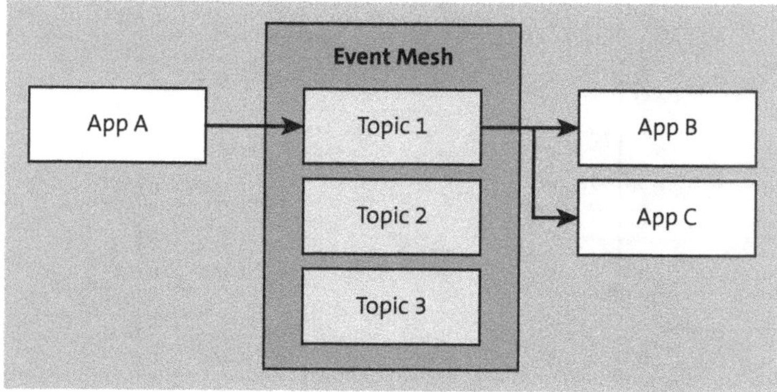

Figure 8.10 SAP Event Mesh Topics

As your organization's usage of event-based integration grows, it will be important to catalog the availability of application events through a catalog and portal. Figure 8.11

depicts the latest API Management developer portal, called API Business Hub Enterprise, which provides a mechanism for publishing SAP and third-party APIs in a common hub. You can check the SAP roadmap for the latest information regarding the ability to publish events, business objects, notifications, and actions.

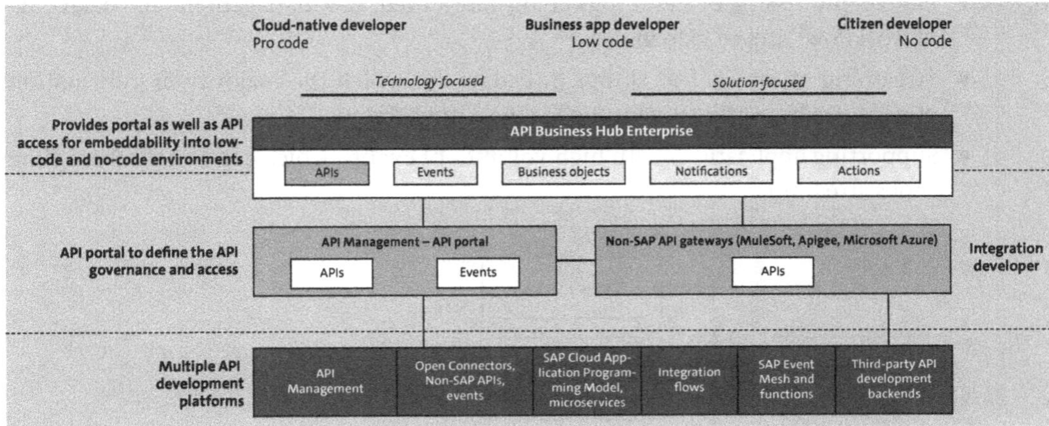

Figure 8.11 API Business Hub Enterprise

8.2 Decoupling Integration Scenarios Using Event Patterns

In terms of event patterns, there are three main types: publish-subscribe, point-to-point, and request-response. We'll discuss each in this section.

The *publish-subscribe* structure, shown in Figure 8.12, supports the asynchronous event-based architecture we have been discussing so far in this chapter. In the publish-subscribe event pattern, an agent application or system publishes an update to the event broker. There can be one or many subscribing applications or systems that are notified of an event. The event can therefore be processed multiple times, and each recipient receives their own copy of the event.

Under the publish-subscribe event pattern, another subpattern is *message replay*, in which the subscribing application or system can request the event broker to resend the event.

There are a lot of common integration scenarios that can be supported by this event pattern:

- Supporting applications or systems that need to update many others, thereby avoiding the burden of an application to maintain a recipient list
- Providing a simpler approach across applications that consume the updates at different rates or periods
- Supporting serverless concepts such as functions as a service, in which the need for dynamic resource allocation is present

- For generally avoiding the burden of different systems polling each other for updates, which in turn improves response times and reduces latency
- Allowing for applications to be developed and scaled separately, with no ripple effects when changes are made
- Parallel processing of alerts, logs being distributed, and management and balancing of workflows and workloads
- Streaming of Internet of Things (IoT) data, for which the majority of information may be irrelevant for consuming business applications
- Supporting applications with high volumes of events, which can benefit from time series streaming

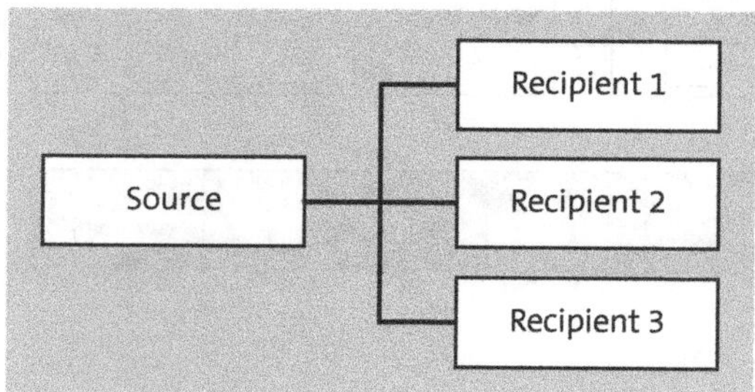

Figure 8.12 Publish-Subscribe Event Pattern

The second event pattern is *point-to-point*, as shown in Figure 8.13. This type of event pattern facilitates traditional synchronous communication between applications and systems. In this event pattern, events are sent to a single receiver. The message delivery is typically guaranteed.

Figure 8.13 Point-to-Point Event Pattern

A subset of the point-to-point event pattern is a *nonexclusive arrangement*, in which consuming applications or systems are grouped together, as shown in Figure 8.14. The events are sent over a single channel to a single end point. The consuming applications therefore share the channel and the arriving events. One of the common integration scenarios that can be supported by this event pattern is applications that need to be tightly coupled—for example, one application or system that needs to consistently map features from another source. Master data integration is a commonly cited use case by SAP customers for point-to-point event patterns. Another example is a stateless application that doesn't save or persist information, relying on a source system for data. This enables the stateless application to scale consumption, with inherent queuing of messages.

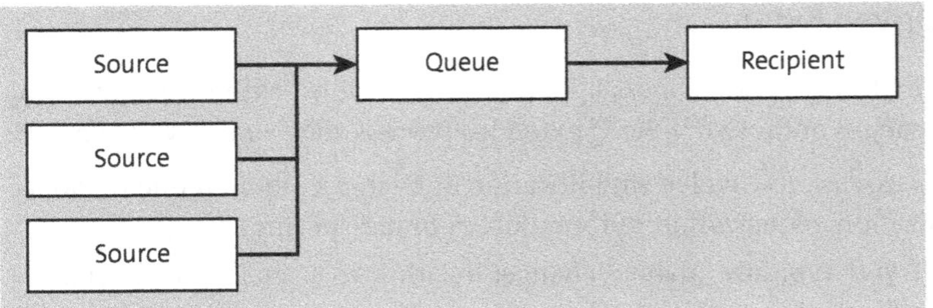

Figure 8.14 Point-to-Point Message-Based Queuing

The third event pattern is *request-response*, as shown in Figure 8.15. This type of event patten involves two-way communication between two applications or systems, with separate channels for requests on one and a second dedicated to replies on the other. Message delivery is therefore typically guaranteed. A typical integration scenario in which request-response is needed is when an interaction with an external system is required prior to an operation being completed. A common example is rendering information for an end user that is sourced from multiple applications or systems—for example, determining which training courses and what content are available to an employee based on their profile when the training information is maintained in a separate application from the user interface the employee is using. The request-response event pattern could therefore involve read or write type actions to the source system (see the available training courses, make changes to the training courses that are available).

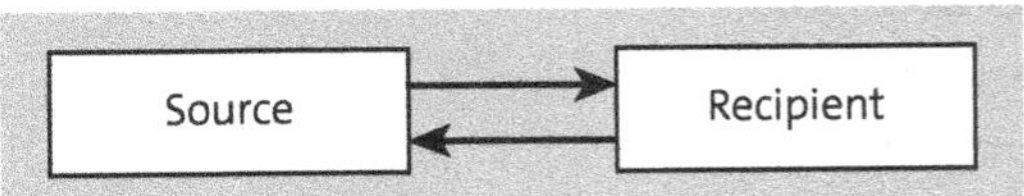

Figure 8.15 Request-Response Event Pattern

A direct request-response communication profile is fast, simple, and has a small number of failure responses. However, a direct communication profile has poor error tracing and can benefit from being leveraged within the context of SAP Integration Suite. SAP Integration Suite can provide support in a number of beneficial ways.

First, SAP Integration Suite can provide a mediation role, by which event messages can be modified or enriched in terms of both request and response messages. SAP Integration Suite also can provide a system-monitoring capability, checking for problems in the event streams with event logging, error management, and credential management. SAP Integration Suite is also able to provide a centralized caching mechanism to help improve the reliability of communication between the two applications. We'll go into more detail on these capabilities in Section 8.4.

8.3 Event-Driven Extensions

There are four main types of business application extensions, extensions involving user interface changes or those involving extended process, data, or ecosystem needs:

1. *User interface extensions* involve simplification or changes required due to globalization/localization, mobilization, automation, or brand-specific changes.
2. *Process extensions* typically involve changes relating to user action, workflows, checkpoints, custom processes, automation, or output summaries.
3. *Data-driven extensions* typically enhance data models, enable self-service, combine data sources, process predefined jobs, and provide insights based on predefined algorithms and predictions that discover and simulate trends, thus facilitating analysis by consolidating data from multiple sources.
4. *Ecosystem extensions* help extend a central software as a service (SaaS) application for usage by multiple customers and/or partners.

Figure 8.16 summarizes how these different types of extensions align with each other in a side-by-side extensibility framework.

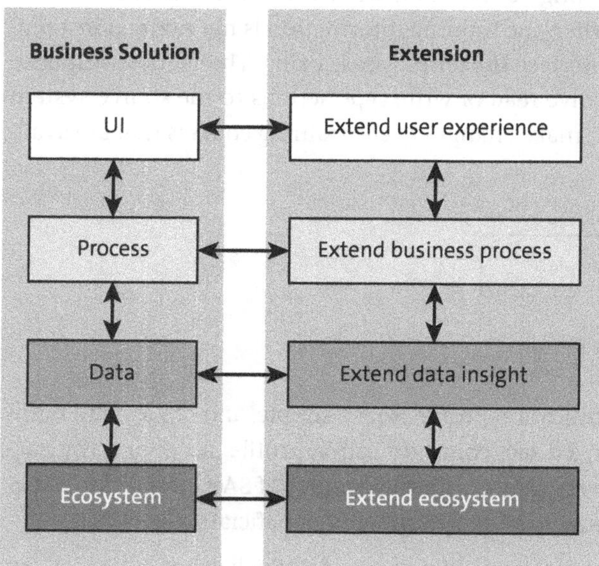

Figure 8.16 Extension Types

Before we go further, let's quickly discuss why extension types are relevant to event-driven integration. Event-driven integration can help drive a decoupled extension approach within an organization. In other words, event-driven integration enables an asynchronous approach that benefits the core application in a number of ways:

- It simplifies the development of the extension by reducing the number of source-application considerations, which simplifies ongoing development and maintenance of the extension.
- It reduces the impact on the source application, keeping the source application standard (or close to it), which in turns simplifies ongoing maintenance and application upgrades.
- A decoupled extension is cleaner for leveraging additional elements such as intelligent technologies (IoT, artificial intelligence [AI], etc.) within the extension's functional scope.
- The agility and flexibility to make changes reduce the total cost of development, which in turn enables the development of extensions to serve stakeholders and audiences that would be become unwieldy in an in-app approach.

Extensions are typically viewed by the business as required to enable a competitive advantage or provide the business with missing capabilities via custom specific innovations of business processes. Extensions provide a competitive advantage by providing insights, intelligent automation, and/or industry-specific functionality.

IT typically approaches extensions through the lens of pursuing a decoupled architecture, keeping the core of major applications clean or close to standard, which simplifies ongoing upgrades and maintenance while enabling rapid iteration and optimization of extensions through a bimodal approach.

The loosely coupled approach associated with extensions contrasts with the traditional or legacy approach of application customizations (tightly coupled application changes that restrict and complicate application upgrades and maintenance). Figure 8.17 highlights which extension scenarios are compatible with a tightly coupled approach and which ones are not in the context of SAP S/4HANA.

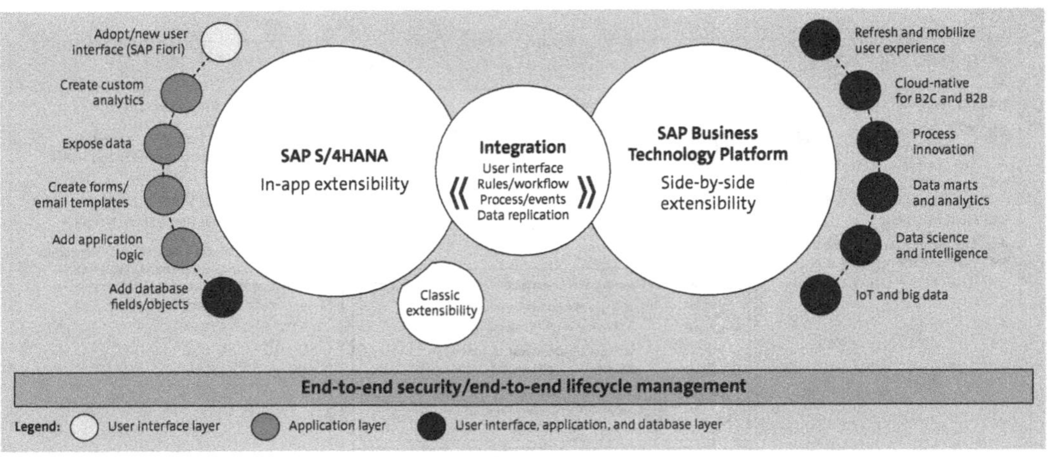

Figure 8.17 Tightly versus Loosely Coupled Extensions

Event-driven extensions are a big part of cloud-native/serverless concepts that leverage functions and application extensions to bring dynamic resourcing into play. With cloud-based approaches, cloud providers take on responsibilities associated with scalability (no usage; no resources; pay for what you need). Functions are invoked that call on database resources, call other applications for data, and make updates to other applications as needed in response to events.

These dynamic computing resources are hidden to the developer that calls the function, so this simplifies the development and maintenance of the code calling the function. Typically, functions are temporary in nature, so the associated resources that are spun up and utilized go back to sleep after use. This temporary nature reduces operating costs and facilitates a pay-per-use type of model.

Functions are atomic, stateless pieces of custom code (commonly written in Node.js) invoked on-demand by *triggers* (e.g., events), which define how the function code implementation will be invoked and the output returned. Function-as-a-service characteristics include the following:

- Enablement of autoscaling in a serverless environment
- Can be triggered by events (event-driven)
- Facilitate usage-based billing (pay-per-use)
- Ensure efficient resource utilization

Extensions can be built in different languages, providing organizations with development flexibility. Extensions no longer burden source systems.

Revisiting the setup of SAP Event Mesh with SAP S/4HANA that we discussed in Section 8.1.5, Figure 8.18 summarizes the steps associated with setting up SAP Extension Suite in association with the SAP Event Mesh event broker configured to receive events from SAP S/4HANA.

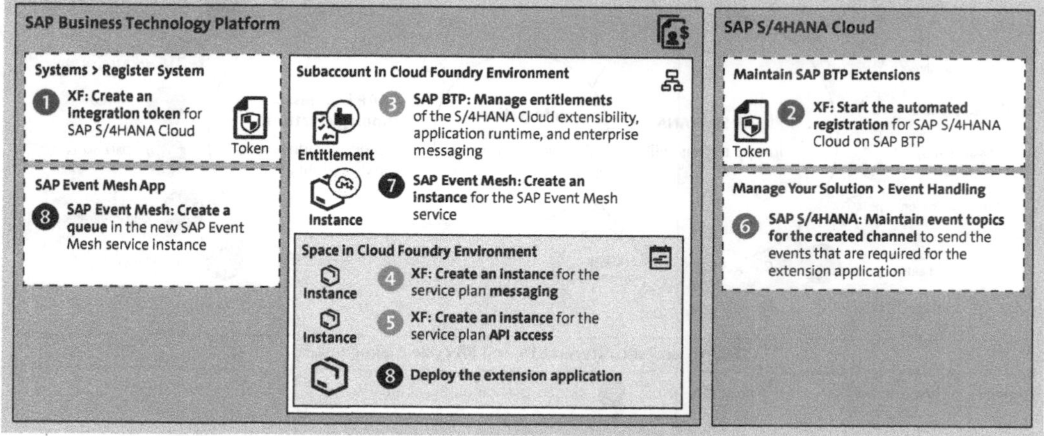

Figure 8.18 SAP BTP-Based Extension with SAP S/4HANA Cloud Example

Figure 8.18 highlights a side-by-side SAP BTP-based extension. In this case, SAP Extension Suite leverages a serverless runtime approach to use functions. SAP BTP extensions can support loose coupling extension formats, built around event-driven frameworks with core applications, as well as tightly integrated based on process-based integrations powered by SAP Integration Suite. SAP BTP's SAP Extension Suite provides templates to make things easier for creating a function-based extension.

Figure 8.19 highlights how extensions can be built leveraging SAP Extension Suite concepts or extensions can be built leveraging alternative approaches, such as an ABAP-based runtime or another application runtime type (e.g., Java, Python, or Node.js). Contrasting with the SAP Extension Suite dynamic approach, these other runtimes tend to take a reserved and dedicated resources approach. A dedicated and reserved approach requires consistent use and budgeting of resources. As mentioned previously, a functions-based approach facilitates a usage-based and pay-per-use approach to extensions, which can often prove financially attractive to a development team.

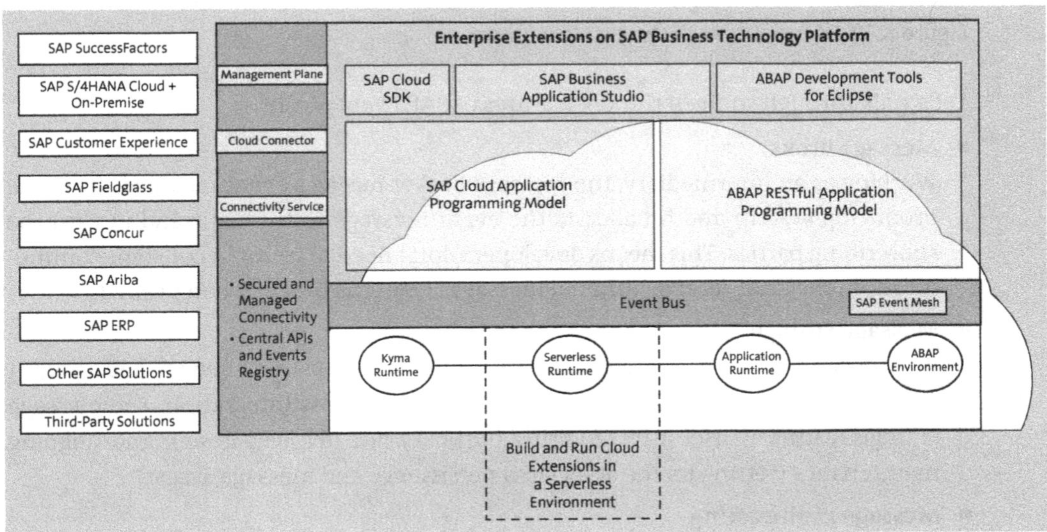

Figure 8.19 Building Enterprise Extensions Architecture

Event-based extensions don't have to leverage functions, with standard extensions able to consume and respond to events as they come in through an event bus. It's also common for extensions to be event producers in their own right; the extension just needs to be capable of issuing a rest message.

8.4 SAP Integration Suite with SAP Event Mesh

As discussed so far in this chapter, there are a lot of benefits to consider with regard to the capabilities of SAP Event Mesh. Next, let's discuss how SAP Event Mesh is being leveraged to enhance SAP Integration Suite.

SAP Event Mesh can help SAP Integration Suite users in a variety of ways. First, SAP Event Mesh represents a great way to persist data. SAP Event Mesh makes it easy to temporarily persist messages in queues until they have been delivered to the receiving clients, as shown in Figure 8.20.

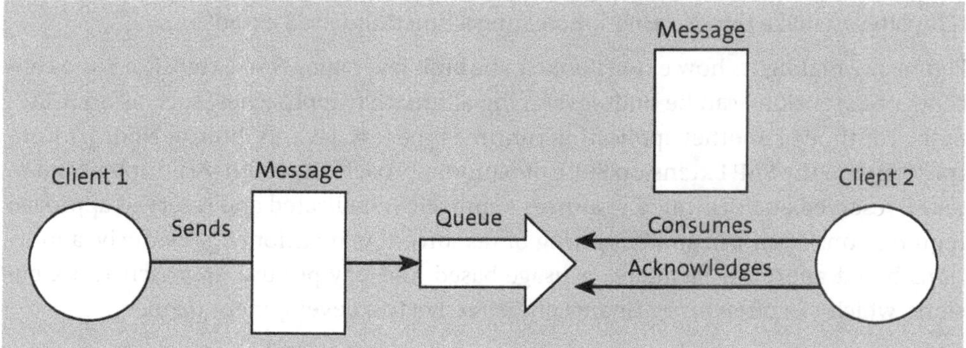

Figure 8.20 SAP Event Mesh Queue Mechanism

Let's walk through some of the key features of SAP Event Mesh:

- **Message broker**
  Working as an intermediary, the message broker receives event messages from the producing system and broadcasts the event message to the topic and in turn the subscribing parties. This means developers don't need to factor in constant communication; they can receive buffered messages from the broker when needed.
- **Message storage**
  By contrast, with an SAP Integration Suite integration flow, you need to add a persist step to store a message at a specific point in the process integration. The message storage feature is useful for auditing purposes but involves design and ongoing maintenance (setting up the associated persistence and message access).
- **Message multicasting**
  Another area where SAP Event Mesh improves SAP Integration Suite is in its ability to distribute messages to multiple recipients using SAP Event Mesh topics. By contrast, when using SAP Integration Suite in such a scenario, you need to setup a multicast (either parallel or sequential). With sequential multicast, you can perform multiple operations on the same message in a single integration process.
- **Message management**
  When building multiple receivers in SAP Integration Suite, you'll need to complete a bespoke development that covers different scenarios. What happens when a message is successfully delivered to a first receiver but fails for a second receiver? Does the integration flow need to restart to deliver the message to the second receiver? What happens to the first receiver in terms of duplication? You could do workarounds to build separate integration flows for different receiving systems, but this requires

ongoing maintenance and involves system load. It's therefore much better to leverage SAP Event Mesh topics and queues that make ongoing management easier.

Leveraging SAP Integration Suite and SAP Event Mesh together can lead to some enhanced capabilities. For example, SAP Integration Suite through its capabilities can enrich a message received through SAP Event Mesh before returning it for distribution to additional subscribers. SAP Integration Suite can also provide powerful authentication capabilities, which can be leveraged to request additional information for information received within an event by SAP Event Mesh.

SAP Event Mesh has also been formally combined within SAP Integration Suite to provide Java Message Service (JMS), Applicability Statement 2 (AS2), and XI support. JMS provides a generic messaging model built around a producer-consumer setup between systems. JMS supports both point-to-point and publish-subscribe message models. AS2 is a data transport specification used in business-to-business (B2B) communications. With AS2, files are encoded as attachments in a standardized format (such as XML or EDIFACT). AS2 provides a standard way for communicating between entities with the use of certificates and specific names before the data transfer is possible.

Figure 8.21 shows an SAP Integration Suite integration flow relating to SAP SuccessFactors and the use of AS2 with a third-party human resources-specific application.

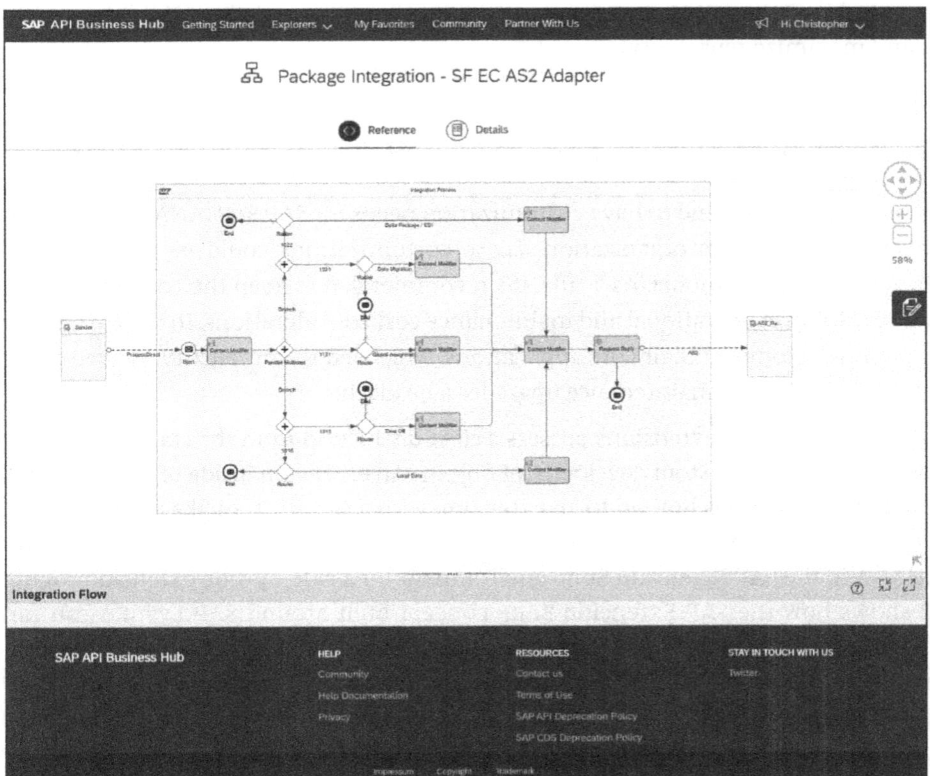

Figure 8.21 SAP Integration Suite Integration Reference for SAP SuccessFactors and HR Application Leveraging AS2 (Payroll)

The XI adapter allows customers to connect their on-premise system to SAP Integration Suite leveraging the XI 3.0 protocol. The XI adapter includes both sender and receiver capabilities.

With the JMS adapter (sender or receiver), a new or existing SAP Integration Suite queue can be created when an integration flow is deployed. Similarly, an integration message queue can be created when an integration flow is built to leverage the AS2 and XI adapters.

The sender and receiver message processing can be decoupled to ensure that a retry is done from the integration system rather than the sender system. This can be achieved in this case with the JMS queue feature temporarily storing the messages. If the receiver system can't be reached, a retry can be attempted. To decouple inbound and outbound message processing, two processes are configured. One process receives the inbound message and stores it in the JMS queue; the second process triggers the message from the JMS queue to the receiver backend.

It's likely that an organization will be seeking to solve multiple integration scenarios simultaneously. As a result, some integration use cases may leverage SAP Integration Suite by itself. Where synchronous integration is required, this could be between SAP and third-party applications or cloud and on-premise applications. For other scenarios, asynchronous integration may be desired, and SAP Integration Suite along with SAP Event Mesh and the API Management capability will be desirable (to reduce dependencies and maximize reusability).

8.5 Event-Driven Integration of SAP S/4HANA with Non-SAP Apps

Typically, customers tend to have customization needs for SAP S/4HANA to serve their individual needs as an organization. These customizations could be specific to an industry or business model. As a rule, it's recommended to keep the core application clean to minimize operational and maintenance cost considerations. In the past, many companies customized their core applications, which led to complexity, slowed down operations, and made maintenance upgrades a headache.

In-app or side-by-side extensions possess a clear distinction from the standard applications and associated custom development objects. An extension inside of SAP S/4HANA (in-app extensibility) is limited to an extension with a specific tool like a *key user tool*. It's therefore recommended to place custom development objects outside of SAP S/4HANA in the SAP Extension Suite environment for a side-by-side extension. Figure 8.22 shows how the SAP Extension Suite concept built around SAP Event Mesh can serve as a central repository for application extensions, with various source applications registered and connected. You can see how the extension is built leveraging SAP BTP's SAP Extension Suite. SAP Extension Suite provides a comprehensive and flexible approach to building extensions. The extension could be built to provide functionality specific to an industry or role, to share data outside an organization, or to provide a user with information stemming from multiple applications. In Figure 8.22, an example is shown in which application data is coming from multiple SAP and third-party applications. An example of this scenario could be an employee-facing application in which the employee is fed a 360-degree view of their customer, with sales and order information flowing from SAP Customer Experience and SAP S/4HANA respectively. This application data could in turn be combined with third-party specialist applications such as ServiceNow, Adobe Creative Cloud, and DocuSign to further enhance the depth and breadth of the customer status being presented to the employee. In terms of SAP Event Mesh, corresponding events from the source applications (leveraging associated application APIs) can be managed asynchronously for the employee to receive the latest and greatest view of a customer's interactions with the business.

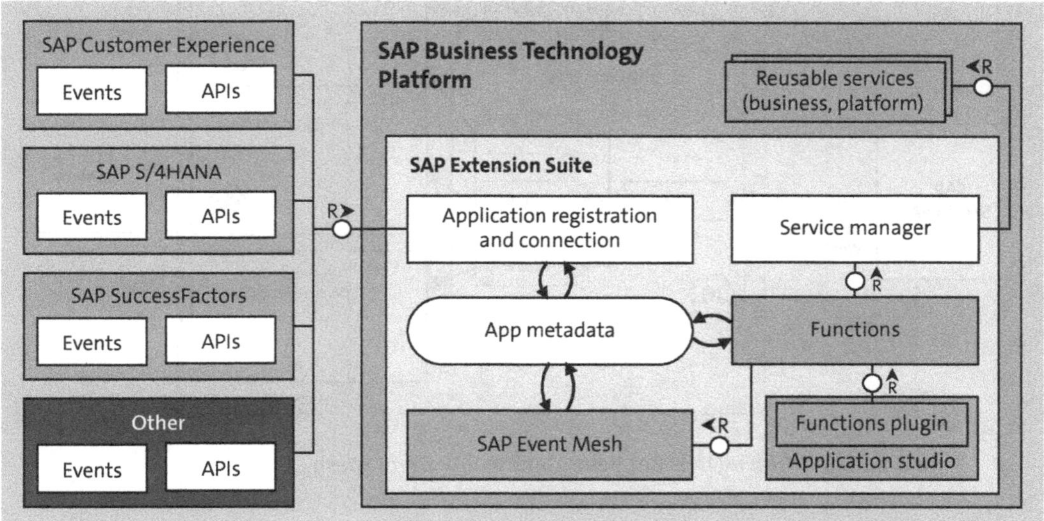

Figure 8.22 Side-by-Side versus In-App SAP S/4HANA Extensions

As discussed in Section 8.1, it's important to weigh the merits of synchronous and asynchronous integration methods when considering how to build extensions for your enterprise. APIs are an extremely powerful and lightweight method for request-response data requests, but asynchronous events also present great benefits in terms of simplifying complex communication patterns and achieving high levels of efficiency for network communication.

An example of an SAP S/4HANA extension that is built in a side-by-side fashion is a project involving a new user interface. This might be a functional subset of SAP S/4HANA that makes interaction easier for a specific user type, or the extension may provide specific functionality that isn't part of the core SAP S/4HANA business process. In the context of this chapter, an SAP S/4HANA event can serve to initiate business logic within the side-by-side extension via SAP Event Mesh. SAP Integration Suite could also be used to pull additional information from SAP S/4HANA (or another application) in response to an SAP S/4HANA event occurring.

In Figure 8.23, we show an event occurring in SAP S/4HANA that is sent to SAP Event Mesh. In turn, the event, which relates to a topic, is sent to multiple queues. In this example, the use cases are distinct. The first is a serverless extension with dynamic resourcing; the second is relating to SAP Integration Suite and the need to forward the information on to an external application (e.g., a third-party CRM application); the third use case is sending the information to a web application (for business usage).

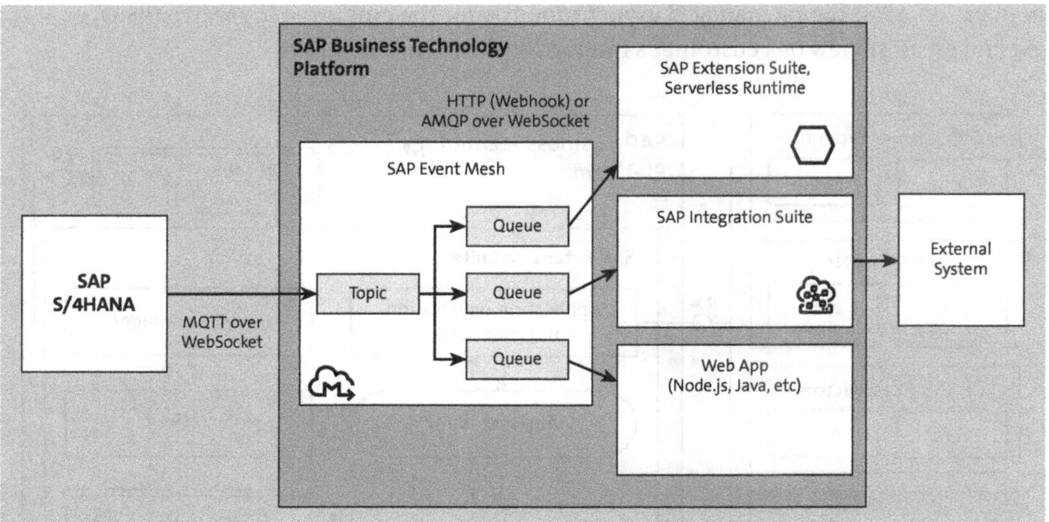

Figure 8.23 SAP S/4HANA Event Being Sent to SAP Event Mesh

8.6 Summary

In summary, SAP Event Mesh has a multitude of benefits for enabling your organization in terms of an event-based architecture. SAP Event Mesh is ideal for loosely coupled applications where one-to-many communication of events is needed. SAP Event Mesh is great for scenarios in which the reuse of events is common (i.e., the data is of interest to multiple groups).

SAP Event Mesh should be considered for use cases involving high-volume and transactional types of updates when subscribing systems can be aware of updates and choose the changes in state they are interested in. Conversely, SAP Event Mesh is not a good candidate for real-time on-demand data access, data virtualization, and gathering cross-functional data for analytical use cases.

In the next (and final) chapter, we'll take a look at recent innovations and upcoming developments with SAP Integration Suite.

Chapter 9
Future Developments

In this book so far, you've learned about SAP Integration Suite, an enterprise integration platform as a service (iPaaS) that allows you to connect applications, processes, and people with speed and ease. In this last chapter, we'll move ahead, looking at recent innovations and the future direction of SAP Integration Suite.

Technology and innovation go hand in hand to embrace the journey of intelligent transformation. Integration is the key ingredient to achieve end-to-end process excellence. To confront the challenges and opportunities ahead, organizations are looking for the right integration platform. SAP Integration Suite is the enterprise iPaaS offering from SAP that can cater to your integration needs holistically. When it comes to aligning products with a company's direct business needs and its feedback, SAP offers workshops, customer engagement initiatives, voice of the customer sessions, and more, as we'll explain in Appendix B. Continuous product improvements and innovations become part of each cycle of product delivery. Now, let's explore what's new and what's ahead for SAP Integration Suite.

9.1 Recent Innovations

In this section, we'll discuss recent innovations and developments made possible by applying advanced technologies and best practices in order to evolve SAP Integration Suite as an open, enterprise-grade integration platform. This has simplified and accelerated integrations for both SAP and non-SAP integration scenarios. The following are the top five recent innovations:

1. A harmonized user experience, achieved via a single launchpad for SAP Integration Suite
2. Low-code application programming interface (API) development
3. An enhanced experience via SAP API Business Hub
4. Centralized monitoring and reporting using SAP Cloud ALM and SAP Analytics Cloud
5. An central API catalog for application developers with API Business Hub Enterprise

Let's discuss each one in detail.

9.1.1 Harmonized User Experience

To provide integration developers with a unified interface, as shown in Figure 9.1, and to easily navigate across different capabilities within SAP Integration Suite, there is a single launchpad for SAP Integration Suite. The following are its benefits:

- Simplified onboarding and role assignment
- Harmonized provisioning
- Ease of collaboration with integration developer, project manager, account manager, and IT

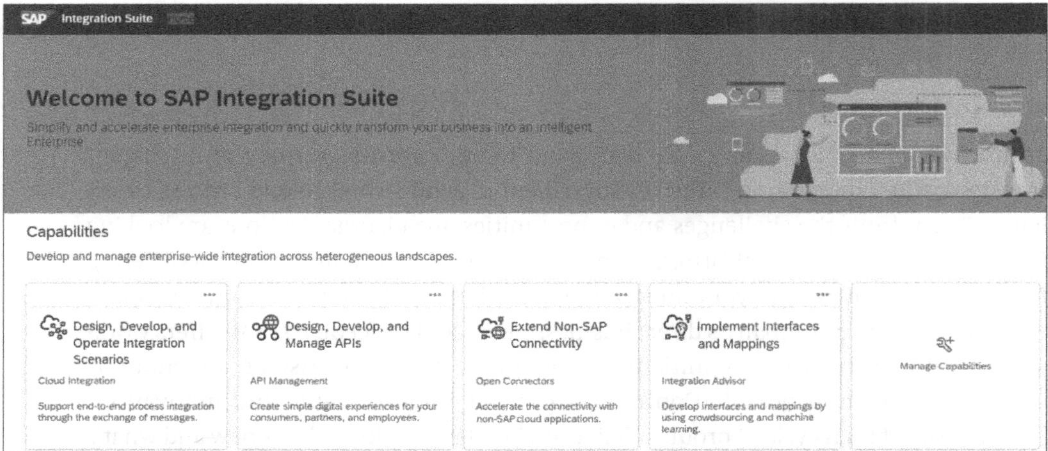

Figure 9.1 SAP Integration Suite Launchpad

9.1.2 Low-Code API Development

Low-code or no-code integration platforms dramatically improve how customers build, deploy, and manage integrations. Having reusable APIs and API-led integration helps application developers to build real-time integrations at scale. Building well-defined APIs that are easy to use, well-documented, error-free, and performant is an art. The process starts by designing the APIs and then following best practices and guidelines for developing these endpoints, as shown in Figure 9.2. Once the API implementation phase for setting the right business logic is over, the application developer can deploy and test the APIs. This is followed by setting the monitoring and monetizing aspects of the APIs by making them available to the developer ecosystem.

SAP is constantly focusing on the following questions:

- Can the complete lifecycle of APIs happen in a low-code fashion?
- Can customers build API-led integration scenarios by using a graphical interface?
- Can business colleagues from an organization get involved in the initial phase of API design?

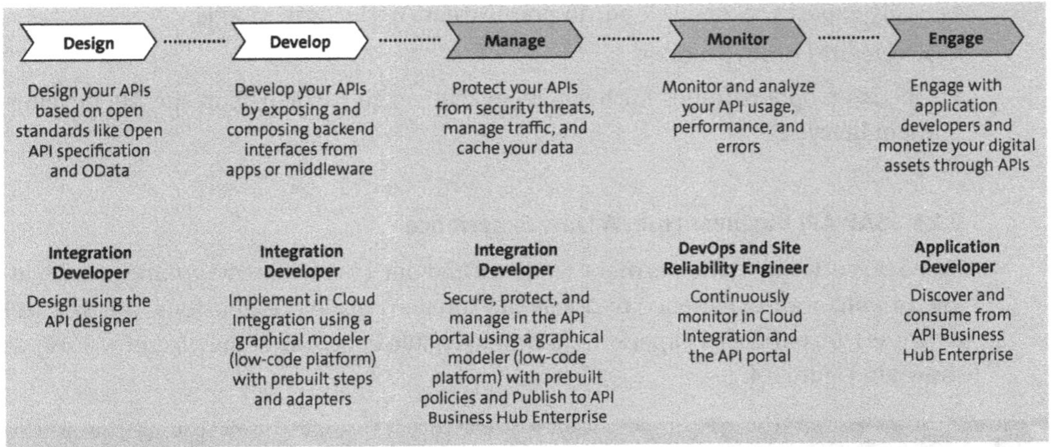

Figure 9.2 API Lifecycle Management

Now let's see how SAP Integration Suite helps you develop APIs in a low-code or no-code approach, as shown in Figure 9.3. It provides the following benefits:

- Makes available graphical and visual interfaces for developing APIs
- Helps integration developers create API endpoints of different types, like OData, REST, and SOAP, in a low-code manner

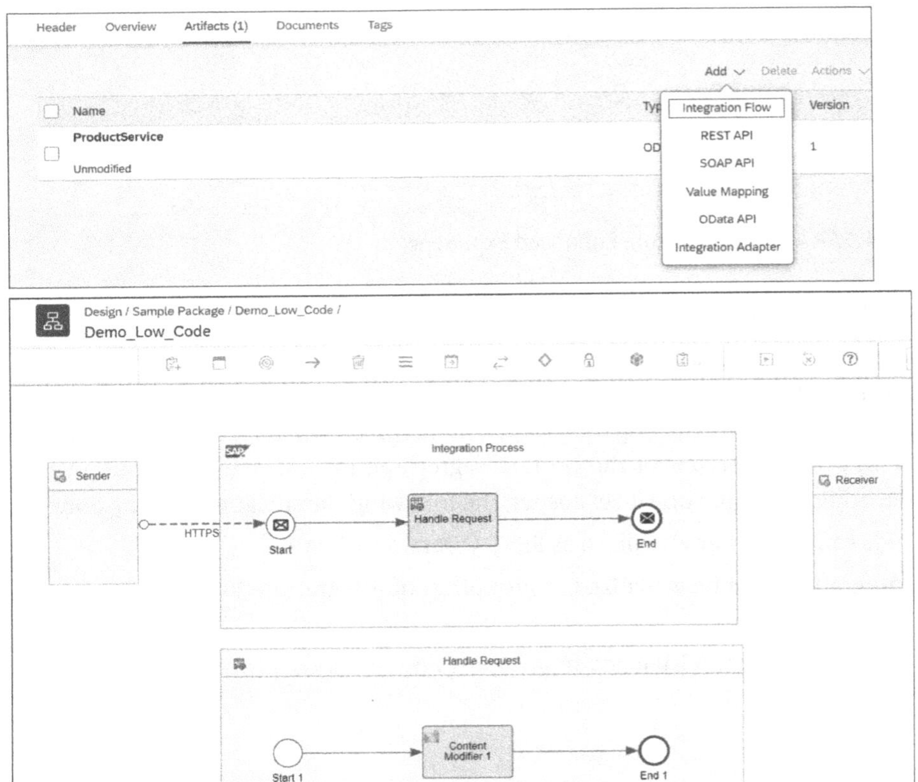

Figure 9.3 SAP Integration Suite as Low-Code API Development Platform

- Easily exposes backend cloud/on-premise data in the form of APIs
- Accelerates time to market
- Opens up opportunities for business executives to participate in the API development lifecycle

9.1.3 SAP API Business Hub: A New Experience

Based on customers' and partners' feedback and due to constantly working on providing an enhanced experience to accelerate integrations and extensions, the SAP API Business Hub enhanced experience is now available in beta as of the time of writing, as shown in Figure 9.4.

Figure 9.4 SAP API Business Hub: Enhanced Experience

Key new features include the following:

- **Discover integrations across products**
  With the enhanced design, prepackaged content associated with products can be visualized quickly by clicking the product name, as shown in Figure 9.5. It's also possible to get an overview of the content segregated into APIs, integrations, events, and workflows. Going one level deeper, the following classifications can be done:
  - APIs can be further visualized as REST, OData, and SOAP.
  - Integrations can be visualized in groupings of integration flows and value mappings.
  - Workflows can be visualized in groupings of decisions, processes, and visibility scenarios.

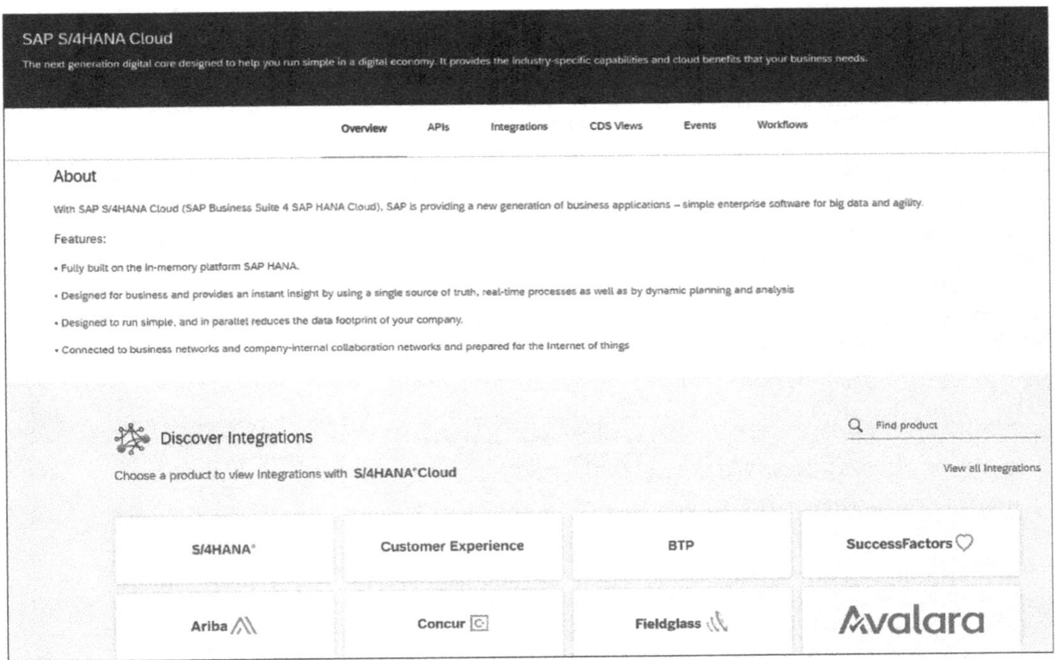

Figure 9.5 Discover Integrations across Products

- **Discover by partners**
  The rich set of APIs and integrations from SAP partners can now be visualized easily by selecting the **Partner Content** tab, as shown in Figure 9.6.

Figure 9.6 Partner Content on SAP API Business Hub

- **Discover reference business processes**
  End-to-end business processes and associated scenario implementations, value

flows, and data flow diagrams can be discovered with visual richness by selecting the **Business Process** tab, as shown in Figure 9.7.

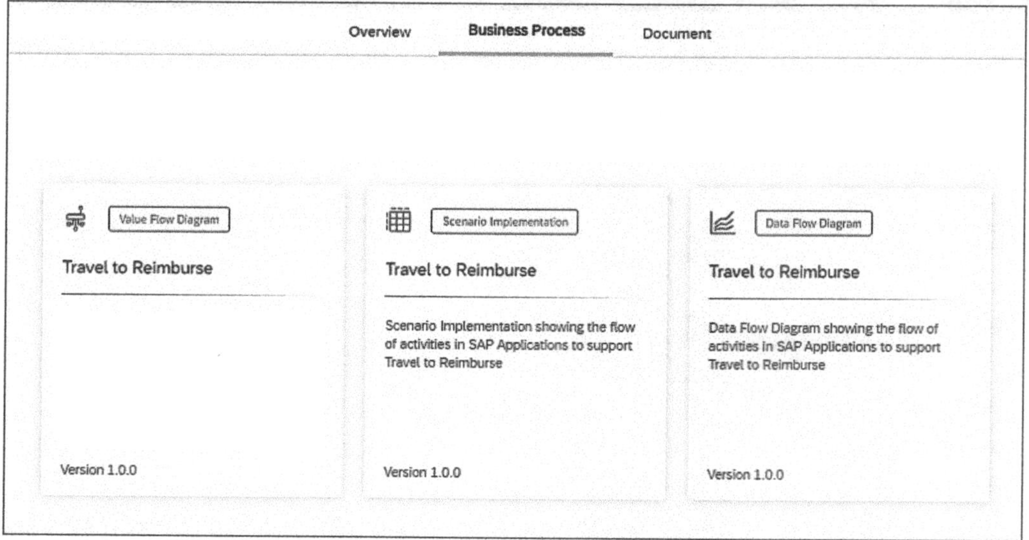

Figure 9.7 Business Processes on SAP API Business Hub

- **Live process content packages**
  To increase the digital process automation of your workflows, there are now live processes offered as prepackaged content on SAP API Business Hub, as shown in Figure 9.8. These readily available packages include process templates, user interfaces, workflows, decision logic, and visibility scenarios.

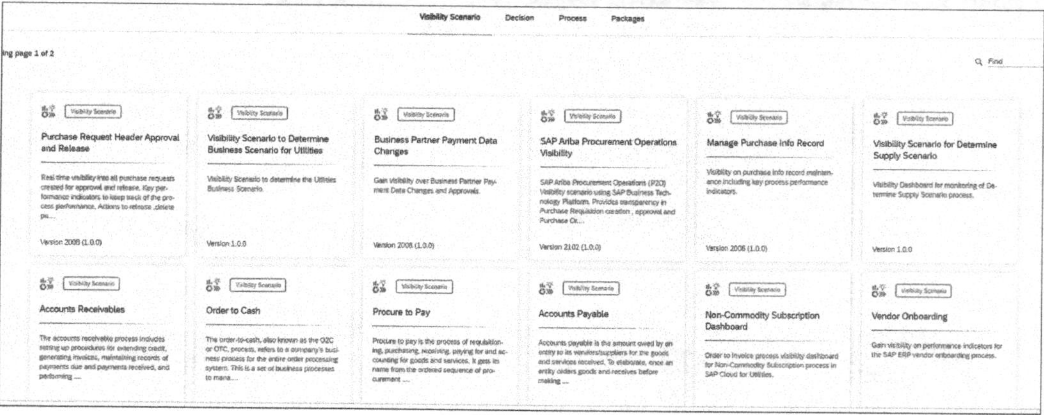

Figure 9.8 Prepackaged Content for Live Processes

- **Consume prepackaged integration content**
  You can discover and consume the prepackaged content seamlessly by following two main steps, as shown in Figure 9.9:
  - Registering the SAP Integration Suite workspace
  - Copying it to the Cloud Integration workspace by clicking the **Add to Workspace** button.

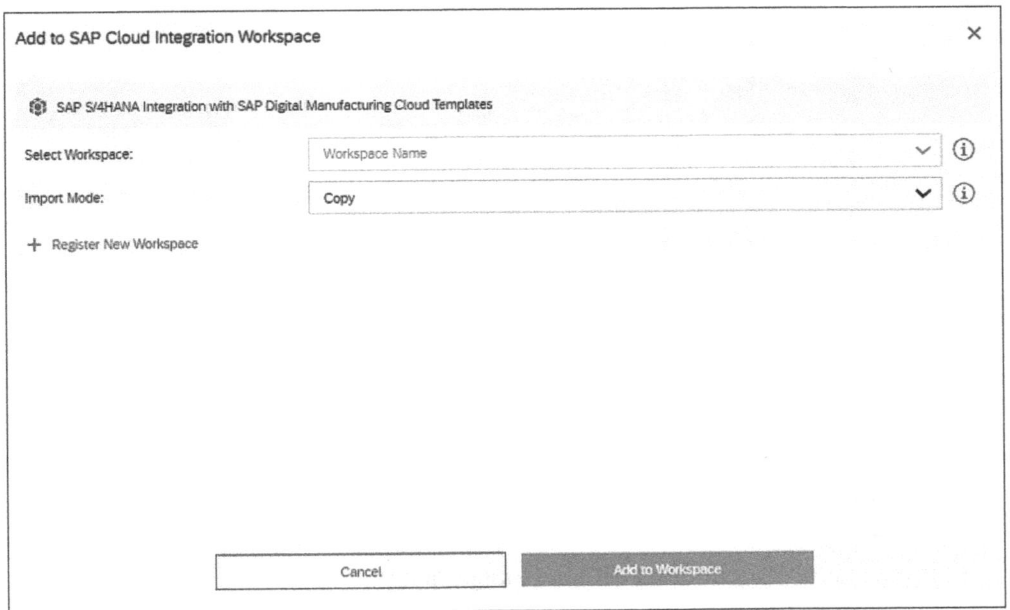

Figure 9.9 Consume Prepackaged Integration Content

9.1.4 Centralized Monitoring and Reporting

The Cloud Integration reporting dashboard is developed in SAP Analytics Cloud. It's a simple and intuitive widget-based analytics dashboard that provides a view of relevant key performance indicators (KPIs) for the SAP Cloud Integration tenant. The widgets in the dashboard display data in a simple metric that helps you visualize the context information with slicing and dicing capabilities, as shown in Figure 9.10.

It enables customers to get responsive analytics information about their Cloud Integration tenants, providing the following views at a glance:

- Message processing view
- Deployed artifacts view
- Service endpoints view

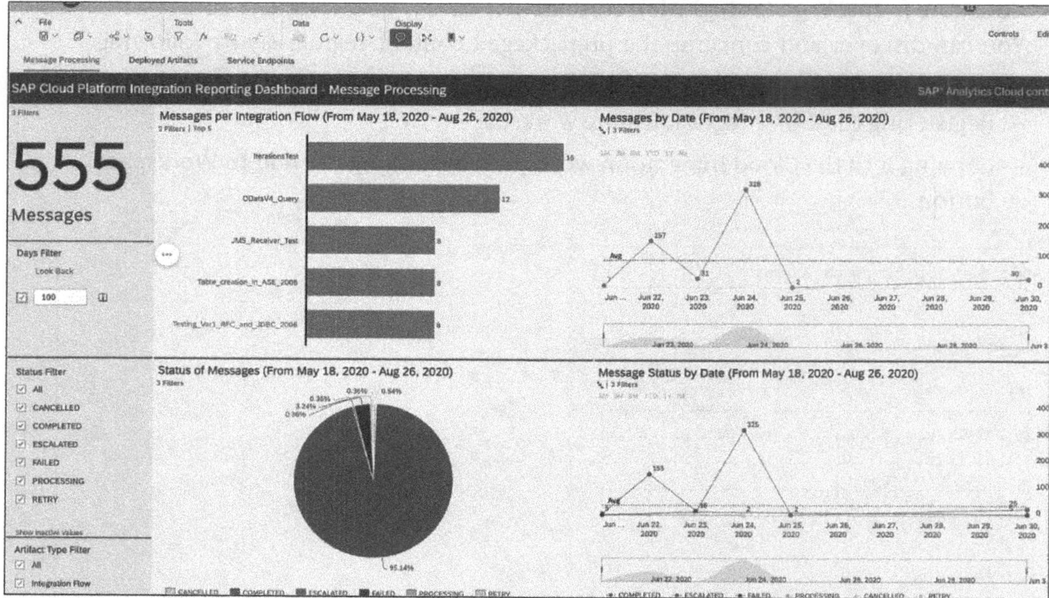

Figure 9.10 Reporting Dashboard Developed Using SAP Analytics Cloud

The following significant improvements have also been made in terms of providing the monitoring capability for the Cloud Integration tenant's health status within SAP Cloud ALM:

- Improve health monitoring of Cloud Integration tenants
- Assure business continuity for customer-critical business processes

9.1.5 API Business Hub Enterprise

To overcome the challenges of supporting specific business scenarios in which the same API should be available in two different portals, and to reduce the redundancy of managing the same API in two API gateways, SAP created API Business Hub Enterprise as part of the API Management capability of SAP Integration Suite, as shown in Figure 9.11. In short, the API Business Hub Enterprise offering is a rebranding of the developer portal, with the following enhanced capabilities:

- Provides a central API catalog for application developers
- Lists APIs from SAP and non-SAP API gateways (i.e., is API gateway agnostic)
- Allows publishing of APIs from multiple instances of the API Management capability to the central API Business Hub Enterprise
- Enables the easy import of externally managed APIs
- Supports seamless search, exploration, and discovery of the listed APIs

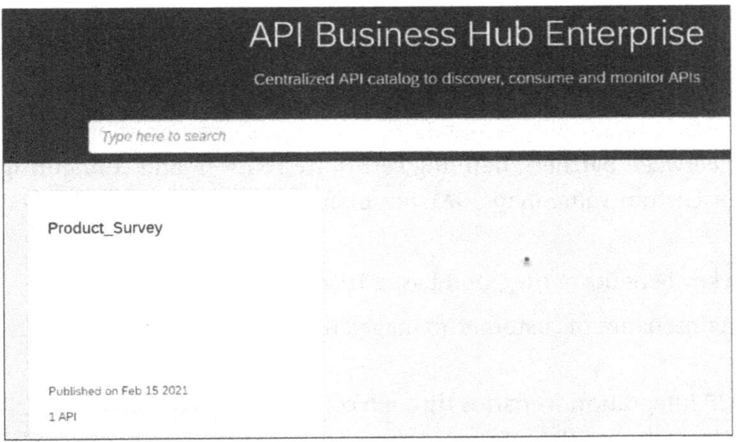

Figure 9.11 API Business Hub Enterprise

9.2 Future Roadmap

In this section, we'll map out the vision and direction for three areas in SAP Integration Suite over time. We'll also explain the rationale behind the following three key focus areas:

1. Cloud-based business-to-business (B2B) integration with trading partner management (TPM) and data archiving
2. Improved high-availability support
3. Hybrid deployment option

> **Note**
> The future roadmap items discussed in this section are subject to change. We'll cover SAP's key focus areas for SAP Integration Suite as of the time of writing.

9.2.1 Cloud-Based B2B Integration

As part of B2B/EDI operations, storing and finding EDI messages becomes a repeated task for the integration developer. As of the time of writing, you can use the following to store files and messages:

- Message processing logs
- Persisting messages in SAP Integration Suite
- Persisting messages outside of SAP Integration Suite—for example, on an SFTP server

To constantly evolve and meet customer needs, a data archiving feature is in the works to fulfill data-retention requirements, typically for B2B and business-to-government (B2G) integration scenarios with SAP Integration Suite.

Further, with the intention to simplify maintaining centralized partner information, creating agreements between partners, defining certificate rollover, and configuring functional profiles for custom value maps, SAP has plans to introduce a cloud-based TPM offering.

The following are the key benefits of the cloud-based TPM offering:

- Onboarding and maintenance of customer-managed B2B integration scenarios with trading partners
- Improve TPM in B2B integration scenarios through collaborative agreement on B2B scenarios directly through the TPM offering
- A one-stop shop, enabling lift-and-shift scenarios in transformation projects to move to the cloud
- Ensure quality and consistency of integration content
- Leverage content provided by customers and partners

9.2.2 Improved High-Availability Support

High availability protects your integration landscape in the event of a service failure. To offer this, SAP plans to leverage the concept of *availability zones* and make services available via data centers with at least three such zones. SAP Integration Suite is hosted in multiple locations worldwide, called *regions*. Regions consist of several isolated locations or availability zones. An availability zone is usually an isolated building within a region that is served by a common set of network devices, power supplies, and so on. Synchronous replication of data is possible because latency between two availability zones in a region is low. For multiple availability zones, the instances of a service are located at different availability zones; that is, downtime in one availability zone doesn't impact the availability of the service.

For improved high availability and scaling of runtime resources and worker nodes, a multiple availability zone approach has the following features:

- Redundant power, networking, and connectivity, housed in several facilities
- Connected with a low-latency, high-bandwidth communication link
- Ability to replicate resources between the availability zones

With the availability zone concept, SAP will provide improved availability of SAP Integration Suite to customers to mitigate risk and make sure businesses can function as usual.

9.2.3 Hybrid Deployment Options

A need for flexibility and changing circumstances are driving companies to accelerate their move to the cloud. This results in a complex landscape with a mix of cloud and on-premise systems to run a business. To leverage SAP Integration Suite in the public cloud of your choice for cloud-to-cloud or cloud-to-ground scenarios, SAP is working toward providing a more lightweight runtime with a hybrid deployment option for ground-to-ground integration scenarios. This means that for users in on-premise environments, a local hybrid deployment of the SAP Integration Suite will be available. As a further add-on, there will be a migration tool to accelerate the move of SAP Process Orchestration integration scenarios to SAP Integration Suite. The following are the working areas to support the hybrid deployment:

- Design, configure, and monitor hybrid integration scenarios, including on-premise–to–on-premise integration.
- Provide a containerized integration runtime as an optional extension of SAP Integration Suite cloud tenants.
- Offer a lightweight runtime software lifecycle in sync with the cloud delivery speed.

9.3 Summary

In this chapter, we explained the recent and future product developments and the rationale behind these investments. With a clear vision of providing seamless integration across heterogeneous and hybrid landscapes, SAP is proactively working on providing an end-to-end, digitally connected experience and making organizations more intelligent. By gathering customer use cases, prospective scenarios, and future opportunities, SAP is focused on building a modular integration platform—on establishing SAP Integration Suite as the modern, versatile enterprise iPaaS.

Appendices

Appendix A
Solution Blueprints

The focus of this appendix is to describe overarching trends that are shaping enterprise landscapes. This appendix will discuss the migration of traditional on-premise workloads to the cloud (infrastructure as a service [IaaS], platform as a service [PaaS], and software as a service [SaaS]). This appendix will also review the role of custom microservices working in harmony with the best-of-breed applications that provide prepackaged value.

A.1 Customer Transformation Journey

According to Boston Consulting Group (*http://s-prs.co/v532619*), a customer transformation journey can be defined as "reconfiguring the way a company develops, delivers, and improves its products and processes in order to deliver the best possible experience to customers."

For a company to achieve this objective, it needs to digitize and optimize its day-to-day operations to enable seamless real-time data access and use across its business, both inside and outside its organization. A company needs to be a master of its day-to-day operations, with the ability to react immediately to customer interactions and business events. Ultimately, a company needs to provide a fully personalized experience that drives high customer satisfaction and loyalty.

SAP Integration Suite plays a vital role in a company's desires to achieve these objectives. SAP Integration Suite provides the technology to enable real-time data access and use. Further, the SAP Integration Solution Advisory Methodology provides the means for a company to optimize its integration resources and provide many services to the rest of the organization as it seeks to optimize its business processes, customer interactions, and support of its strategic objectives. In this section, we'll explore some specific examples in which SAP Integration Suite can help companies achieve their transformation objectives.

A healthcare services provider in the Asia Pacific region supports over 300,000 facilities across 13 markets. It viewed its market dominance and 100 years of history as a potential impediment to achieving the goal of establishing itself as a technology and digital innovation hub for its customers. As part of its SAP Integration Suite deployment, the company was able to do the following:

- Conduct data processing, reporting, and analytics in a consolidated, consistent manner that achieved efficiencies.
- Propel its sales force into the digital age through real-time data access on the go, increasing decision-making autonomy and creating a competitive edge on the ground.
- Combat counterfeiting and quality incidents by using blockchain technology to facilitate coordination across the supply chain.
- Leverage SAP integration services to help manage the proliferation of digital healthcare applications both within and outside of the company.
- Transform untraceable manual processes into digitized value-added business processes that span multiple systems.
- Optimize efficiency and productivity at customer service centers with coordinated task management and 360-degree visibility.
- Improve customer satisfaction throughout the end-to-end experience and reduce processing times.

Figure A.1 shows this customer transformation example organized by technology component. The data layer at the bottom comprises on-premise applications and databases.

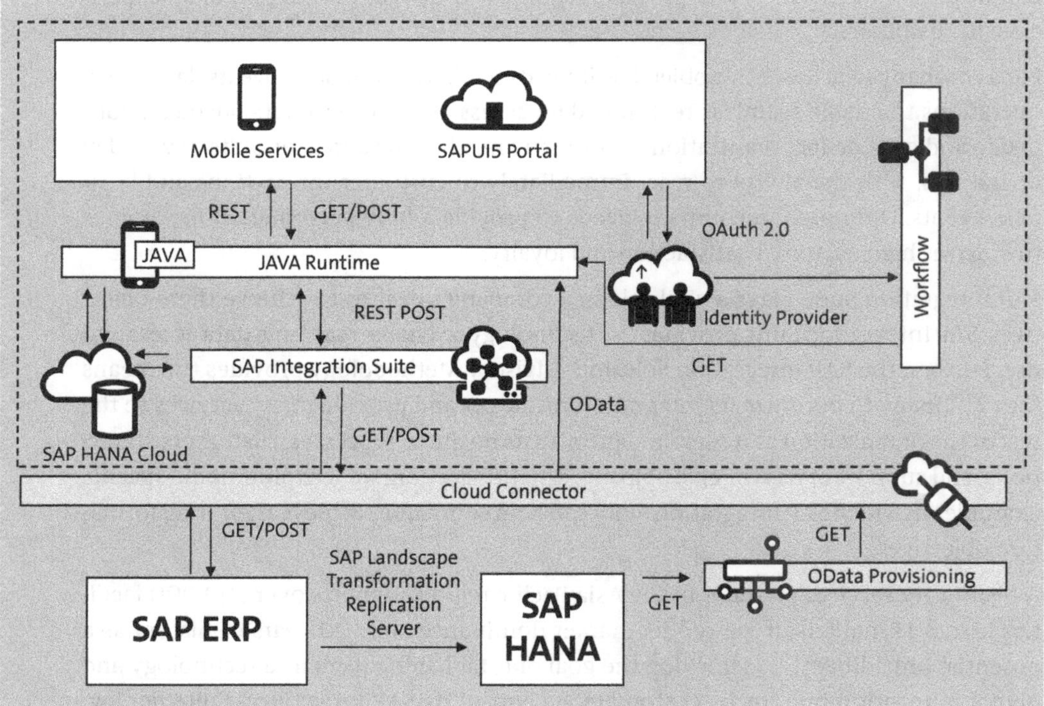

Figure A.1 Customer Example of Leveraging SAP Integration Suite for Customer Transformation Journeys

The cloud connector is a secure reverse proxy agent that provides connectivity to the cloud. SAP Integration Suite is the receiver of the on-premise data, which in turn is stored in an SAP HANA Cloud sidecar (persisting the data for improved responsiveness and efficiency). The Java runtime is the application runtime running in the cloud along with a mobile and web-based user interface. Finally, the solution includes identity and workflow capabilities to ensure appropriate data access and role-specific functionality.

A.2 Accelerated Cloud Agility with SAP Integration Suite

As the name implies, SAP Integration Suite is a comprehensive set of integration-related services. SAP Integration Suite has been built specifically to support cloud-first landscapes and embraces modern approaches to integration. As covered in this book, SAP Integration Suite supports business processes, APIs, event-based integration, and data replication. However, the focus of SAP Integration Suite is not to provide a broad selection of tools; the aim of SAP Integration Suite is to provide a set of capabilities that help make your life easier. SAP approaches this goal by providing a big inventory of prepackaged integration content for both SAP and third-party applications. SAP API Business Hub, covered in Chapter 3, provides integration content for process integration, APIs, events, and other aspects, such as eDocument templates for electronic invoicing and more. Figure A.2 highlights the time and effort savings associated with the use of prepackaged integration content. Templates can represent a significant benefit for companies that leverage this approach to integration.

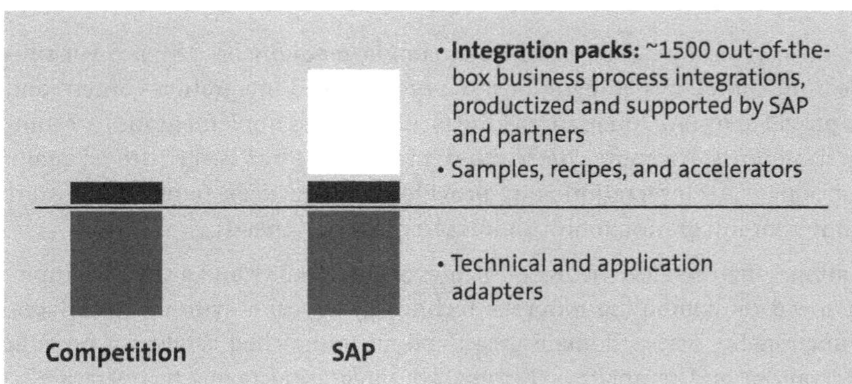

Figure A.2 Value of SAP Integration Suite Prepackaged Integration Content

Multiple companies have been able to witness great improvements in their overall performance by leveraging SAP Integration Suite, compared to legacy on-premise solutions and third-party middleware that takes a developer persona-orientated approach. On many occasions, not only have companies been able to reduce their operating costs by moving to SAP Integration Suite, they have also been able to do so with small, agile teams that can implement integration initiatives in a matter of days or weeks. The fact that SAP maintains and updates a lot of the associated integration content frees up integration personnel to work on more value-adding activities. As Figure A.3 shows, SAP has complemented integration technology with modern capabilities such as the Integration Advisor and SAP Graph, which provide machine learning recommendations and harmonized API access, respectively.

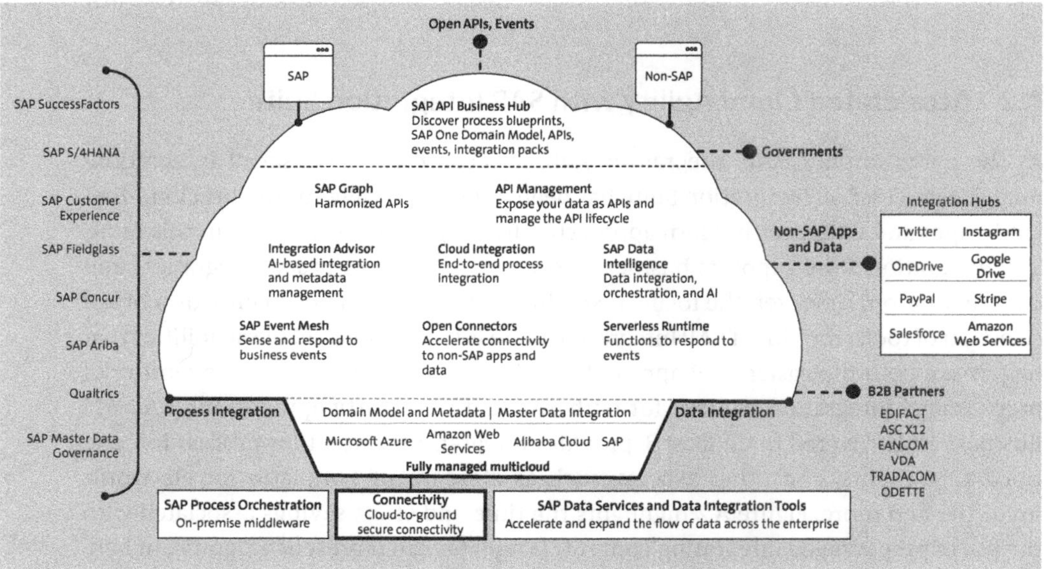

Figure A.3 SAP Integration Suite Offering

Integration flows typically can be classified as simple, medium, or complex. For simple and medium complexity integrations, the prepackaged integration content and other tools provided by SAP Integration Suite facilitate rapid implementation of common integration flows. For more complicated integration flows (which are normally company-specific), SAP Integration Suite provides best practices, templates to start from, and different integration tools to address a company's needs.

SAP Integration Suite therefore helps companies achieve agility and a scalable implementation speed by combining different technology types, easy-to-use interfaces, seamless functionality across domain areas, and an unmatched catalog of prebuilt integration content and templates. This enables integration teams to operate efficiently and predictably and ultimately enables an organization to adopt best practices for an enterprise seeking seamless data access and usage across the organization.

A global brand retailer in the UK was able to leverage SAP Integration Suite to move from using a single, complex, nonresilient middleware product to operating via a cloud-first integration strategy in three months, running over three million transactions per week. It was able to achieve this quick transformation by using the following features:

- Prepackaged content
- Backward-compatibility with its legacy service repository
- Cloud-hosted approach
- Easy-to-use interface

This enabled the retailer to manage its cloud integration solution with a team of less than four people. As a direct result, it was able to focus more time and effort on value-adding activities such as the following:

- Servicing customers through a high-end fashion marketplace online
- Addressing extended markets, especially in China and Saudi Arabia
- Ensuring stability for business-critical processes
- Eliminating the need for patching windows and associated downtime thanks to the cloud-based approach
- Out-of-the-box support for standard functions and policies to accelerate the implementation multifold

As shown in Figure A.4, the retailer leverages SAP Integration Suite to seamlessly connect both SAP and third-party applications, across on-premise and cloud-based environments. For the on-premise applications, the retailer is leveraging a reverse proxy agent to provide secure communication with SAP Integration Suite. SAP Integration Suite is also leveraged for direct connectivity with Microsoft Azure. The API Management capability is leveraged for connectivity with several third-party SaaS solutions. Finally, the data is consumed by an application runtime to provide the high-end online marketplace.

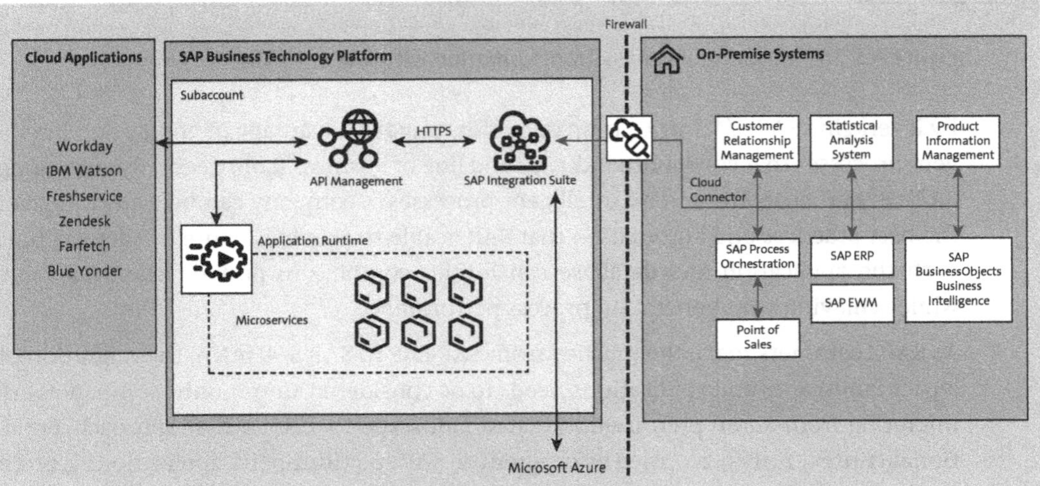

Figure A.4 Customer Example of Cloud-Based Agility with SAP Integration Suite

A.3 SAP ERP to SAP S/4HANA Migrations

A lot of SAP customers leverage or have leveraged SAP ERP. SAP ERP is the previous generation of SAP's enterprise resource planning software. SAP ERP provides modules covering a full range of industry applications, including finance, logistics, HR, product planning, and customer service, linked together into a single, customizable system run on a database of the user's choice. SAP ERP was succeeded by SAP S/4HANA in 2015. As shown in Figure A.5, the goal of SAP S/4HANA is to transform the business operations of SAP users by minimizing the cycles it takes to run the business on a day-to-day basis, providing more time and resources to focus on value-adding activities that provide both real-time and predictive insights.

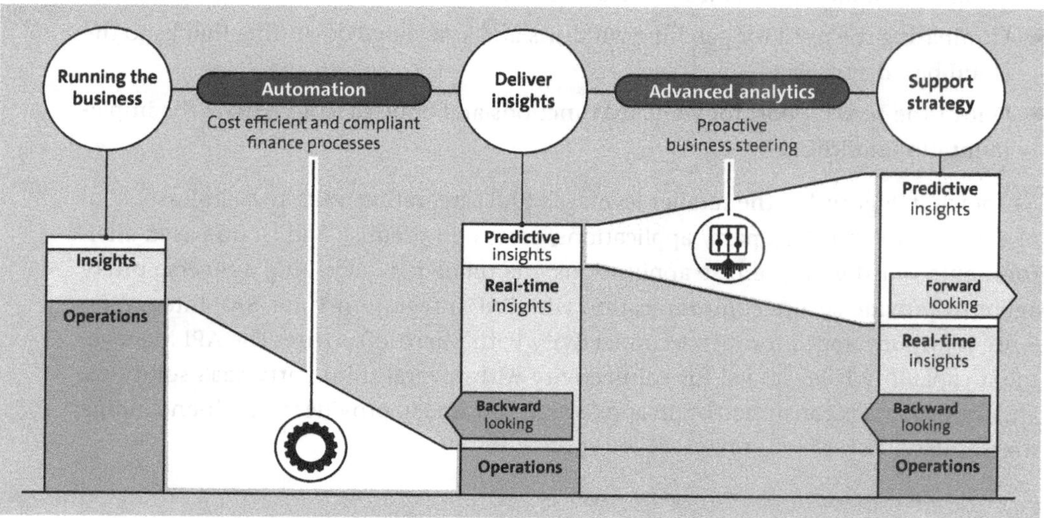

Figure A.5 SAP Vision for Business Transformation with SAP S/4HANA

By leveraging SAP S/4HANA, companies also can take advantage of end-to-end intelligent processes that cover both industry and line of business (LoB) needs. By leveraging SAP's expertise in end-to-end intelligent processes, a company can take advantage of the best practices and automation that SAP is able to provide. Figure A.6 depicts how both the application and database capabilities combine to provide seamless value while achieving next-generation process performance.

As SAP customers make the journey from SAP ERP to SAP S/4HANA, the migration of legacy landscapes and applications needs to be considered. Commonly, companies are migrating from an on-premise-dominated landscape to a cloud-first approach. Functionality previously performed by on-premise SAP and third-party applications is often replaced with cloud-based SaaS applications that provide customers with increased agility, reduced costs, and, ultimately, improved performance.

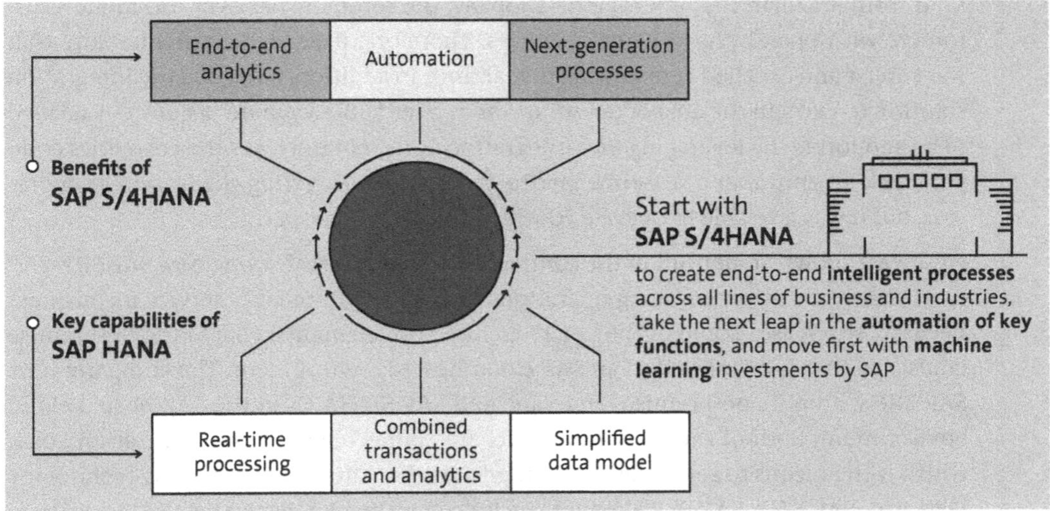

Figure A.6 SAP S/4HANA and the Intelligent Enterprise

Very often, an on-premise integration layer or architecture needs to be replaced to fit the new cloud-first based philosophy. It's common for these on-premise SAP ERP users to have leveraged SAP Process Integration or SAP Process Orchestration. Moving to SAP Integration Suite is the natural choice for SAP Process Integration and SAP Process Orchestration users given that SAP Integration Suite was built with cloud-first architectures in mind, was built by the same team that built the on-premise integration offerings, and includes over 1,800 prepackaged integration packs that are maintained and updated by SAP, freeing up integration specialists to pursue value-adding tasks that transform businesses.

The guidance from SAP once you have an instance of SAP Integration Suite in place is to do the following:

1. Update your SAP Process Integration/SAP Process Orchestration instance to version 7.50.
2. Connect your enterprise service repository via cloud connector.
3. Evaluate which mappings make sense to import and which ones can be adopted in new ways (e.g., by leveraging prepackaged integration packs from SAP).

SAP Integration Suite supports integrating with both SAP ERP and SAP S/4HANA, so current and future integration needs can be supported regardless of the status of a company's digital core.

An example of a company that has benefited from moving its digital core from SAP ERP to SAP S/4HANA is a Brazilian cosmetics company with global operations. The cosmetic company processes over 24 million sales orders per year, with deliveries across over 5,000 cities in Brazil alone. This business volume breaks down to over 3,000 orders per hour from Brazilian customers alone. Globally, the company has over 100 million customers with operations in eight countries. Therefore, it needed an architecture that was robust and excelled at operating in real time. In addition, it needed the integration solution to excel in the area of tax solutions as over 1,800 separate tax entities needed to be supported. By leveraging SAP Integration Suite, not only was the cosmetics company able to empower a real-time end-to-end business covering the lead-to-cash process, but it was also able to operate 70% faster and 40% cheaper.

Figure A.7 shows a blueprint of the company's architecture, showing how SAP ERP, SAP S/4HANA, and customer-facing operations fit together to reliably service its business needs. It showcases some of the key IT architecture elements. First is the on-premise landscape that is connected to an SAP cloud-hosted environment. The company runs SAP ERP within its on-premise landscape and SAP S/4HANA in the SAP-hosted cloud environment. Both of these environments are connected with SAP Integration Suite, which is the means to serve the other extension elements of SAP Business Technology Platform (SAP BTP). SAP BTP elements include identity elements, an SAP HANA Cloud sidecar, and a user experience layer. Together, the technology elements provide customer-facing role access to all the data and order-related activities the company needs to access to be successful.

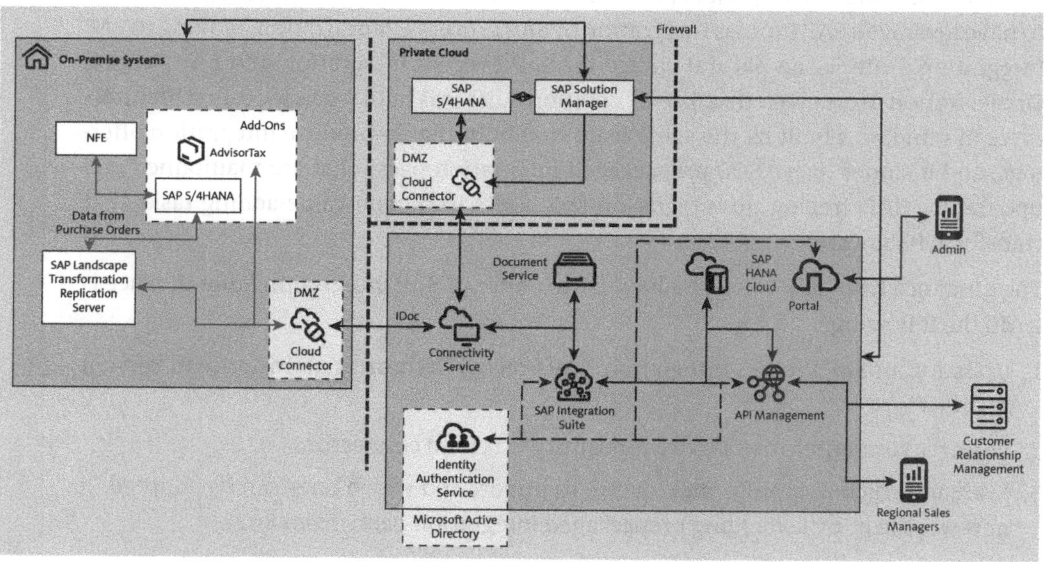

Figure A.7 SAP ERP and SAP S/4HANA Illustrative Customer Landscape

A.4 Digital Integration Hub

A popular approach for modern enterprise landscapes is the usage of a digital integration hub. Gartner defined a *digital integration hub* in 2018 as "an advanced application architecture that aggregates multiple back-end systems of record data sources into a low-latency and scale-out, high-performance data store" (*http://s-prs.co/v532620*).

The digital integration hub is a response to the challenges that are associated with an API-led strategy. An API-led strategy can on occasion be limited by the capabilities of the underlying applications. For example:

- Large enterprise applications are difficult and costly to scale to handle massive API calls.
- It's difficult to leverage cloud-native, high-availability APIs over legacy business systems.
- Fragmented data across multiple applications and data sources makes integration complex.

In response, digital integration is an emerging architecture for implementing large-scale, high-throughput APIs by inserting a high-performance in-memory data store layer between the API service layer and the system of record.

Figure A.8 depicts the working elements of a digital integration hub.

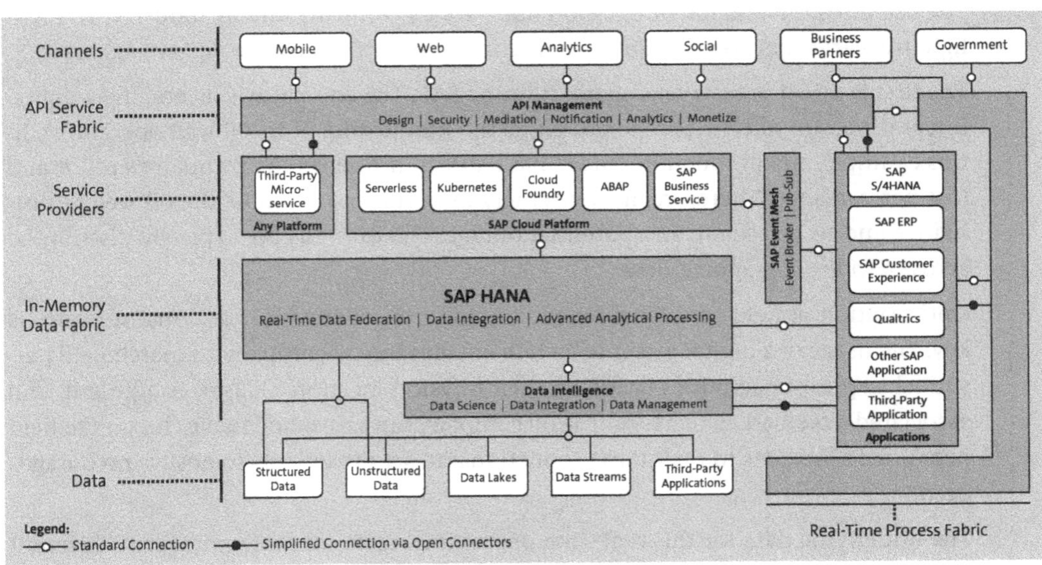

Figure A.8 Anatomy of a Digital Integration Hub

In Figure A.8, different elements of a digital integration hub are highlighted. The combined solution helps the landscape surface the right information to the right user at the right time, leveraging information from a disparate range of source applications.

API Management is shown at the top, acting as an abstraction layer for associated users and stakeholders. The API Management capability provides the following:

- Low-latency frontend API services connect to a high-performance cache for read queries.
- Update calls can be performed directly on the backend systems (command query responsibility segregation).

The high-performance data store is represented in the middle—in this case, SAP HANA, which does the following:

- Stores relevant subsets of the backend data in cache
- Provides a single consolidated view of entities, the data for which is stored in one or multiple systems of record

The integration layer is depicted on the right-hand side. In this layer, SAP Integration Suite does the following:

- Keeps the system of record sources and high-performance data store in sync
- Leverages event-driven integration
- Uses open connectors for third-party SaaS connectivity

The use cases for a digital integration hub vary by company and by industry. To highlight the varied nature of a digital integration hub, let's discuss a couple of examples.

The first is an oil industry company running SAP. The company is leveraging a digital integration hub to centralize and digitize the monitoring of its oil well operations. In this example, any production outage has a direct impact on the company's profit and loss performance. Therefore, it's extremely important to monitor the oil well performance and to anticipate any potential outage. The oil wells are typically distributed across a wide geographical area.

Optimization of field personnel scheduling in terms of site visits and maintenance is key. The usage of a digital integration hub enables the oil company to match field personnel to priority activities based on their physical location, skillset, equipment, and associated workload. In real time, the oil company can assess and make changes to field personnel schedules to maintain production and act proactively to ensure no outages occur.

The underlying data for this real-time operational view comes from over 14 different applications. The data is combined seamlessly to provide the operational control center a single consolidated view. The associated data is persisted in an agile data mart (high-performance data store). The integration with the underlying applications is managed by SAP Integration Suite and SAP Event Mesh. The API Management capability is leveraged as an abstraction layer to provide smooth user experiences.

A second digital integration hub example is a sports team in North America that operates a large stadium, which can hold over 40,000 fans. Prior to the digital integration hub implementation, it was very difficult for the sports franchise to optimize the performance of its stadium on game day. In fact, corresponding reports were not available until two days after the sporting event. With the implementation of the digital integration hub, the sports team was able to achieve a real-time view into stadium performance as it related to the fan experience.

With the digital integration hub, the stadium management team was able to receive real-time visibility into parking, gate entry, point-of-sale activity, bathroom facility status, and more. The real-time views into stadium performance provided the management team with the capability to react in real time to performance data. If the score of the game was close, for example, more personnel could be assigned to seat concessions; if the score was lopsided or the weather changed, management could reallocate personnel to other areas of the stadium. If a bathroom facility reported low scores from fans, again, real-time action could be taken.

The use cases for a digital integration hub are unlimited, and with the capabilities provided by SAP Integration Suite, along with other areas of SAP BTP, comprehensive solutions can be implemented quickly.

A.5 Enterprise-Wide Integration Landscape

SAP has been supporting enterprises as they optimize their business operations for over forty years. A key part of SAP's value proposition is its industry and LoB expertise and focus. Industry and LoB expertise can be grouped together in terms of end-to-end business processes that span multiple steps and frequently involve different stakeholders. Figure A.9 highlights four major processes common in most companies: lead-to-cash, source-to-pay, design-to-operate, and recruit-to-retire. These end-to-end processes in turn are made up of many substeps, as shown in Figure A.9.

In some cases, companies will be able to operate standard out-of-the-box business processes, but it's highly likely that many companies will need to adjust or change these processes and steps based on their unique business operations or circumstances. This is where SAP Integration Suite as part of the broader SAP BTP can help. SAP Integration Suite, as we've seen throughout this book, is a very powerful set of capabilities for achieving connectivity and data access needs. SAP Integration Suite can also be leveraged in combination with other elements of SAP BTP to achieve intelligent enterprise goals in the context of unique company needs and expectations. For example, key processes can be extended to include custom workflows, usage of automation technologies, and custom user experiences (simplified, branded, and/or mobilized).

SAP Integration Suite has prepackaged integration content and templates that cover all areas of the intelligent enterprise. Therefore, SAP Integration Suite can be leveraged across your entire enterprise landscape with prepackaged content covering all business areas, plus SAP and third-party applications across cloud and on-premise elements.

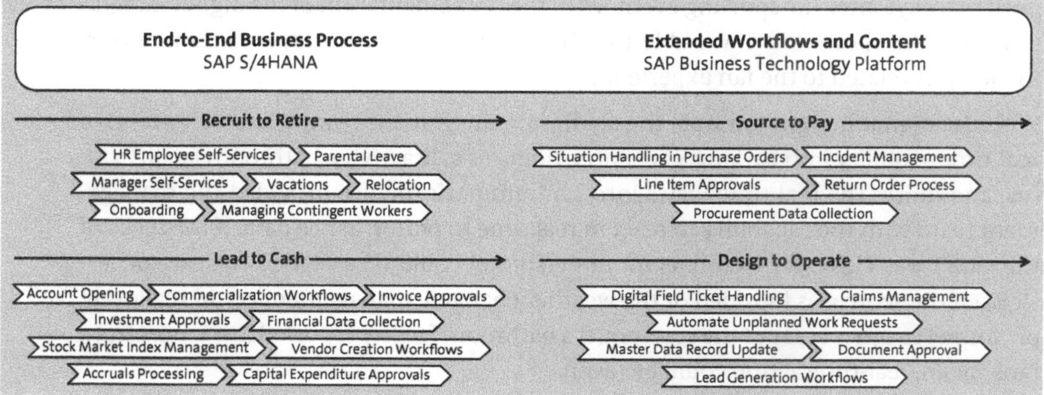

Figure A.9 Optimization of Business Processes

A European sensor company has built a big data and integration hub for standardizing its enterprise-wide data access (B2B, application-to-application [A2A], Internet of Things [IoT], and databases) in response to competitive pressures that require an effort to differentiate and establish customer loyalty. As a result, the company's customers are now able to perform self-service analytics and product configuration, which has improved customer satisfaction. Internally, stakeholders are also able to take advantage of the big data and integration hub.

It was important for this company to move at speed, and as a direct result of leveraging SAP integration Suite, it was able to quickly realize its digitalization goals. For example, it was able to roll out a new digital sales channel that generated €50 million in sales and improved the company's overall project speed by fivefold. This company saw integration, data science, and innovation as a single related item that enabled it to achieve data-driven intelligence. This company was able to do the following:

- Use SAP Integration Suite prepackaged content to accelerate access across the organization.
- Eliminate data silos by establishing a uniform, standardized approach to data.
- Build standard APIs for data access (clean data model for everyone).
- Establish a business data lake for a single source of truth.
- Leverage SAP Integration Suite for A2A and B2B and SAP Data Intelligence for database data.
- Boost machine learning, B2B customer interactions, and use of data science with its business data hub.
- Use SAP Integration Suite to make it easier to comply with legal integration requirements.
- Empower the transformation of end-to-end business processes (e.g., EDI usage in the order-to-cash process) with integration and digitization.

As shown in Figure A.10, the company has implemented SAP Integration Suite in combination with a business data lake. The business data lake is depicted by the shaded box. SAP Analytics Cloud is the main choice for leveraging this data. SAP Data Intelligence and SAP HANA smart data integration are the cloud services responsible for ingesting on-premise information. SAP Integration Suite comes into play for sharing information in the business data hub—serving internal and external users via a website, a portal, and SAP Asset Intelligence Network.

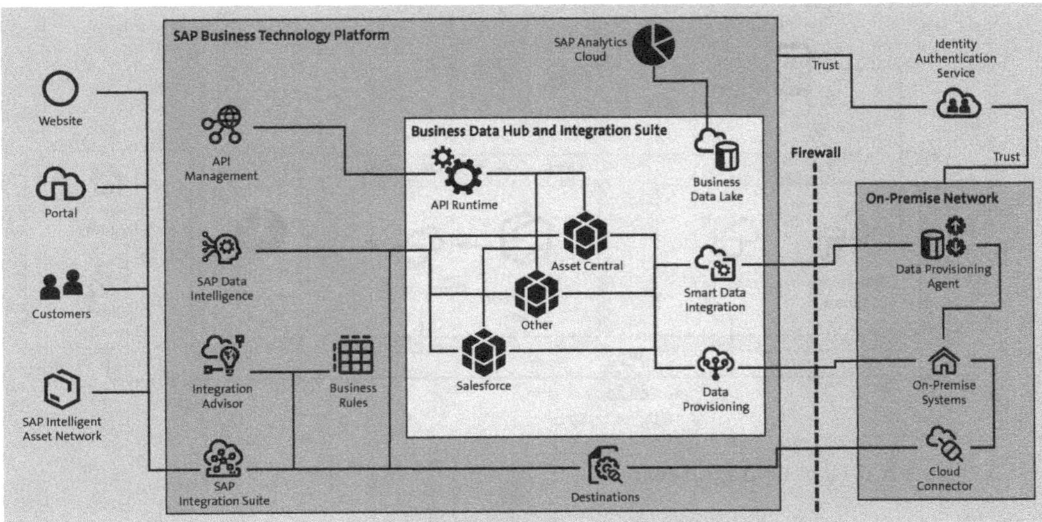

Figure A.10 Example of Enterprise-Wide Integration

A.6 Industry Transformation

The fourth generation of industry transformation is upon us. The utilization of technologies such as advanced robotics, IoT, artificial intelligence (AI), and big data are considered the realization of intelligent, flexible, and distributed commercial capabilities. This stage of industrial maturity is responsible for the generation of unprecedented amounts of information and data.

A great example of industry transformation is a global light bulb manufacturer that's leveraging SAP Integration Suite to support a "light bulb as a service" business model. Through real-time data usage, the light bulb manufacturer can offer lighting on a variable usage basis, moving away from a fixed-cost contractual model. Usage charges can be broken down by building, floor, or room. This helps inform the manufacturer's customers, which can then make better decisions and drive energy efficiencies. Moving from a capital expenditures (CapEx) to an operational expenditures (OpEx) model is expected to save customers approximately 40%, with 33% of customers seeking to participate.

As shown in Figure A.11, SAP Integration Suite acts as a central element in connecting information from a variety of sources—in this case, the interaction platform and SAP Business Suite. In turn, the information is stored in an SAP HANA Cloud database to improve user responsiveness. Finally, SAP Integration Suite provides data to SAP Subscription Billing and SAP Analytics Cloud.

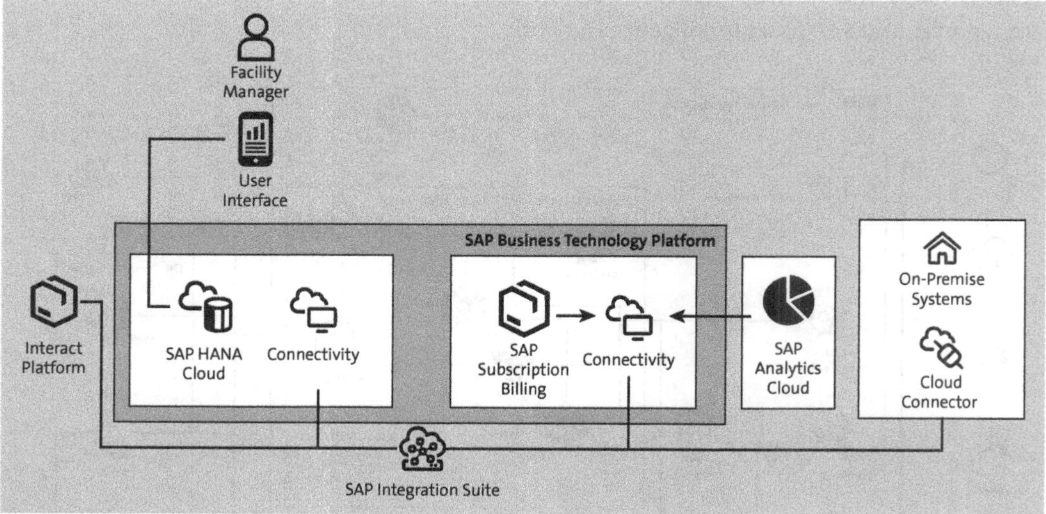

Figure A.11 Light Bulb as a Service

Another example is a US utility that rolled out a self-service digital customer engagement platform that has transformed the utility's relationship with its customers by allowing convenient, easy, real-time access to utility services and information. The utility's mobile app, powered by SAP Integration Suite, connects to relevant systems and applications and allows the utility's customers to manage their accounts, view and pay bills, view and compare their water usage to stay informed about their consumption habits, find water savings tips and apply for rebates, find authorized payment locations, view service outages, report water leaks, and request service or help. The business results have been drastic, with a 20% reduction in water usage and a 32% reduction in operational expenses stemming from a 25% reduction in call center volume.

In Figure A.12, the mobile phone icon represents the user experience for the water management application. The cubes represent data sources connected by SAP Integration Suite that are in turn provided to the water consumers in a seamless and comprehensive manner.

The last example is a European gas company with operations in multiple countries and serving over 20 million customers. Like many enterprises, this company had a hybrid heterogenous landscape consisting of SAP and third-party applications across on-premise and cloud frameworks. The gas company sought to improve its consumers' experience by adopting an API-first strategy to transform its mobile and digital customer service.

Ultimately, it was able to reduce portal response times from over two seconds to less than 200 milliseconds by leveraging a digital integration hub approach at scale.

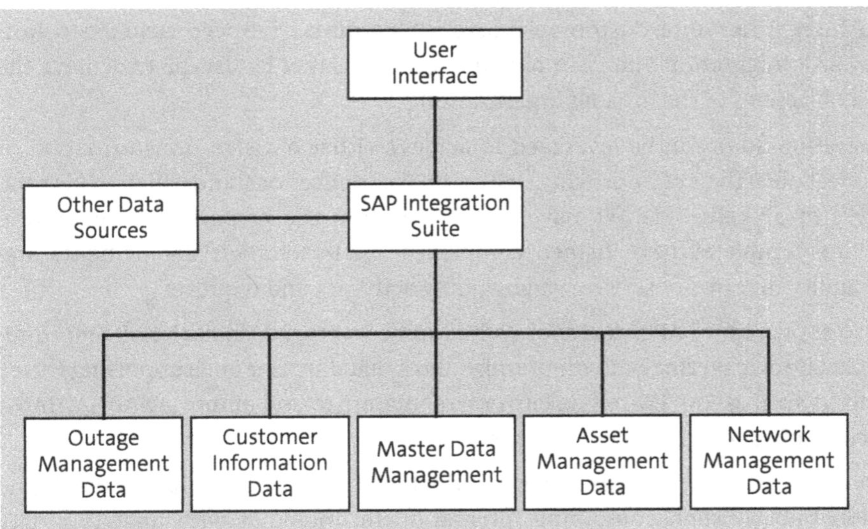

Figure A.12 SAP Integration Suite, Empowering Water Conservation

Figure A.13 shows a blueprint of the customer's landscape. On the left-hand side is the user experience, covering internal and external personas. After being authenticated, a user can access the application. The application leverages SAP Integration Suite for connecting to data sourced from SAP and third-party applications, including those in both on-premise and cloud environments. Multiple elements of SAP Integration Suite are leveraged, including Open Connectors, API Management, and Cloud Integration. A cloud database layer is also leveraged to optimize user responsiveness.

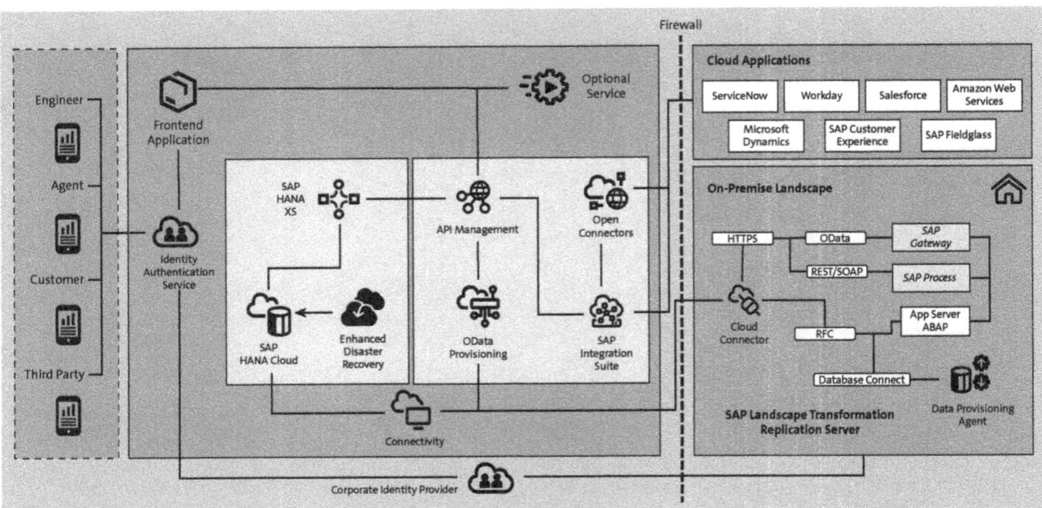

Figure A.13 Improve Customer Experiences with Hybrid Heterogenous Landscapes

A.7 Summary

In this appendix, we covered several exciting areas in which customers are benefiting from SAP Integration Suite. Customer landscape blueprints have been included to illustrate how SAP Integration Suite can plug into any customer landscape to achieve the goals and objectives of the broader organization.

SAP Integration Suite can be leveraged to achieve either massive transformation or more modest objectives of improving customer experiences or standardizing information access by stakeholders. We have highlighted how the prepackaged integration content and templates have helped companies accelerate their deployments and increase agility in terms of serving value-adding activities and requests.

We've also shown how SAP Integration Suite can be leveraged via both SAP and third-party applications, covering both on-premise and cloud domains and supporting legacy core systems such as SAP ERP while customers migrate to and embrace SAP S/4HANA and the concepts of the intelligent enterprise.

Finally, we showed a couple of examples in which customers are leveraging the digital integration hub concept, combining integration technologies with analytics, data management, and user experience to achieve real-time visibility into and digitization of core business operations and processes.

The use of SAP Integration Suite is unlimited, but hopefully these concrete examples will give you an initial idea of how it can benefit your organization. Good luck!

Appendix B
Advisory Program

In this fast-moving digital era, organizations are looking to reap the benefits of digital transformation. Integration being the bedrock of digital transformation, it becomes even more important for SAP to foster a close relationship with its customers and partners. Collaboration, engagement, and communication become business differentiators in winning a company's trust. SAP helps organizations across industries and different market segments to leverage its products, the latest technology, and its platforms to build innovations easily and accelerate time to market. The SAP community can opt in to be part of the advisory program, which is offered at different levels. Finally, SAP customers are part of the feedback process. They also become part of the journey, taking the first looks at new designs and features before they go to market.

The benefits of advisory programs include the following:

- Facilitate easier communication
- Resolve challenges by providing proven practices
- Provide real-world implementation experiences
- Fast access to integration content and a team of experts across the globe
- Insight into emerging trends, opportunities, and practices

In the following sections, we'll see how SAP delivers tools and services to help customers and partners at all maturity levels to educate, engage in, and strengthen the community. We'll look at SAP's high-touch and low-touch engagement channels. You'll also see how global integration evangelists help enterprises define integration strategies and guidelines to win in this digital economy.

B.1 High-Touch and Low-Touch Engagement with SAP

For a customer-centric organization like SAP, it's very important to build a process around a customer engagement model. Engagement models help SAP to collaborate with customers and partners, build teams, and provide them with the right set of information. SAP classifies its engagement models broadly into two categories, high-touch and low-touch engagement, as shown in Figure B.1.

We'll walk through both categories in the following sections.

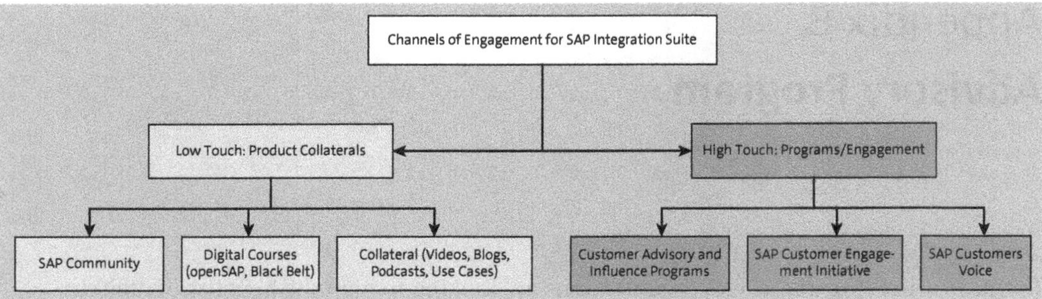

Figure B.1 Channels of Engagement

B.1.1 High-Touch Engagement

Global integration and application programming interface (API) evangelists have been advising SAP customers across industries on accelerating their enterprise integrations and crafting and executing their enterprise API strategies. To help customers leverage the power of SAP Integration Suite, SAP offers four programs as part of high-touch engagements:

1. **Customer Influence program**
   The scope of integration in hybrid and heterogeneous landscapes is constantly increasing. Getting customers' input about a product is extremely valuable. The Customer Influence program, as shown in Figure B.2, focuses primarily on having an in-depth discussion about the product roadmap and exploring the unique needs of large enterprises. There is an influence portal where SAP customers and partners can submit their requests for new features or feature improvements. Other customers can vote for the feature requests that already exist and, based on the top votes, SAP will accept requests and address them in the next development cycle.

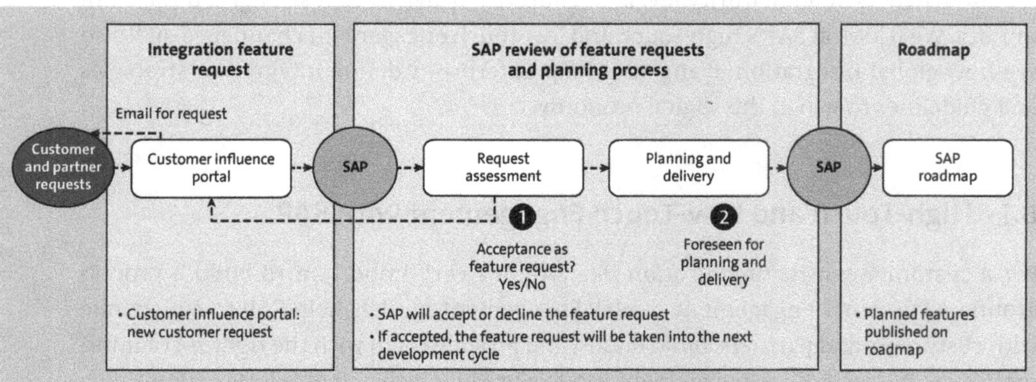

Figure B.2 Customer Influence Program

- **Custom Advisory program**
  The Customer Advisory program, on the other hand, aims at engaging with key customers on their integration use cases and helping them to realize different integration

patterns, as shown in Figure B.3, in a design-thinking-led engagement with business and technical teams. SAP shares solution possibilities and supports enterprise architects in overcoming the integration challenges faced by offering the following:

- The SAP Integration Solution Advisory Methodology simplifies integration via a system integration approach. Refer to Section B.2 for more details.

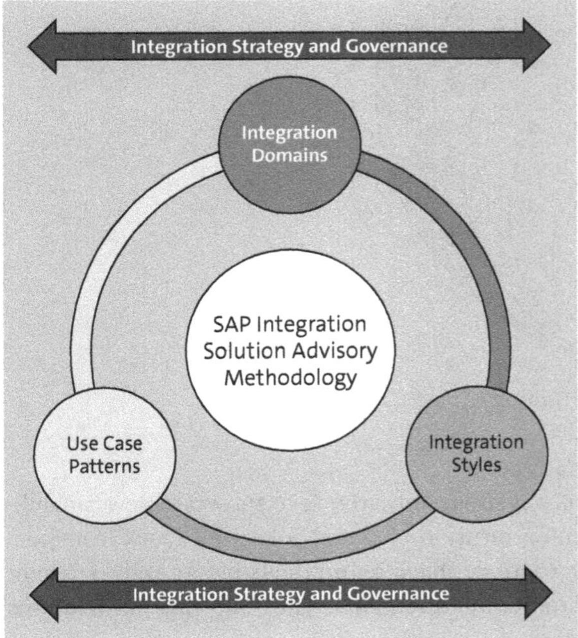

Figure B.3 Customer Advisory Program: SAP Integration Solution Advisory Methodology

- Craft and execute your API strategy to simplify your API experience. APIs enable multispeed IT, foster digital interactions with partners and supply chain business networks, and enable omnichannel customer experiences. Given the importance of APIs in enterprises, a systematic API strategy, architecture, and design are necessary to benefit from the enormous potential of APIs. As part of the Craft and Execute Your Enterprise API Strategy advisory program, as shown in Figure B.4, integration and API evangelists can help you with the following:
  - Align your enterprise strategy to an organizational strategy to gain business benefits.
  - Set up an API program with executive sponsorship to execute an API strategy successfully.
  - Understand and engage with your ecosystem to grow partnerships.
  - Create an API makers community to help ensure high-quality APIs.
  - Use the full-lifecycle API Management solution on SAP Business Technology Platform (SAP BTP) to reduce time to win.

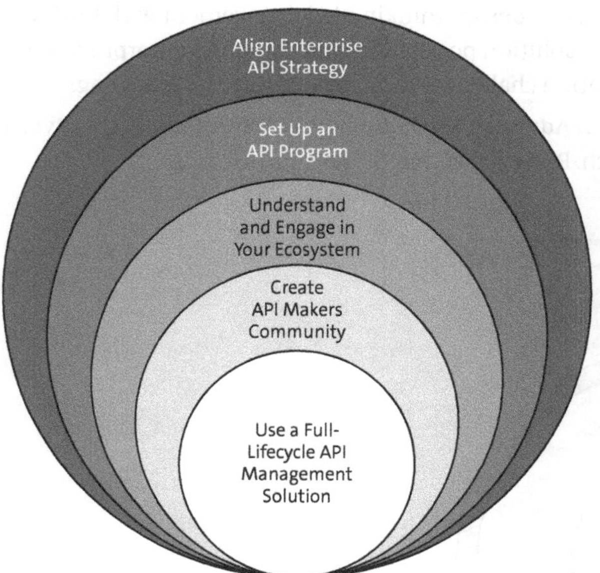

Figure B.4 Craft and Execute Your API Strategy

- **SAP Customer Engagement Initiative**
  The main goal of SAP Customer Engagement Initiative is to showcase new capabilities, discuss proposals for upcoming future topics, and collect feedback from customers and partners. This enables SAP to shape its products better and to deliver features in alignment with customer requirements. The following motivations make this a useful option for customers:
  - Gain early insights into SAP's product visions/ideas/research projects.
  - Get an opportunity to influence the product roadmap and its prioritization.
  - Have an opportunity to share ideas, solutions, and best practices within an ecosystem of similar SAP customers and partners.
  - Establish a direct line of communication to the product development organization.
- **SAP Customer Voice**
  Companies can leverage *voice of the customer* sessions via SAP Customer Voice to share their experience of and feedback about SAP Integration Suite with SAP's development organization. The voice of the customer sessions aim at describing the following:
  - Overview of customers' landscapes, integration scenarios, and use cases
  - Sharing statistics information, such as number of API calls per day/hour, number of productive interfaces/scenarios, number of messages per day/month
  - Technical details about leveraging SAP Integration Suite—such as API policies, prepackaged integration content, adapter types, mapping types, and so on
  - Roadmap for SAP Integration Suite

  - What works well in SAP Integration Suite and where is SAP better than its competition
  - Implementation of any workarounds because of product limitations
  - Overview of requirements for and feedback about SAP Integration Suite

B.1.2 Low-Touch Engagement

The low-touch approach focuses on crafting collateral to support and simplify a company's journey toward adopting SAP Integration Suite. SAP offers blueprints, use cases, missions, learning journeys, digital courses, and interaction through the community to ensure companies have all the knowledge and collateral needed to execute their goals and get started with SAP Integration Suite.

Let's look at the low-touch engagement options in detail:

- **Collateral**
  Collateral helps companies understand in depth how SAP's product features can be used to build business scenarios around hire-to-retire, procure-to-pay, lead-to-cash, and travel-to-reimburse. SAP also provides interactive guides and step-by-step tutorials at *https://developer.sap.com* and missions in the Discovery Center at *https://discovery-center.cloud.sap/*, as shown in Figure B.5.

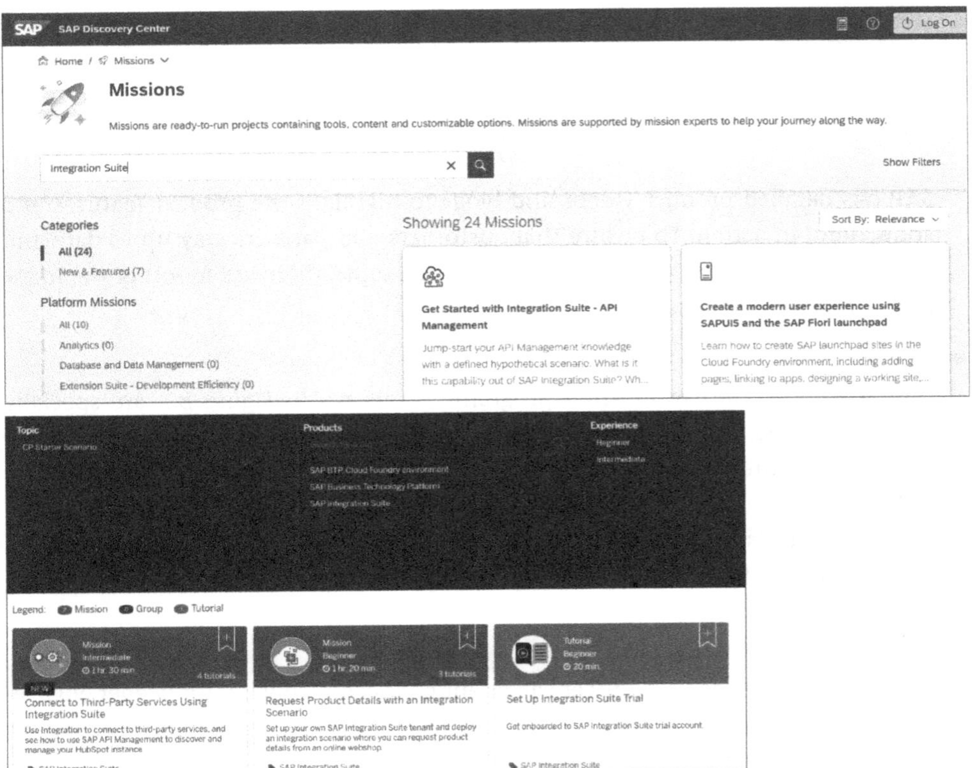

Figure B.5 SAP Integration Suite: Missions

You can also look at the SAP Integration Suite learning journey, designed to help learners understand and navigate the complete path to becoming fully competent with the product, as shown in Figure B.6.

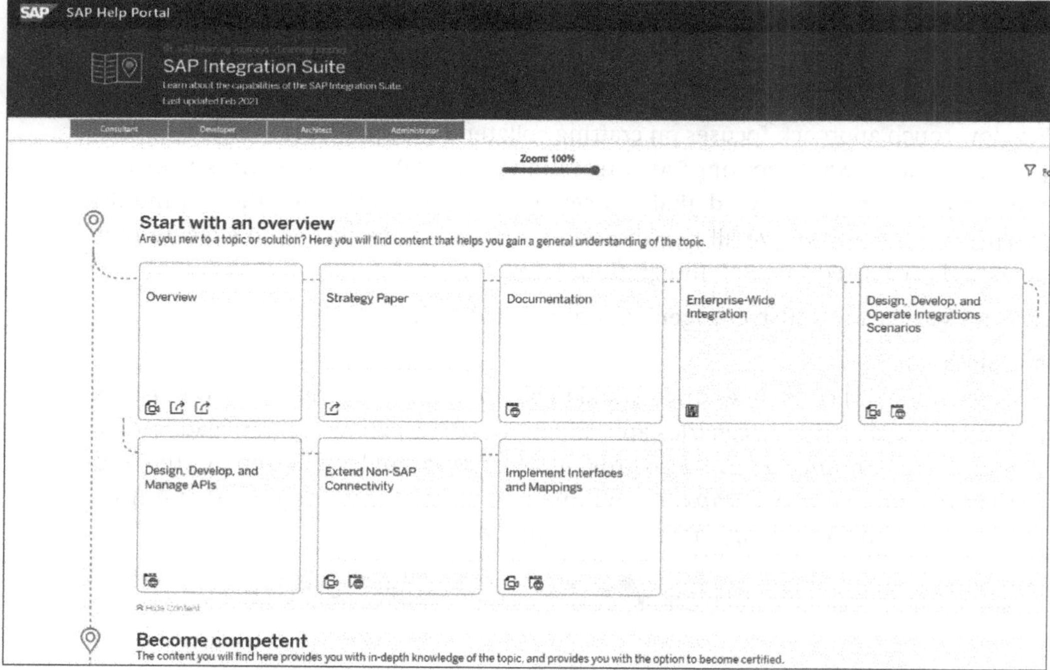

Figure B.6 SAP Integration Suite: Learning Journey

SAP has detailed product videos and blogs to highlight the product features and show them in action. To ensure that customers and partners stay up to date and maintain a sufficient level of knowledge on the topic, there are monthly webinars, podcasts, and code talks.

- **Digital courses**
  Digital courses and certification programs, as shown in Figure B.7, are specially planned, crafted, and executed to give you the experience of real-time customer use cases. Led by integration experts across the globe, digital courses like openSAP and Black Belt certification programs offer different learning experiences and different levels of expertise. The courses cover allied skills, starting from getting started with SAP Integration Suite and continuing through hands-on experience with building business scenarios.
- **SAP Community**
  The importance of community engagement has become pivotal for deeper engagements. It gives the flexibility to collaborate, share ideas, and learn together. You can read more on the SAP Integration Suite community in Section B.3.

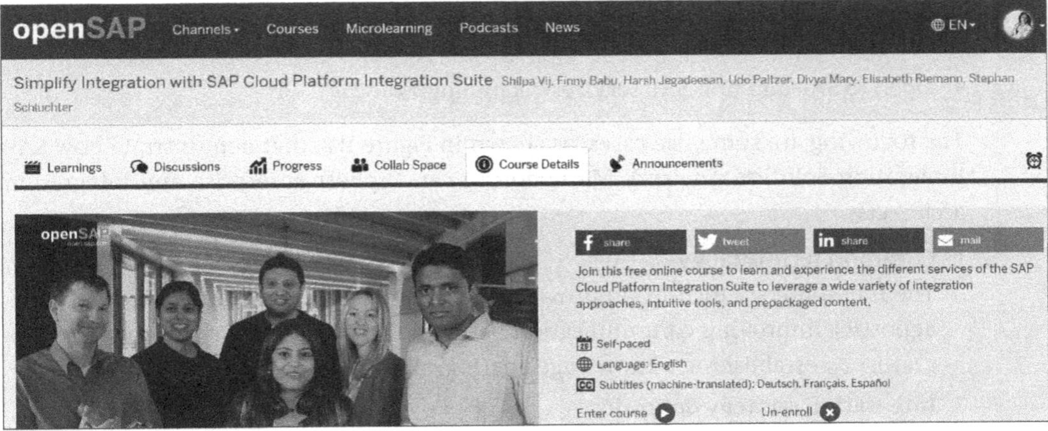

Figure B.7 openSAP Course on SAP Integration Suite

B.2 Defining Integration Guidelines for an Organization

Nowadays, experts agree that digital transformation is not an option, but rather a necessity. In this fast-changing market, a solid integration strategy is of the utmost importance for providing agility and making an informed decision based on real-time data. Companies are taking cloud-first and API-led approaches to adopt intelligent technologies and provide seamless business processes and integration solutions.

The following are some key integration challenges for organizations:

- How to define an integration strategy and guidelines for the organization
- How to select the right tool and integration pattern for the business
- How to approach new integration domains
- How to enable business users for integration
- How to deliver integration solutions rapidly

These are the fundamental questions an organization has to consider before kick-starting its integration journey. To address integration-related questions holistically, SAP provides technology-agnostic integration patterns and styles. These can be leveraged by integration architects and consultants to define a scalable and flexible integration strategy that includes an extensible set of use cases to suit customers' hybrid landscapes.

The goal of SAP Integration Solution Advisory Methodology is to simplify integration and provide a seamless experience for enterprise architects to implement integration use cases in hybrid and heterogeneous environments. SAP Integration Solution Advisory Methodology supports integration architects in the following ways:

- Helps them to shape their integration strategy
- Provides a collection of typical integration styles and patterns

- Maps to integration services (based on customer context)
- Offers architecture blueprints for integration, including decision criteria
- Open to SAP and non-SAP services/technologies

The following are some use cases, as shown in Figure B.8, that demonstrate how SAP Integration Solution Advisory Methodology can support enterprise and integration architects:

1. **Common terminology**
   The methodology focuses on providing a holistic view of integration (technology-agnostic), improving communication between the project teams and system integrators by establishing a knowledge-sharing culture.
2. **Integration strategy definition**
   The methodology can be used to assess and review your existing integration architecture. It helps to define your target integration use cases and identify future building blocks.
3. **Integration governance and hybrid integration platform**
   The methodology can be used as a starting point to define integration guidelines and blueprints for your hybrid integration platform, thereby moving from an an ad hoc integration practice to a more systematic method of integration.
4. **Self-service integration**
   The methodology helps to define areas for self-service integration by enabling citizen integrators to implement simple integration scenarios on their own.

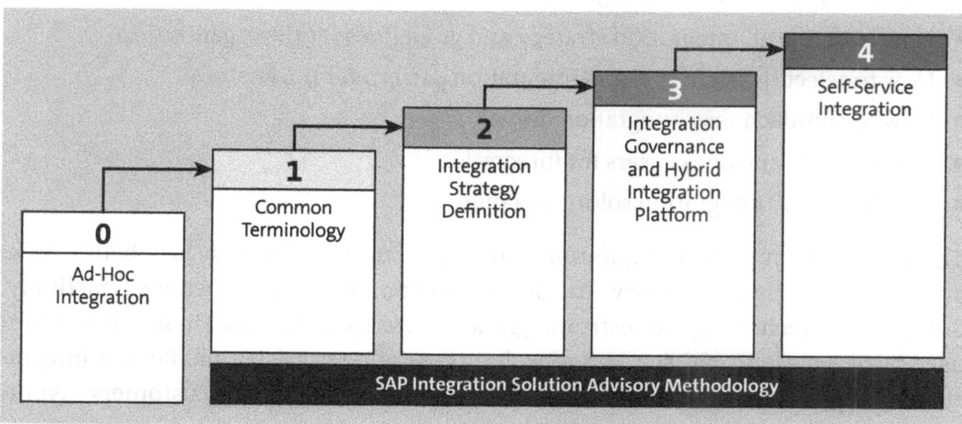

Figure B.8 SAP Integration Solution Advisory Methodology Use Cases

B.3 Integration Community

To build momentum, prompt seamless onboarding, and get your company started quickly with its digital transformation, SAP has a vibrant community to engage with you for your integration projects, as shown in Figure B.9 (*https://community.sap.com/topics/btp-integration-suite*). The integration community at SAP is open-ended, and it encourages peer-to-peer discussions to collaborate and learn together. SAP shares information, important announcements, collateral, product roadmaps, solution blueprints, and customer use cases as part of a deeper engagement with its members.

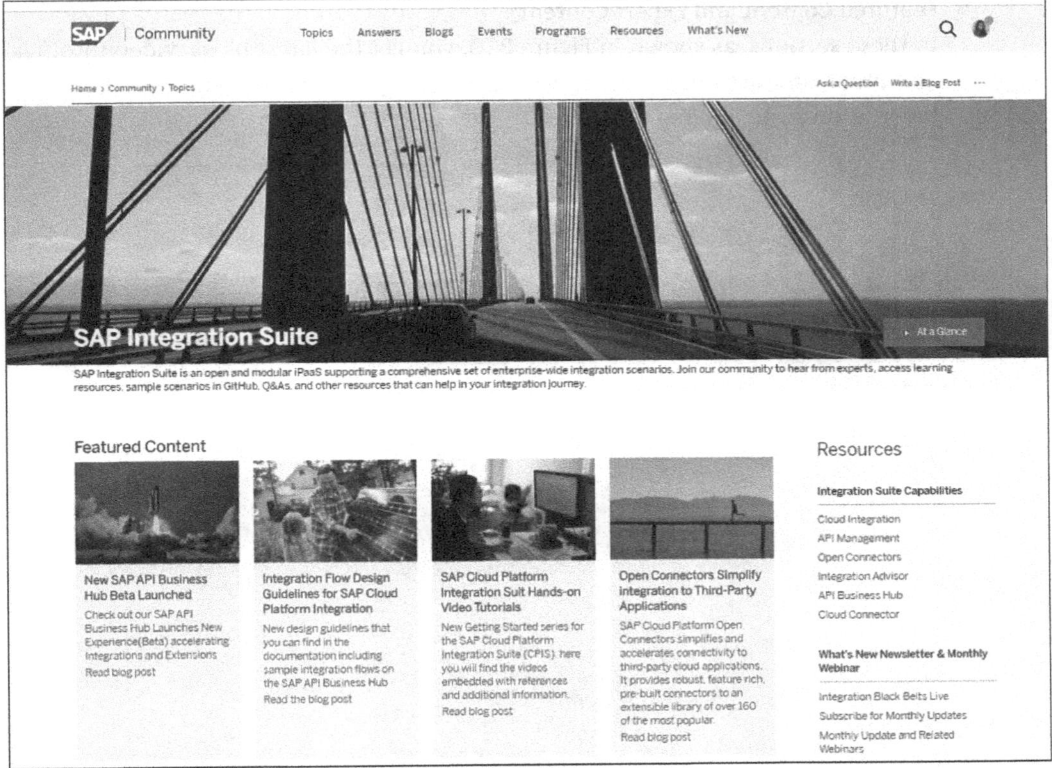

Figure B.9 SAP Integration Suite Community

The following are the benefits of being engaged with experts in the SAP Integration Suite community:

- Get the latest and the greatest updates and build stronger trust with community members.
- Discuss ideas, use cases, blueprints, and landscapes from an early stage of your project.
- Address your concerns and make better decisions.
- Become better informed about opportunities, webinars, and other business decisions.
- Make your decisions robust, starting from the planning phase of your integration journey.

With integration as the foundation of SAP BTP, SAP sees the value of customers, partners, and stakeholders coming together and representing a strong integration community.

To deliver a seamless experience and the latest updates, the SAP Integration Suite community has the following sections:

- **Featured Content and Expert Content**
  In these sections, as shown in Figure B.10, you get the latest blogs, video tutorials, and other product highlights, as well as recent questions and answers about SAP Integration Suite.

Featured Content

New SAP API Business Hub Beta Launched

Check out our SAP API Business Hub Launches New Experience(Beta) accelerating Integrations and Extensions

Read blog post

Integration Flow Design Guidelines for SAP Cloud Platform Integration

New design guidelines that you can find in the documentation including sample integration flows on the SAP API Business Hub

Read the blog post

SAP Cloud Platform Integration Suit Hands-on Video Tutorials

New Getting Started series for the SAP Cloud Platform Integration Suite (CPIS). here you will find the videos embedded with references and additional information.

Read blog post

Open Connectors Simplify Integration to Third-Party Applications

SAP Cloud Platform Open Connectors simplifies and accelerates connectivity to third-party cloud applications. It provides robust, feature rich, pre-built connectors to an extensible library of over 160 of the most popular.

Read blog post

Expert Content

Ask your questions and get answers from experts

Get answers to your queries on various enterprise-wide integration scenarios supported by SAP Cloud Platform Integration Suite.

Ask Question

Figure B.10 Featured Content in SAP Integration Suite Community

- **Resources**
  To help guide customers, partners, and the developer ecosystem, there is a **Resources** section that contains links to the following areas:
  - Help documentation for product capabilities
  - Latest and greatest product updates via newsletters and monthly webinars
  - SAP Integration Suite roadmap to outline SAP's vision, direction, and priorities for each quarter

  - Video tutorials to ease learning and get started with SAP Integration Suite
  - Groupings of key material in the form of a learning journey

- **Latest Blog Posts and Latest Questions**
  To experience the latest product updates and get step-by-step explanations, SAP Integration Suite blogs are a handy part of the SAP Integration Suite community page, as shown in Figure B.11. In addition to the collateral, discussions and knowledge sharing are encouraged in the questions and answers section.

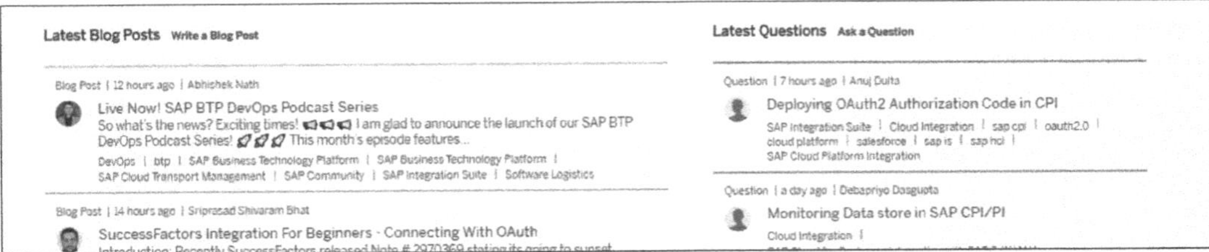

Figure B.11 Latest Blogs and Questions

- **Recipes in GitHub**
  Accelerate your integration projects with recipes available in the GitHub repository at *https://github.com/SAP/apibusinesshub-integration-recipes*. This repository contains recipes (best practices), samples, and reusable code to serve as starting points for your integration projects. This is a community-driven repository and can help you in the following ways:
  - Discover: Browse for the right integration accelerators to suit your business needs.
  - Contribute: Share best practices and give feedback on the recipes.
  - Lead: With sufficient contributions, you have an opportunity to lead, manage, and maintain this repository with your expertise.

B.4 Summary

In this appendix, you learned about the low-touch and high-touch engagements SAP offers to drive omnichannel customer engagements. As part of high-touch programs, you learned how Customer Advisory, Customer Engagement Initiative, and SAP Customer Voice provide the right sets of engagement models at every stage of your journey. Different video tutorials, blogs, missions, step-by-step guides, learning journeys, and certification courses are part of the low-touch engagement and provide easy onboarding and drive value for customers. With ever-changing IT infrastructure, integration is becoming more and more complex; to overcome these challenges, integration architects can leverage SAP Integration Solution Advisory Methodology to manage complexities. It focuses on defining the integration domain in a hybrid landscape, leveraging integration styles and use cases, and finally mapping the technology that should be used in a specific customer context. You can also communicate with and involve yourself in the SAP Integration Suite community to share best practices and exchange ideas.

Appendix C
The Authors

Christopher Aron is vice president, go-to-market and a leading expert for SAP Business Technology Platform (SAP BTP; previously known as SAP Cloud Platform) and has been nurturing and growing SAP Integration Suite's market adoption for the past five years. Chris has more than 20 years of technology and telecommunication product marketing experience. He received his master's degree in business administration from Webster University (USA) and his bachelor's degree in economics from Birmingham City University (UK). Chris holds three US patents from his work in Internet of Things with PassTime.

Day-to-day, Chris works across regions and industries with customers, system integrators, and technology partners to evangelize and improve SAP's broad range of capabilities. Internally, Chris works closely with local sales teams on key deal revenue support and with product teams for roadmap prioritization, launching new capabilities and long-term strategies. Chris leads additional efforts to help align internal and external perspectives, for example leading multiple customer councils designed to optimize product feedback and ultimately aid customer adoption. Chris has often featured as a keynote executive for external events around the world. Originally from England, Chris is currently based in Denver, Colorado.

Piyush Gakhar is director of product management for SAP Integration Suite and leads a team of integration product managers. He has more than 15 years of experience in technology across various roles, including product management, business strategy, consulting, and presales.

Piyush received his master's degree in computer applications and contributed to several successful internal and external projects, working closely with customers, partners, and key stakeholders. He is responsible for the product design, strategy, roadmap, and go-to market for SAP BTP services. Piyush has represented SAP in multiple global platforms/conferences and led key strategic initiatives and partnerships.

Shilpa Vij is a senior product manager for SAP Integration Suite. She started her professional journey at Tata Consultancy Services Pvt Ltd as a C++ developer. She received her bachelor's degree in electronics and communication engineering in 2007 and then worked towards being a business analyst, pursuing her passion for client interfacing and consultancy roles. Shilpa has worked at SAP Labs India for more than a decade, where she has been a seasoned cloud engineering expert, program lead, product expert, and now works as a product manager.

At SAP, she is dedicated to and contributes to product and customer success. Day-to-day, Shilpa works globally across industries with customers and integration experts for product planning and execution throughout the product lifecycle, including gathering and prioritizing product and customer requirements and defining the product vision. She also works closely with engineering, sales, marketing, and support to ensure revenue and customer satisfaction. She provides enablement sessions and expert-level workshops in the areas of SAP Integration Suite. Currently, her area of interest lies in discovering and projecting product management as an art. She is a coauthor of *SAP API Management* from SAP PRESS.

Index

A

G

H

I

T

- Set up and use SAP Cloud Platform Integration in the Cloud Foundry and Neo environments
- Model synchronous, asynchronous, and B2B integration scenarios
- Monitor, maintain, and secure integration flows

Bilay, Gutsche, Krimmel, Stiehl

SAP Cloud Platform Integration

The Comprehensive Guide

Bring your cloud and on-premise applications together with SAP Cloud Platform Integration! Learn how to integrate processes and data in your system step-by-step by developing and configuring integration flows in the Cloud Foundry and Neo environments. Explore prepackaged integration content in the integration content catalog, debug and secure integration projects, enable B2B integration, develop custom adapters, and more. With this guide, master SAP Cloud Platform Integration!

[illegible] pages, 3rd edition, pub. 09/2020

E-Book: $79.99 | Print: $89.95 | Bundle: $99.[illegible]

www.sap-press.com/5077